the
vanishing
ice

the
vanishing
ice

Diaries of a Scottish snow hunter

Iain Cameron

Vertebrate Publishing, Sheffield
www.v-publishing.co.uk

the
vanishing
ice

Iain Cameron

First published in 2021 by Vertebrate Publishing.
Hardback edition reprinted in 2021. Paperback edition first published in 2021.

 Vertebrate Publishing
Omega Court, 352 Cemetery Road, Sheffield S11 8FT, United Kingdom.
www.v-publishing.co.uk

Front cover: The snow cathedral of Ben Nevis, September 2016. © Murdo MacLeod.
Author photo: The author partially hidden beneath a snowdrift on Ben Nevis in July 2016. © Alistair Todd.
Other photography by Iain Cameron unless otherwise credited.

This book is a work of non-fiction. The author has stated to the publishers that, except in such minor respects not affecting the substantial accuracy of the work, the contents of the book are true.

A CIP catalogue record for this book is available from the British Library.

ISBN: 978-1-83981-108-1 (Hardback)
ISBN: 978-1-83981-087-9 (Paperback)
ISBN: 978-1-83981-088-6 (Ebook)
ISBN: 978-1-83981-089-3 (Audiobook)

Cover design by Jane Beagley, Vertebrate Publishing.
Production by Cameron Bonser, Vertebrate Publishing.
www.v-publishing.co.uk

Vertebrate Publishing is committed to printing on paper from sustainable sources.

MIX
Paper from responsible sources
FSC
www.fsc.org   FSC® C013056

Printed and bound in the UK by TJ Books Limited, Padstow, Cornwall.

To Adam.
Though no longer with us,
his presence is enduring and ubiquitous.

Contents

prologue

In the belly of the beast

I emerged at last, breathless, on to the plateau. The ominous-looking wintry squall that had been chasing me up the hill arrived from the west just as I completed the last few heavy steps of a relentless climb from the col, some 1,200 feet below. As I stood upright once more and filled my lungs with the rarefied Lochaber air, the first outlying snow-flakes of the squall danced down harmlessly enough. The initial flurry tends to do that: a few slivers corkscrew down from the sky and land softly on your sleeve or face, just on the leading edge of the gale that rides invariably behind them. But this is, literally, the calm before the storm. I've been in this situation more times than I can remember, and I know what's coming next.

Overnight snow already lay thick on the plateau, and the forecast promised more for the rest of the day and the next. Taking a minute to orient my internal compass and restore my breathing to something resembling normality, I reached into the side of my rucksack for one of the bottles of flavoured water I carried, only to find that there was ice floating in it. As I tipped my head back to pour the freezing mixture into my mouth, the first wave of the storm hit me – hard.

* * *

The day had started three hours earlier in the deserted glen far below. A man I trusted – a long-time snow observer – had told me that there were two old patches left over from the previous winter in one of the high corries on this hill.[1] This seemed unlikely to me, as the autumn weather had been mild – warm, even. I needed to be sure, though. I *had* to count every last relic of snow from the preceding winter before they were buried for another nine months or so. Missing even one would have felt like failure. But there was a problem: I had never been to this corrie before. It lurked on the edge of my knowledge. An apparently tricky place to reach even in summer, it was going to be doubly so in winter conditions. But, since no one else had volunteered to venture into this neglected corner of the Highlands, I decided there was no alternative but to travel there myself.

The climb from the car park had started memorably, but not for the reasons one might hope. On the upward pull that eventually reaches on to the col, about twenty minutes after starting, I chanced across a dying red deer. He was a handsome old stag and lay about fifteen feet from the stalkers' track. His front-right leg looked badly broken and, clearly, he could not move. His big brown eyes stared at me as I neared him, and he opened his mouth, attempting a roar to scare me off but managing only a hollow, thin rasp. It was a pitiful, tragic sight to see, and I could offer him no solace. I moved off, regretfully, hoping his death would come quickly and that this sad, chance encounter wouldn't amount to a macabre omen for my trip.

About an hour and a half later, I arrived at the foot of the col, where the first impediment presented itself. The overnight snow had been heavier than forecast and my path upwards on to the plateau was unfathomable beneath the fresh falls. All around me the landscape was winter, not autumn. The late-year, lifeless vegetation poked through the snow in an apparent attempt at defiance of its surroundings. Thick and ominous cloud swirled around the higher reaches of the hills to the west, whose summits were coated white.

---

1   A corrie is a glacial basin in the side of a hill, the Scottish Gaelic equivalent of an Alpine 'cirque'.

In the absence of any discernible path, I could do nothing else but one thing: climb. So, picking out the least hazardous and most obvious-looking route, up I went. It was steep. *Really* steep. I gained height quickly as a result, though, and luckily the path became clearer. The snow that had fallen filled the hollows where thousands of boots had trodden, and it snaked up the escarpment at a more forgiving angle – although to call it a path would have been generous. It was no better than an indistinct goat track, and goats would have thought twice about ascending it in these conditions. It became steeper the higher I got, and as I paused halfway up for breath, I turned around to see whence I'd come. The hills that encircled me were splattered with thick snow, and any heather that didn't lie under the white blanket was a dull, lifeless mahogany brown. A dark line of approaching weather, the squall of snow, raced towards me. I reckoned I had fifteen minutes to reach the top before it caught me up and deposited its considerable payload.

But moving upwards in these conditions was not an easy endeavour. The rutted path soon vanished yet again, leaving me to judge the steep slope and where best to place my hands and feet. I cursed myself for not coming a few days previously when the weather had been better.

I made slow progress, but progress nonetheless. At length I managed to negotiate a large bluff and grasped a solid-looking boulder for purchase to heave myself up.[2] To be safe I tested its grip on the ground by means of a stout pull, which was just as well, for it rolled from its berth almost without resistance and disappeared down the hill to the bottom, some 1,000 feet below. My heart beat loudly as I watched it tumble, and the clatter of it smashing off a granite outcrop raced back up the hill and into my ears. In less than ten seconds, the stone had reached the point where I had stood thirty minutes earlier. I cursed myself once again.

But then, I was up. The last big step negotiated, I arrived on the plateau. I was still at least an hour from where the two old lumps of snow from the previous winter were purportedly located. Finding them would not

---

2   A bluff is a steep headland.

be easy. Would they even be visible by now? If they *had* survived, would I know? The new snow might have buried them.

* * *

As I emptied that drink of icy, flavoured water into my mouth, I felt a push on my back as though I'd been hit by a car, and it propelled me forward on to my knees. I laughed a hollow chuckle to myself to try and brush it off, but this was serious. I needed to give the weather the respect it deserved. There didn't seem any point in pretending anything else. To accompany this seemingly tenfold increase in windspeed, the flakes of snow started to fall – if fall is the right word – horizontally. I took a bearing and made for the drop-in point to the corrie on the other side of the hill where the old snow would, hopefully, be lying. But before a minute had passed, I could see less than thirty yards. Another minute later it was half that. 'It'll pass. It'll pass,' I said to myself over and over as I walked across the plateau. By this point the ground and sky had fused together into a maelstrom of choking white. I checked my bearing again and trusted in it. I had to. This weather was as bad as any I'd witnessed on the hills of Scotland.

The walk that day across the plateau, which reaches almost 4,000 feet, can best be described as a penance. The buffeting by the gale was intense, easily above storm force. It once threatened to send me airborne in its ferocity. The snow, too, could scarcely have been thicker. With visibility this bad, it would be all too easy to walk off the edge of a cliff, so I stopped every few minutes to check my location. So far, so good.

But, despite the grim conditions, I made good progress across the plateau, and quicker than I anticipated, doubtless due to the wind paddling me on. At length I saw the rim of the other side of the summit ridge and, to my relief, it now looked as though the falling snow was moderating. I took the map out once more and checked my position. By good fortune, perhaps, I had arrived just near the drop-in point of the corrie. The sky now turned from white back to grey, and the wind eased to a playful breeze as quickly as it had started.

Gazing into the abyss below was like staring into a murky cauldron that contained otherworldly ingredients. Great swirls of snow and cloud spun and danced before me. The tail end of the gale whipped eastward, ready to inflict a sting on the next range of hills in its path. For me, at least for now, the ordeal was over. I sat on the snowy grass and put crampons on my boots for grip and waited a minute to see if the snow would lessen further, which – mercifully – it did. I decided, at last, to drop in.

Very deliberately, I eased down the steep grassy slope which itself wore a substantial covering of snow. The wind that had flattened me on the summit had all but disappeared in the shelter of the corrie. I relaxed my body. The absence of noise lent this place an altogether more serene atmosphere than just twenty yards away on the summit ridge. I breathed easier, too. The snow still fell, though, even if only in light, large flakes. It came up to my ankles and I was glad of the crampons on my feet and the ice axe in my hand, though there wasn't enough snow to anchor on if I should fall, so the axe locked into my hand as a mental crutch and very little else.

In the high places of Lochaber, visibility can be so poor, even during the middle of the day, that it is a struggle to see more than twenty yards ahead if the weather is appalling. The cloud and snow can be so thick that you must trust to navigation and instinct. It's an unsettling experience. Half of your mind – the subconscious – is telling you all the while to get out of this place; the other half – the pragmatic – wills you on to complete your task. I had come too close to the prize to let my subconscious win out.

But then, without warning, the snow started to fall thickly again, and I could hear the wind howl off the top of the cliffs above. Time was now against me. I needed to find last winter's snow if it were there and get out of the corrie before the new snow barred my exit. I had good coordinates, but in this decaying place a compass offered little in the way of help. I instead used instinct and judgment as I traversed along the foot of the cliffs that, on my right-hand side, reached up 300 feet into the vortex of wind and snow. On my left there was nothing except cloud and a steep slope that did not break for some 1,000 feet. Slipping here could not be countenanced. 'Steady, Iain,' I repeated to myself. For ten minutes I tottered across broken rock and drifting snow, hoping that at any second I'd stumble across the icy relics from a year ago. But nothing. 'They've probably gone,' I said to myself, just on the cusp of giving up. Mentally I was almost spent. 'Another few steps. Another thirty seconds.'

Just then, however, the ground changed ahead. 'What's that?' I said out loud, as though someone were with me. I took a few more steps forward. I strained my eyes and wiped my glasses. In front of me, not more now than ten yards distant, there they lay. Two big slabs of dirty white snow sat at the foot of the cliff, almost invisible in the camouflage of their surroundings. In form they resembled large white coffins which, considering the environment I was amid, seemed grimly appropriate. 'YES!' I shouted, as though I'd discovered some long-lost, priceless treasure in one of the deep places of the world. I was ecstatic that they endured and that I had found them. They were larger than I'd expected, too, and as I crouched beside them, I knocked on their surface: hard as the rock that they sat on. Months and months of compression, melting and refreezing had cast them like concrete. Their tops were like dragon skin, scaly and scalloped – and just as tough. There was no doubt that, at about fifteen feet long each, they'd survive now. The snow that had already fallen half buried them, and the heavier snow that was due in the next day would cover them up for the season. This would be the last time I, or anyone else, would see them for many months.

But how had it come to this? The snow that lingered on this hill never used to melt. Or, at least, it hardly ever did. Travellers in previous centuries even commented on being able to see it from the valley below all-year round. 'The only snow visible from a British train station in every month of the year,' one excited journal read in the 1930s. No longer could the same thing be said. The snow was a shadow of its former self, reduced to two small lumps of ignominious irrelevance, hidden from, and of no interest to, all but the most dedicated enthusiast. This was the first year in five that any snow had endured the whole year in the corrie. Could it be a sign of things to come? Had climate change stretched its long fingers into even the smallest nooks of the highest Scottish hills?

I had no time to ponder this any further. The snow was deep, and I had to get out of this cauldron of ice and rotting rock. Fresh boulders from rockfalls lay all around, serving as a reminder that this landscape is constantly evolving and not a place to linger.

I retraced my steps back up the steep slope and on to the lip of the plateau. As I came over the edge, the blast of wind that hit my chest forced me to sink down momentarily and grab the vegetation. The snow stung my face so hard it felt like I was being sandblasted. At that point it was obvious that going back down the way I had come would be tantamount to a death wish, so I abandoned the idea. Though the car lay that way I could not risk the descent back down to the col in such atrocious conditions. Poor decision-making on the hill can cost lives and I had no wish to call mountain rescue, nor to become another statistic. An easier route off the north side of the hill would necessitate a taxi journey back to the car, but it would be money well spent.

I descended slowly, and with every downward step I took I cared less about the weather that chased me off the hill. I was beyond pleased I had come to this place today because, had I not, the two patches would have gone unrecorded. When researchers look back in a hundred years' time, they will maybe remark that in 2007 one enthusiast noted that two small patches of snow had survived the year on this hill. This may just be a footnote in a paper they write, and they will not know the human story that lies behind the record. This is of no moment to me. What matters is the snow that was recorded, not the means.

one

The spark

I got up first, as usual. It was early, about 6.30 a.m. The sun streamed in through a gap in the curtains of our bedroom and, for the first time in what seemed like over a week, I hadn't woken up to rain bouncing off the window. Letting my six-year-old brother sleep on, I went upstairs to the living room to put on the television. Prior to pressing the heavy analogue button, which rocked the TV back on its stand, I opened the curtains to let the light of the day in. There were no clouds in the azure blue sky that morning and on the hills, spring's first flush of growth had turned the grass a verdant green. The warmth of the sun could be felt through the single-paned window. Even then, just nine years old on that early May day in 1983, I was in awe of the views our new house had. It sat right at the top of a housing estate in Port Glasgow, literally the highest street in town, 500 feet above the Firth of Clyde. We'd moved there just a few months back, when my dad started a new job locally. The vistas from the front of the house were panoramic and stretched over the estuary from the Erskine Bridge in the east to Dunoon to the west. The high hills on the opposite side of the Firth looked impossibly far away and seemingly unclimbable. Normally they poked into the clouds and were invisible from the house. Some of the hills were rolling and green, but others were grey and craggy. How could anyone, even a child, not be drawn in by the allure of these new visions?

One hill, many miles distant, towered higher than the others. It rose up from behind two smaller hills and the summit cone sat atop two large shoulders which jutted out horizontally in pleasing symmetry. On that beautiful May morning, sitting on this distant peak's south-facing slope, gleaming prominently in the early sunshine, was a large patch of what seemed to be brilliant-white snow. I couldn't understand how it had endured. I stood for a moment and looked at it. 'It can't be,' I said to myself, doing a double take. No snow had fallen at my house for many weeks. Surely it had all melted? The last remnant I'd seen that year had been a long, streaky drift against a lichen-covered drystone wall, just a hundred yards or so behind our garden. But that was ages ago. Yet here on this muscular, high hill there remained an apparent relic of winter just past, clinging on despite the warm weather. Why, when it had melted away everywhere else, did it persist here? Though old enough to understand that there must be a reason for it, I had not yet enough wit to work out why.

For reasons that I cannot adequately convey, my curiosity had been piqued to the extent that I wanted – needed – to identify which hill's chest this white medallion hung on. To do that I was going to have to conduct my own detective work, as I knew nobody else in my family would be able to help me. They weren't much interested in the outdoors, so any attempt to enlist them as helpers would have been a futile exercise. That said, I remembered that somewhere in the house we had a tatty old AA road atlas that my parents kept, despite not having a car. I unearthed it from the ragged old ottoman that the phone sat on, only to find the atlas in equally poor repair. There was now no front or rear cover, and the staples were fighting a hopeless, rearguard action to hold the whole thing together. It was in miserably large scale, and covered in coffee stains to boot, but I figured it might contain some clues. Even then I was fascinated by maps and geography so, standing with the atlas open at the window, I oriented it in the direction I faced. Glancing between the open atlas and the view, as though trying to crack a secret code, I finally surmised that the hill where the snow reposed could *only* be Ben Lomond. According to the atlas it rose to an impressive 3,194 feet. I knew that people in

Scotland called hills over 3,000 feet 'Munros'. I'd no idea why this should be the case, but I was pretty sure that I was gazing upon Scotland's most southerly Munro.

I do not recall that morning whether I mentioned this revelation to my parents. I suspect not. It was pointless asking them in any case why snow would still be there on Ben Lomond. They'd barely set foot on a hill, never mind studied the vicissitudes of mountain weather. But, that morning, a light bulb flicked on in my young mind. A spark. As for why the snow hung on, that was something I needed to figure out for myself.

For the next few days the pattern remained the same: up at first light, run upstairs and peer through the window to see if the snow was still there, go to school, then look again in the evening before I went to bed. It had turned into a strange fascination. But, at length the weather turned again and the hills were obscured by cloud and rain, so for three or four days I saw nothing. 'Surely it's melted,' I said to myself. Then, one day after returning from school, the cloud blew off Ben Lomond and its cohorts. The snow, alas, had gone. At least, I couldn't see it. I was disappointed by its demise. Sad, even. I knew well enough that it eventually *had* to vanish, but this didn't lessen the feeling of disappointment. It felt like saying goodbye to a relative you'd probably not see again for ages. I was determined that, one day, I would get to the bottom of the mystery and visit this snowy place.

Because of my parents' lack of interest in anything to do with hills, or the general outdoors, and because independent motorised transport was beyond our financial means, my interest in patches of snow remained confined for the next eight years to what I could see through the living room window and from the Ordnance Survey trig point about half a mile from our house (the views from there were even more wide-ranging and took in hills that were over forty miles distant). The option of what I could glean from very primitive technology presented itself, too. Chief amongst these was page 421 of Ceefax. Every morning in winter I would load up this highly unwieldy and slow proto-internet. Page 421 showed all five of the Scottish ski resorts at that time, and how many of the lifts were running. A teenage me wanted to know how much snow lay on the hills, even though I had never skied in my life, nor climbed a snowy hill. It was completely illogical. I remember at the age of thirteen being castigated by my parents for dialling a premium-rate phone number that gave a real-time overview of which ski lifts were spinning at Scotland's resorts. I think, actually, they were quite relieved when they found out what I had used the phone for, and that their eldest son wasn't calling an altogether less wholesome premium-rate service.

At secondary school I recall disagreeing with my geography teacher on the subject of snow. I didn't rate him, mostly because he used to enjoy demeaning his pupils by showing them how poor their knowledge of his precious subject was. It wasn't just that, though. He seemed dis-interested in our learning, often teaching in what I thought was an overly perfunctory manner. I wouldn't have it, and he disliked me from day one. The snow disagreement centred around a road. He said to the class one day that the A939 Cockbridge to Tomintoul had been closed in the past by snow in every calendar month of the year. I knew this to be nonsense (it had never closed in July or August) and told him as much, though obviously not in such blunt terms. It was a nail in my educational coffin. Looking back on episodes like this, I wish I'd learnt to keep my mouth shut.

The school that I attended, St Stephen's High, could often be found propping up every other school in the annual exam results league tables for Strathclyde Regional Council, as it was back then. Many people speak of their time at secondary school as the happiest of their lives, but for me they were the worst and I despised it, as I did every subject bar English and geography. As a result of this disinterest and apathy, I decided to leave school at sixteen with only a handful of mediocre qualifications, much to my parents' chagrin. I had no clear idea of what I wanted to do next, or how to do it; I was just thankful to be out of that awful place. However, by an unusual turn of good fortune, I managed to secure an apprenticeship to be an electrician with a local, family-run company. For the next four years, this provided me with welcome cash, and with the discipline, learning and practical skills that I cherish to this day. In con-trast to the gloom of my secondary education, I look upon these four years fondly as some of the best I've ever had. They were the making of me. (Though my siblings and I were brought up with an abundance of love, discipline and a good sense of right and wrong in the home, an apprenticeship among older, skilled men instilled in me a sense of diligence and hard work that can hardly be overstated. The pride with which these ordinary men worked still resonates with me thirty years after I was first exposed to it.)

As soon as I was able to, I learnt to drive and got a car. I wanted to explore my country, something that I had never been able to do independently. I wanted to climb hills and to go to see some snow in summer. However, having lived a relatively sheltered life thus far, I was much too nervous to do anything ambitious. My first walks were done on easy hills with obvious paths and tracks. I recall one such trip, with my brother. Aged seventeen and fourteen, we climbed up Ben More near Crianlarich in early June. Mum and Dad, perhaps blissfully unaware of the potential risks of letting two children loose on the hills unsupervised, or perhaps just trusting in our ability, let us get on with it. Being Britain's sixteenth-highest hill, Ben More felt like a big adventure – a huge and hulking mass of steep grass and rock. Just below the summit, we came across fourteen ptarmigan, which were the first I'd ever seen. I was transfixed and followed them as they deftly scuttled across the rocks, always just too far away for a decent camera shot on the Olympus Trip I had bought from a charity shop. At the summit itself, patches of snow on Ben More's neighbour Stob Binnein were visible. What a treat. A whole world of new possibilities and excitement. I remember it as yesterday.

Around this time, interest in the outdoors was growing, though it was very much a minority pastime. TV programmes appeared which hitherto would have struggled to get a commission. This was to a significant extent the result of a hugely popular TV show called *Weir's Way*, presented by the irrepressible Tom Weir. A Scottish staple during the 1970s and 1980s, *Weir's Way* showcased the bobble-hatted Tom visiting parts of the country that were not well known to the average viewer. Weir remains a legendary figure to the outdoors community in Scotland; a pioneer of opening up the outdoors to a wider audience via his written and TV output.

Of the TV shows that appeared in the early 1990s, the most popular was Muriel Gray's *The Munro Show*. This programme had a big effect on me and many more folk my age. It was a window to view parts of Scotland that I scarcely knew existed, save in the maps that I pored over as a teenager. Gray had an infectious enthusiasm too, which came across well on screen. The format of the show was original and highly engaging. It started with each walk being introduced by the extremely eccentric Gaelic poet Sorley MacLean, sitting on an armchair. Next to him was a blackboard with the hill's name written on it. MacLean then translated the often unpronounceable and impenetrable name, sometimes imparting an anecdote that he'd either heard or experienced himself. Watching as a young man I thought him a bit odd but, viewing the programme back on YouTube many years later, I completely underestimated just how wonderfully over-the-top he was.

Once the meaning and pronunciation had been delivered by MacLean, the rest of the show contained splendid scenic shots of Gray wandering up whichever hill she and her team had decided to climb.

One episode of *The Munro Show* stood out. In it Gray walked up Braeriach on a beautiful late summer's day, to judge by the height of the sun and the colour of the vegetation. This hill is the second-highest in the Cairngorms and third-highest in the UK. As she walked across the vast, Arctic-like plateau, the camera cut to her standing on a precarious rocky spur, with what looked like drops on all sides of hundreds of feet. It then panned downwards into the black abyss, stopping when it framed four large patches of snow.

'In this corrie,' Gray continued, 'is Scotland's most permanent snowfield. In fact, it's only melted twice this century, in 1933 and 1959.'

Upon excitedly hearing this gem of information, I sat bolt upright in my seat. Several things leapt into my mind. Firstly, how did she know they were the most permanent snowfields in Scotland? Secondly, who on earth was monitoring long-lasting patches of snow on Scotland's hills? Finally, why did patches of snow persist all year? This information was a revelation to me. Needless to say, I recorded all the episodes of this show on VHS and would come back time and time again to the Braeriach episode just to see those massive banks of snow in the shaded corrie below. At that time, I thought them impossible to reach, but I longed to see them nonetheless.

Inspired by the revelatory information that people must be monitoring long-lasting snow in Scotland, I vowed to find out more. This was a much taller order in the early 1990s than nowadays, with no internet to draw on. Finding out such things required visiting libraries and bookshops. Just finding the latter in the area where I grew up was hopeless because establishments that sold books were often confined to the larger John Menzies or WHSmith newsagents. Apart from the odd Ordnance Survey map or glossy 'Scotland is Beautiful'-type tartan shortbread books, there was nothing remotely like the sort of tome I yearned for.

The first few sorties to the local public libraries yielded little either, with librarians knowing virtually nothing about the outdoors, never mind this esoteric subject. But, because diligence is the mother of good fortune, one day in a bookshop in Glasgow I came across a publication that gave me my eureka moment.

The outdoor section of Waterstones bookstore in Glasgow wasn't particularly well-frequented at that time. I recall perusing it at leisure, with no one else in close proximity. Because it contained the largest section of its type in Glasgow, I figured it would be the best option for locating any literature that might conceivably cover this odd interest. Sure enough, it was, and I stood rooted to the spot as I read a particular chapter of a book I found. Many people will have experienced a similar sensation when they read something that electrifies them or informs them of something that makes them think, 'Yes! Yes, that's it!' It is a rare occurrence but, because of that, one is inclined to remember it for the rest of one's life.

Martin Moran's *Scotland's Winter Mountains* was a revelation to me.[1] Intended as a guide for mountaineers, its pages were packed full of technical information and instruction on winter climbing. Most of the content didn't interest me. At that point I had little or no desire to indulge in rock

---

1   Moran, Martin (1988), *Scotland's Winter Mountains*, Pynes Hill, Exeter: David & Charles Ltd.

or ice climbing, but what grabbed my attention was a brief section that specifically dealt with snow. The chapter in question discussed the last ice age, many thousands of years ago, and the more recent so-called 'Little Ice Age' of the period 1600–1850 (or thereabouts). On one page it posed the hypothetical question, 'Are today's snow patches dying relics of a bygone era or the harbingers of new glaciers?' This was what I'd been searching for! This question accompanied a full-page table that had some golden nuggets of information, such as a list of hills and long-lying snow patches, the elevation they sat at, their aspect and some commentary on them. These observations on snow patches were wonderful. 'Never known to wholly disappear.' 'Not closely observed but occasional survival has been noted.' It was fascinating. Names that nowadays I am familiar with to the point of saturation, I read for the first time – Ciste Mhearad, Aonach Mòr, Ben Wyvis and many more. Proof in print, at last, that these patches were being monitored. There was also a photograph of the same snow patches that Muriel Gray showed us on the Braeriach episode of *The Munro Show*, with massive drifts visible in a picture from October 1984. I simply had to have this book and would have bought it on the spot, but for one small problem: its price tag.

Even these days, shelling out £25 on a book would make many people think twice. Spending that amount in 1990, when my weekly take-home pay from my apprenticeship was about £50, meant plenty of patience and discipline. So, for the next five weeks I carefully saved £5 each week, setting it aside and putting it behind a poster of Madonna that hung by Blu Tack on my bedroom wall. It meant I couldn't buy any of the nice clothes or fancy trainers that I had developed a penchant for when I started earning my own money. This small sacrifice for the pleasure of what was sure to be a fascinating chronicle seemed worth it. Indeed, it became one of my most prized possessions.

Going back to Waterstones to buy the book was exciting. I even eschewed a gathering of the other apprentice electricians from my college who were going into Greenock to play pool and then on to the amusement arcade. I didn't dare tell them the reason for doing this. To do so would have meant inviting some unwanted adolescent abuse.

The woman who served me that day smiled as I handed over the five £5 notes. I put the book in a shiny Waterstones carrier bag and ran back to the station to catch the train home. I seldom put it down for weeks thereafter.

After finishing my electrical apprenticeship in October 1993, I took some time out and worked, briefly, as a volunteer for the National Trust for Scotland at Glen Coe. In between (unpaid) dogsbody duties such as red deer counting and digging footpaths, I used to note, for my own interest, the patches of snow still clinging on below the highest peaks. October is extremely late for snow to persist from the previous winter in Glen Coe, so I made sure to record the exact dates. I was certain that in the future this would be of interest to someone. How unusually prophetic that turned out to be.

Many years later, in 2005, I read a journal that featured an esteemed Scottish scientist. The paper he authored in the journal focused on the subject of snow patches. By now I was aware of some of the people who engaged in this study, and I wanted to get involved. In the paper the scientist wrote that 'snow has never been known to persist from one winter to the next at Glen Coe'.

Upon reading this statement I cast my mind back twelve years to the diary entry I'd made at Glen Coe. The large patches high on the slopes of the unpronounceable Stob Coire Sgreamhach were visible in October, which is normally a critical month for old snow patches. They are usually terminally small at that point and susceptible to vanishing. There was a very good chance these large patches I saw would have survived through to winter, given the lateness of the year and their size. I couldn't be sure, as I left Glen Coe before the lasting falls of the new winter came. So, with some trepidation, I penned a note to the author of the paper, whose email address appeared at the bottom. The reason for my nervousness was not that I doubted my observations, but because of the reputation and clout of the person who wrote the paper.

Adam Watson, holder of quadruple doctorates and possibly Scotland's pre-eminent outdoors scientist at that time, authored the paper. Here was I: a grunt; an electrician with zero academic pedigree or academic gravitas. I hovered over the 'send' button, about to fire off a note to one of the country's most respected scientists telling him the information contained in his paper might not be 100 per cent accurate. What was I thinking? I shelved the note. But something nagged at me over the next forty-eight hours. I told myself that the diary entry had been correct, and that I'd been looking at these things long enough to know that there was a decent chance those Glen Coe patches survived until winter.

'Send the bloody email, Iain,' I said to myself. So I did, and to my surprise I received a reply the very next day. I sat for a while looking at it unread in my inbox, wondering what it contained. I opened it nervously, through semi-closed eyes. My relief was palpable as I read its contents, which I still have, as they were highly encouraging and warm. A couple of lines stood out:

I am interested to come across a snow enthusiast, as there are so few of us! Best wishes and keep in touch,
Adam Watson

Phew. I was immensely relieved. Here was a scientist who was not precious about the content of his life's work. If data existed that added to the collective knowledge of the study, then he wanted to hear about it, irrespective of whether it meant revisions to his own.

From that point onwards Adam and I corresponded a good deal. We also became friends. Whenever I went out on to the hills, I would seek out places that I knew or thought there would be old snow on, and I shared my observations with Adam. Some of these he would collate and insert into an annual scientific paper of which he was the lead author. The paper, which appeared in the Royal Meteorological Society's *Weather* journal, chronicled how many patches of snow lasted from one winter to the next on Scotland's hills, as well as how long into the year others on lesser hills lasted. This paper was a treasure trove of information to me. In 2008, after I'd been contributing for a few years, I whooped with delight when Adam asked me to become a co-author of the paper. Having my name in a peer-reviewed scientific journal is something that

I never thought I'd achieve. Just a few years later, in 2011, Adam suggested that I have a stab at writing the paper with a view to becoming its lead author. Perhaps sensing that he might not be around for ever, he appeared keen to hand over the authorship to someone who could keep it going for many years to come. I was humbled, and in truth probably more chuffed by this than by anything I'd ever been asked to do before. When my first paper was published, in late 2012, I framed it. A number of years later, in 2017, the Royal Meteorological Society presented us with the annual Gordon Manley Weather Prize for 'outstanding contribution to *Weather* in the preceding five years'. It felt like a real honour to receive recognition from our peers and I attended the ceremony in London with John Pottie, another longstanding expert and co-author who had contributed much to the paper over the years.

Honours like these are very special to us. This is because all the work we do on snow-patch research is done as amateurs. In other words, we do it as a hobby. Every pair of boots, every waterproof jacket, tape measure, rucksack, tank of fuel, even down to our sandwiches – every penny we spend comes from our own pockets. For a living I am employed in construction, specialising in safety and health. I advise the company I work for on what course of action it ought to take to mitigate risk in terms of legislation. It could not be more different from the esoteric world of snow-patch spotting.

Besides, doing it as amateurs has its own advantages. For a start we are not beholden to an institution to spin the results of our research a certain way. In today's cut-throat world of academic funding, leaning towards a favoured position can be crucial to securing financial backing for future studies. Not having to account for our time, or explain to people where we are going, is another distinct advantage. The Scottish weather obeys no calendar. We go to look for snow when conditions are most suitable and, wherever we can, most agreeable. Having to justify ourselves to a supervisor or funding body is something that none of us would feel comfortable with.

Occupying such a small niche does have fascinating consequences. Given the lack of genuine experts in this field, coupled with the increased interest in the subject over the last few years, the level of exposure our work receives has hit unexpected highs. Snow-patch stories have appeared in virtually every national newspaper, as well as on countless radio and TV programmes. This sort of exposure is very helpful for raising the profile of our work, but not just for its own sake. The more people who walk on the hills in summer and autumn, the more eyes there are to report back on the state of any remaining snow. Consequently, the data that we compile and submit to the Royal Meteorological Society becomes that bit more valuable. Without doubt, the number of people who now express an interest in this topic is remarkable. Back in the early and mid-noughties we thought there were only a handful. The upsurge of awareness, not only in enthusiasts but the general public, is most heartening. Long may it continue.

When I first became absorbed in snow patches aged nine, I could have had no idea that it would eventually lead to me being the author of over twenty papers on the subject. It seems to me that you are either born with a love for something or you are not. For sure, interests can be fostered, nurtured and encouraged by parents, guardians or teachers. Generally, though, these tend to be 'normal' or prosaic pastimes like playing an instrument, football or reading. I have yet to come across a parent who actively encourages their offspring to take a tape measure on to the back end of a high cliff or steep gully to measure an eight-month-old patch of snow.

Perhaps that day is coming. Who knows?

two

With friends like these

Some people go to priests; others to poetry; I to my friends.
— Virginia Woolf[1]

The question is *why?* What makes someone become so interested in snow and, more specifically, patches of snow, that they are prepared to go to some of the remotest parts of the country to record relics that are sometimes smaller than the average dinner table? Moreover, what is it that made a young boy become fascinated by the subject?

Childhood interests are often fragile things, being both whimsical and ephemeral. Here today, gone tomorrow curiosities picked up by infants are many but, just occasionally, something sticks. An idea or experience attaches itself to a child, where – allowed to blossom – it can remain for the rest of his or her life. There are an *ad nauseam* number of accounts where respected scientists or naturalists have read or seen something inspirational as children. In their memoirs many of these people seem to be able to recall the event which has gone on to inform their life's study with almost perfect clarity, despite it sometimes being many decades ago.

One beautiful day in August 2012 I set off on a trip to a remote part

---

1   Woolf, Virginia (1931), *The Waves*, London: Hogarth.

of the Central Highlands of Scotland. The ostensible aim of my outing was to check on the status of a large patch of extant snow that I heard lay on one of the area's high hills. When I arrived at the starting point of the walk, some ninety minutes after setting out by bike from my car, I found – somewhat surprisingly, given it was a weekday – a small group of people packing their camping gear away and getting ready for a round of six Munros. They'd come the previous evening, setting up their tents and sleeping under canvas, because they fancied a leisurely start to the day rather than attempting the ten-mile cycle that I had just completed. By the time I'd stowed and locked my bike for the return journey they'd already set off on the path that my route followed. I quickly caught them up and engaged in some conversation typically indulged in by walkers who bump into each other in remote areas. At length one of the walkers noticed the large tape measure strapped to the rear of my rucksack.

'What's the tape measure for?' she asked. I had forgotten it was visible. Normally I kept it in the rucksack so that people didn't ask what I found to be awkward questions. But because I was in the wilds this day, and it was midweek, I hadn't expected to meet anyone. I fumbled around for an adequate answer to the question, but none was forthcoming. It wasn't that I was embarrassed or self-conscious *per se* (though there was a touch of that), but rather that I liked to go about my business quietly, without having to explain my rather individual hobby.

'Oh, just some scientific research,' I replied, hoping that this would deter them from further enquiry. But, of course, it didn't – it only served to make them even more curious.

'What sort of scientific enquiry requires such a large tape measure?' one of the other walkers asked. I knew at this point the game was up and I had better just answer transparently.

'I'm going up to measure the last patch of snow in these hills,' I said. 'It's one of the very few that has been known to last all year outside the Cairngorms or the Nevis range.'

'Oh, how interesting,' another of the walkers said. 'I saw on the news or some other TV programme last year that there were people going out into the hills and doing this. Are you one of these people, then? Well done!'

Aside from the slight self-consciousness that I felt (a relic of childhood), the other reason I was reluctant back in 2012 about what I was doing was that I didn't – and, to a degree, don't – fully understand *why* I found it so interesting. For me it is a deeply profound question and one that I have struggled to answer ever since I was a boy. Even now, after all these years of being immersed in the subject and asked about it frequently, I don't think I have ever given a wholly satisfactory or comprehensive reason.

On the specific encounter that day, my main memory is that of relief. The camping party's reaction of interest and congratulations was heartening to hear. That chance meeting on the hill that day was, for me, a watershed moment, and the point at which I realised I needn't worry what people thought, as most of them seemed to be genuinely interested anyway.

Being called a 'geek' used to be a term of derision, attributed to people who were socially awkward or lacking in personable skills. When I was growing up, the word carried baggage that conjured up the image of a lone boy in a bedroom working on a science project or reading books about computer coding. It seems, though, that over the last ten years the word has been largely divested of its negative connotations and is now used far more positively. We geeks have been liberated.

* * *

In the previous chapter I mentioned that the nine-year-old me was sad to see that final fragment of snow disappear on Ben Lomond all those years ago, as though I were waving cheerio to an old friend.

The notion that a particular patch of snow is somehow imbued with a personality, or is in any way anthropomorphised, is patently absurd. And yet, despite this, I have on many occasions set off on a journey to one of my regular long-lying snow haunts, acutely concerned with the condition I would find it in. Why should this be? What does it matter? In the grand scheme of things, it hardly does. It appears, though, that this seemingly odd sentiment is more prevalent than I first imagined.

In his lovely book *Among the Summer Snows*, Christopher Nicholson relays the feeling better than any account I've read. There is a chapter which describes one occasion the author went to the Cairngorms in search of summer snows. A longstanding chionophile, he knew where to look for these patches.[2] But as he stood waiting, the cloud refused to lift, thereby prohibiting a clear view over the valley to the snow. He waited and waited, and then – at last – he was rewarded when the cloud lifted suddenly:

> So here I was, on my knees, late one day in the summer of 2016. My heart in my mouth. Absurd. Is this what it had come to? It: me, the world, human history. The winter would bring more snow, so why did it matter, it did not matter, not really, not at all, the old argument raged away in my head, but at that point it mattered at least enough to keep me there. I needed to know if the snow had survived.[3]

So, why *does* it matter to spotters like Nicholson and me whether snow survives or not? Why do people like us lament its passing and ascribe almost human-like properties to something that is about as far away from being human as it is possible to be? My own view, arrived at after a good deal of reading and listening to psychologists, is that when one develops an interest in something that can border on an obsession, irrational attachments arise. This is especially true for me when visiting the same locations time after time, year after year. One gets to know what a particular patch of snow will look like before one even arrives at it. They tend to take the same shape every year due to the topography of the location they inhabit, so there is a familiarity for the observer. Then there is the fact that the snow is not permanent. It is liable, indeed sometimes likely, to disappear entirely. In this regard the analogy to the human condition is, I believe, apt: the snow is engaged in an existential struggle against time and the elements. It is being chipped away at for every second of every day of its existence until, ultimately, it loses its fight. In other words, from the moment it comes into being, it is always dying.

---

2 *Chionophile* is from the Greek words *chion* meaning 'snow' and *phile* meaning 'lover'.
3 Nicholson, Christopher (2018), *Among the Summer Snows*, Tewkesbury: September Publishing.

To extend the analogy, if one is used to visiting a specific location and the snow is not as large as you expected it to be, then you are invariably disappointed. Sad, even. In the past I have likened this sensation to visiting an elderly relative: you hope that things will be good when you get there, but if the person turns out to be in poorer health than you anticipated, then your own feelings are correspondingly negatively affected.

The patches that Nicholson described earlier as those that caused him to be on his knees are the ones at Garbh Choire Mòr on Braeriach, which I describe fully in chapter 4. I have a longstanding interest in those ones as well, so I can sympathise entirely. But there are others. Ben Nevis's upper reaches possess such drifts, as do its near-neighbours Aonach Mòr and Aonach Beag – both 4,000-feet-plus hills. All these patches are the very longest lasting in Scotland, and the ones that are most visited. Perhaps, therefore, it is no surprise that these are the snows which have been invested with so much emotional capital.

Whatever the reasons are for people attaching themselves to these long-lying wreaths, I for one am glad of them. They help to engender an interest that sees us go back year in, year out. These elderly relatives have family that cares for them.

* * *

There are a multitude of reasons why a hill can cling on to a patch of snow well into summer and sometimes beyond, but it is a multifaceted issue and not always consistent. Even on the same hill, with identical weather patterns occurring over the years, different locations can and often do swap with each other as to which will hold the last drift in any given year. And with temperatures during summer sometimes exceeding 20° Celsius at the locations where the patches sit, people often wonder how *any* snow can possibly resist weather this warm for so long. How can they endure for such protracted periods?

By far the most important reason that affects a snow patch's longevity is the amount of snow that falls in winter and spring at that locality. This may seem like a strikingly obvious thing to say, but it does bear repeating.

If a hollow or gully contains snow that is in the region of thirty feet deep, then it's going to take longer for an 'average' Scottish summer to melt it than if the depth were only fifteen feet. Fifteen feet, or approximately four and half metres if you prefer metric, is in no way unusual for many hills at the end of winter. I have seen depths of old snow in excess of thirty feet in *September* on Ben Nevis's north-east face. Even on relatively modestly sized hills, the right winter conditions can result in extreme depth. Allied to the *amount* of snow that falls, the *direction* that it comes from is hugely important too, as is the *strength* of the wind that propels it. You could quite reasonably say that the quantity of snow is directly dependent on these other two elements.

The second factor that allows snow to persist so long is the physical geography of the hill on which it reposes. In Scotland, as with many other countries in the northern hemisphere, glacial action over the millennia has gouged out corries from the hillsides, many of which face north-east. In form they resemble a large bite being taken out of a cake, especially in the Cairngorms and other areas where rolling hills are prevalent. These bites into the hill are characterised by sheer cliffs and shattered rocks below them. Moisture-laden storms arriving from the west and south-west in winter can, and often do, drop massive amounts of snow above 3,000 feet, with a good deal of it being carried into the corries. This is done, as already mentioned, by the direction and strength of the wind. If there's nothing for the snow to stick to when it falls from the sky, it will be blown across a hill's summit, where it is unrestrained by any obstacles. As it dances along the tops of these hills it is constantly searching for a place to stick and settle. Invariably it will be driven over the edge of the north-east-facing corries and, when it tumbles downwards, it will settle at the foot of the tall cliffs. Each individual storm can add many feet of depth to a single location, and this process repeats numerous times over winter. The very biggest storms in winter can literally add over twenty feet of depth to a single location.

A simple analogy: when one walks in the countryside after a blizzard, fields can appear covered. When, inevitably, the temperature rises by a degree or two during the day, the fields are stripped bare, but the surrounding ditches often contain quite deep drifts. It is exactly the same phenomenon. The fields have nothing on them for the snow to stick to when the wind blows. When it is driven over the edge, it gathers in the ditches and is trapped by the bank. This is why wind is such an important factor in snow's durability. If snow fell vertically on a breathless day it would lie evenly and melt at a uniform rate.

Another reason why north-east-facing patches do so well is that they are, to a greater degree than the rest, protected from the glare of the sun, and the worst of the warm rain, for a good deal of summer. Patches that are tucked in against cliffs that face north-east may only see direct sunlight for the first few hours of the longest days of the summer months. For the rest of the year, the sun traverses the southern sky, its azimuth insufficiently high to reach over the cliffs to trouble the snow. By the same token, the equivalent Atlantic weather systems that deposit so much snow in winter can do the opposite in summer. Southwesterly 'hairdryers', as we call them, from June through to August can wreak enormous damage on the snowpack, but the patches in the north-east corries are protected from the worst of this weather by the cliffs above them.

The third important factor that contributes to snow's longevity is the height the patch sits at. A good rule of thumb is that the higher up something is, the colder the temperature will be. It needs to be said that the higher patches tend to be the ones that survive longest. Generally speaking, if the temperature at sea level is 8° Celsius, it'll be around 0° Celsius at the summit of Ben Nevis. Or, roughly two degrees lower for every 1,000 feet of altitude.

The final aspect that contributes towards a snow patch's durability and is worth discussing at length is a fascinating process called freeze/thaw. It works something like this. When snow falls and settles on the hills in winter, air will be present between the crystals when they land to a greater or lesser degree. Soft, puffy snow is often encountered when temperatures are extremely low and, if further falls of this type occur, then they will simply heap on top of the ones that already exist. While this may make for superb powder skiing and winter pursuits in general, it is of limited value to the snow-patch enthusiast. This is because this type of snow is unconsolidated, or to put it another way, it is fragile and prone to rapid thawing.

For snow to survive long into summer and beyond, a sharp rise in temperature soon after there has been a big deposit of winter snow is desirable. This will have the effect of vanquishing much of the air that is trapped between the individual crystals by the introduction of melt-water and/or rain that invariably accompanies an increase in tempera-ture. What this does is saturate the snowpack, making it heavy and dense. Then, when the air temperature falls to below zero once again (as it is bound to do in winter), the dense, wet pack freezes. This turns fragile snow into resilient snow. Skiers call this kind of snow 'bulletproof' or 'bombproof', and for good reason. When subsequent fresh winter falls cover this 'bulletproof' snow, the process repeats, often many times over the course of a season. The snow will be in a much better condition to resist the summer heat when it arrives as a result of the freeze/thaw process. Consider this example: anyone who has ever tried making a snowball from crystals that have fallen in very cold temperatures will find that they won't bind together very well. No matter how hard you try to pack the snow between your hands, the snowball disintegrates as soon as it's thrown. Make a snowball when the temperature's a bit milder, though, and you'll pack a fair bit of water in with it, binding it together and ensuring maximum effectiveness when throwing it.

Winter 2010 was a very good example of when freeze/thaw played a decisive role in snow endurance. That year, sub-zero temperatures on the Cairngorms persisted for weeks and weeks. Fresh falls of snow were encountered regularly, borne on northerly winds. Soft, puffy snow lay thick and uniform across great swathes of upland Britain. But, as quickly as it came, it vanished when the temperature warmed up. This was purely because the lying snow was unconsolidated. Between the falls of fresh snow there had been very few – if any – rises in temperature, so the pack remained fragile. Warm weather removed it rapidly and what might have been a vintage year for snow-patch enthusiasts turned into a poor one, with very few patches surviving the whole year.

What of the depths, though? Just how deep can snow get in Scotland? As alluded to previously, if the right conditions are met then accumulations that most would find almost unbelievable can build up. The very deepest that I've seen in this country exist, unsurprisingly, on Britain's tallest hill, Ben Nevis. It has certain attributes that confer on it advantages other hills cannot compete with. Firstly, and obviously, is its height of 4,411 feet. Where rain or sleet falls on a hill 1,000 feet lower, most of the precipitation that lands on Ben Nevis between November and April will be snow. Secondly, its proximity to the west coast of the country ensures that it gets both barrels of the Atlantic storms that can drop thousands of tonnes of water or snow every hour. Thirdly, the plateau of the hill allows for the phenomenon described earlier, whereby snow races off the summit and on to the cliffs and gullies below. And what gullies below! Dozens of them snake from the summit plateau, running sometimes for over a thousand feet. The biggest of them all, Observatory Gully, carves a path from just below the summit cairn all the way down to the floor of Coire Leis, 2,000 feet lower. During the very snowiest of years, depths of 100 feet would not be surprising in the heart of Observatory Gully. For context, 100 feet is the equivalent of seven double-decker buses stacked on top of one another. 1951 and 1994 are two years when accumulations may have reached this remarkable total.

* * *

What is it that I and others *do* when we get to a patch of snow in summer and autumn? What we are primarily concerned with is what size the patch is and, more importantly, what date it is likely to vanish, if at all. If the snow has already gone when we arrive, then an estimate is made as to when we think it disappeared. For the former, there is no hard and fast rule that can be applied, and nor is there a formula. Calculating how long a patch will last is based on its size at the time of the visit, what the weather does afterwards, and applying the human factor of judgment based on experience. Concerning the latter, it is unquestionably disappointing to trudge for miles and climb to 3,500 feet, only to find no snow. However, it is just as important to prove a negative as a positive, so no trip is wasted effort. Often the vegetation surrounding a recently deceased patch can give clues as to when it did finally disappear (flat and dormant grass, waterlogged and semi-frozen terrain). It is important to log the final melt date of a patch so long-term data can be built up. Recording the same one for, say, twenty years allows a broader perspective on any trends, should they exist.

In terms of actual fieldwork and planning which areas to visit, this is now very much a product of experience. Though it is always useful to keep an eye on how much snow falls over winter, my work generally doesn't start in earnest until April. One gets to know certain areas where snow vanishes relatively early (April or May), so attention can be focused there initially. It is possible every year to find snow in April on even modestly sized hills in northern England and southern Scotland. In exceptional years, such as 2013, relics of winter can be found in May on the relatively low hills of the Peak District. The chionophile must always be ready to go somewhere in the briefest of weather windows, or if an unlikely patch has been spotted in a curious location. More recently, social media has proved enormously helpful in this regard. There are now scores of people who actively keep tabs on certain hills, sending me pictures via Facebook or Twitter. Even people who post pictures of hill walks on internet forums (such as *www.walkhighlands.co.uk*) often unwittingly provide invaluable data when the summit photo they have taken shows a patch on some distant hill, over the shoulder of the grinning walker-cum-model. All this allows forward planning, with visits far less haphazard than they used to be. Even fifteen years ago, when I started assisting with Adam Watson's annual surveys, we'd go to a hill not even having a clear idea if we'd find any snow. This may have been part of the charm, but as one grows older, charm is gratefully sacrificed for accuracy.

In the last few years satellite imagery, too, has appeared and now plays a decisive role in deciding where to go. Providing it's clear, which is annoyingly rare in Scotland, the Sentinel satellites give a reasonably good perspective, and can identify individual patches of snow down to about ten metres in length. Without doubt, technology has made the art of snow-patch observation more precise. It is rare now that I go somewhere and am genuinely shocked that the snow has gone or is far larger than I'd anticipated.

Knowing where on a hill to look for the elusive relics of winter, however, is part of the fun. Most of the time this isn't as difficult as it might sound. Because of a hill's topography, patches virtually always form at the same place and take almost exactly the same shape. There are minor variations, of course, but year after year it never ceases to amaze me how uniform their outlines are. But it is not just deciding where and when to go that has been made easier by technology. In terms of the equipment required, no longer do I have to carry a 100-metre tape measure on the outside of my rucksack. These days laser tape measures, which weigh a fraction of the manual equivalent, do a better job. It is also nice not to have to fend off questions from people who wonder why on earth my rucksack is so light that I feel the need to strap a large tape measure to it. (And I will not miss the innumerable times that I weighed down my tape with a large rock at one end of an eighty-metre snow patch and walked across precarious ground using one hand for balance while holding on to the body of the tape measure in the other, only for the tape to ping off just as I was about to reach the other end of the patch.) Predictably, the ubiquitous mobile phone has removed the need for carrying a host of other heavy gear, including notepads, Dictaphones and, to a lesser degree, cameras – though I never go to the hill without a proper camera. As good as modern phones are, the depth of field and lens quality means they are far inferior to a proper camera.

Generally speaking, snow recedes northwards. Therefore, eyes are trained in spring on the Southern Uplands (all the high hills south of Scotland's Central Belt).[4] No hill in this range exceeds 3,000 feet, though, and this combined with their rounded summits and lack of proper cliffs means that winter's snows are quickly diminished in the spring. There is little shelter from sun, wind and rain on offer here. A rare exception to

---

4   I also write a paper each year for England and Wales.

this rule is found on the second-highest hill in the Southern Uplands, Broad Law. On its north-east face sits a rough, rocky escarpment called Polmood Craig. Snow here occasionally reaches proportions that can rival higher hills much farther north in the Highlands.

Once the Southern Uplands' last snows have expired, it is time to focus on the Highlands. This is when it becomes altogether more difficult. In the hills of southern Scotland, the same four or five vie with each other to see which can hold on to the last snow. North of the Highland Fault Line, all bets are off.[5] Eyes are needed absolutely everywhere, so trying to decide where to visit takes a good deal of research. Such is the complexity of the terrain, and such are the sheer *numbers* of locations that can hold snow, that there needs to be an element of circumspection. No person can hope to cover all areas, even if they were employed full-time to do so. It may be possible to do that later in the year, when the number of snow patches is fewer than fifty, but from April to August it is simply not realistic.

In an ideal world I would be in charge of four or five full-time observers who were on the payroll, each keeping tabs on his or her designated area. This would ensure that far more data could be collected and compiled. The data would, ideally, then be analysed by the best minds in the country, the better to build up a wider resource base. Short of a huge lottery win, the chances of this happening are almost zero, but I can dream.

---

5   The Highland Fault Line is a geological fault line stretching across Scotland in a north-east direction, separating the Highlands from the Lowlands.

three

The footsteps of others

If sport like this can on the mountains be,
Where Phoebus flames can never melt the snow;
Then let who list delight in vales below,
Sky-kissing mountains pleasure are for me.

– John Taylor[1]

… the upper and innermost part of the country, where the tops,
or summits of the hills are continually covered with snow, and perhaps
have been so for many ages, so that here if in any place of the world they
may justly add to the description of their country:
Vast wat'ry lakes, which spread below,
And mountains covered with eternal snow.

– Daniel Defoe[2]

---

1 Taylor, John (1618), *The pennyles pilgrimage, or the money-lesse perambulation, of John Taylor, alias the Kings Majesties water-poet*, London: Printed by Edw. Allde, at the charges of the author.
2 Defoe, Daniel (1726), *A tour thro' the whole island of Great Britain, divided into circuits or journeys [etc.]*, London: D. Browne [etc.].

Historical sources for Scottish snow observations, before the era of Victorian rigour, tend to be infuriating and tantalising in equal measure (if no less entertaining for all that). When I was engaged in research for a book I co-authored back in 2010, I used to frequent the reading rooms of the British Library in London. The number of instances where I audibly cursed a writer for their slipshod reportage are too many to remember. There I'd be, on the edge of my seat, reading a wonderful eighteenth-century account of a writer seeing snow on the tops or cliffs of some hill or other, only for them either to omit the day or – unforgivably – the month that they observed it.

The two opening quotations are both classic examples of this. In the first, written during the early seventeenth century, John Taylor wrote about 'Phoebus flames' being unable to melt the snow (i.e. the snow was perennial) that lay on the hills in the locality of Braemar – which is where he was hunting with Lord Erskine when he penned the short poem.[3] It is too easy to dismiss Taylor's flowery description of the snow being permanent as mere romantic embellishment. I'm not at all sure it was.

When Taylor wrote this verse, the country was well in the grip of the Little Ice Age. It is almost certain that many of the high hills around Braemar – the southern Cairngorms – would have carried far more snow in summer than they do now. Maybe he was merely reporting the hills as he saw them, or even recounting a story told to him by his host Lord Erskine, which prophesied the peer's tenure of land being lost if all the snow were to vanish. Echoes of this legend can be heard right to this day.

The major landowners in the Braemar area are now the Farquharsons of Invercauld. One of the highest hills on Farquharson's land is Beinn a' Bhuird. On it, facing south, sits a highly conspicuous snow patch, called for many years by indigenous residents 'the Laird's Tablecloth'. According to local legend, the Farquharson family will lose their lands of Invercauld if the snow ever vanishes. The most recent laird, Captain Alwyne A.C. Farquharson, said during a radio interview in 1982 that:

---

3   Phoebus is another name for the Greek deity Apollo.

so long as the Tablecloth … remains spread, the Farquharsons would hold their lands. And that's the reason why no Farquharson will ever admit that the Tablecloth is not spread. They might go so far as to say that the linen was dirty and needed a wash, but if you take the trouble to climb up and have a look you'll find it spread, summer just the same as winter.[4]

The last sentence is clearly tongue in cheek, because the Laird's Tablecloth almost always vanishes nowadays. But for such a legend to gain a foothold in the minds of Deeside locals there needed to be a basis in fact. So, if we take Taylor at his word, we could easily believe that he was not indulging in a bout of overexaggeration by asserting that 'Phoebus flames can never melt the snow', but, rather, he was reporting on how it was at that time.

Equally, in the second account, Defoe's story of hills' summits being 'continually covered with snow' seems to have a whiff of truth about it. The book in which he describes the countryside is not filled with hyperbole, so it is a reasonable inference that what he portrays is accurate. Having said that, it is unlikely that the summits *per se* would have been 'continually covered'. It is more plausible that it was the cliffs and hollows around them. As outlined in the last chapter, snow is far more attracted to these features than it is to summits.

* * *

There are different types of author that an interested researcher can draw on when he or she digs into the archives of historical snow literature. The first is the one exemplified by Taylor and Defoe: people who compose quite lovely verse but are next to useless at providing reliable dates. (In Taylor's case he didn't even give the *year* he travelled.) Another is the eighteenth- and nineteenth-century travel writer. These authors tend to be far more reliable in what they are reporting on. True to the spirit of the age, where adventure, shooting and having servants were highly fashionable, there is a relative abundance of these.

---

4   The interview was conducted by Adam Watson and passed to the author.

Many writers of that age ventured to the wilds of Scotland from their leafy, west-London villas in search of romantic and heroic characters. Often they'd return south disappointed, having failed to find what they were looking for. Several more complained about the weather and uncouth locals. (At the time many indigenous Highlanders lived in very poor accommodation, sharing their houses with not only a pig or a cow, but associated vermin.) Of the ones that toughed it out, however, their accounts can be fascinating, and not just for the references to snow. Usually there are wonderful cultural asides thrown in, which give a unique window into how everyday folk lived.

Among the most informative of the travel writers of this era was a quite remarkable woman called Mrs Sarah Murray. Having married into upper middle class London society, she was clearly not short of funds nor inclination. This is amply testified to by her ability to spend a start-ling total of five months travelling around the UK with two servants. Clearly very well educated, she provides vivid descriptions in her book of what she found. One of the best of these recounts a trip she made to Cairn Gorm in 1801. After taking a pony up to the summit of the hill on 6 September, she observed that:

> Those who mount that eminence [Cairn Gorm] should walk to the edge of the precipices hanging over the hollow towards Loch Auon [Avon] and to the snow house. The snow house is not far from the Cairn or heap of stones, on the highest part of Cairngorm, and is a hollow, in extent an acre or two. This hollow is filled with snow, and although it faces the south, it is never melted either by the sun or rain … in its bed are large stones standing high and thick, serving for supporters to the roof of snow, which seems to be in some degree petrified … I therefore, by bending my body, walked up the bed of [the] rivulet for three or four yards … but it was so intensely cold under snow and in water, that I was obliged quickly to return.[5]

---

5   Murray, Sarah (1803), *A Companion and Useful Guide to the Beauties of Scotland [etc.]*, London: printed for the author.

This superb snippet is of far more value to researchers than those of Taylor and Defoe. Not only for the detailed description of where it is, but for the accurate date and what was encountered when she visited. Reading it over 200 years later, several things jump out. The location of her trip is, almost certainly, a place that is now called Ciste Mhearad (Margaret's Coffin). It is no longer called 'the snow house', if it ever were called that. This sobriquet may have been a name invented by Murray herself. Ciste Mhearad is situated precisely where she portrays it, close to the summit of Cairn Gorm, the hill that gives its name to the whole range of the Cairngorms. She says it (the 'snow house') faces south, but it is more east than south. This is but a minor quibble. 'It is never melted by sun or rain' is clearly a reference to the snow being permanent. Murray wouldn't have known this, of course, as she was just a visitor. It may be assumed with certainty that she had a guide on this trip, one who may have relayed the information on the snow being permanent. Indeed, for her to venture there in the first place suggests that the local folk knew this location well, and its proclivity to hold snow all-year round. It is invisible from the Spey side of the Cairngorms, which would have been the largest population centre at the time. Clearly, this was a place unusual enough to warrant guides taking people to it on horseback.

The 'rivulet' she walked up unquestionably refers to a snow tunnel of the type commonly found in Scotland in summer and autumn, where stream courses gouge out lying snow, sometimes to a height exceeding fifteen feet. We may be certain that Murray's story is not confected because of the accuracy of her portrayal of this feature. She could not have described it so well without direct experience. For such a tunnel to form at this location in September nowadays would be unthinkable, and it is evident that far less snow is present here nowadays than there was then. Rather than being permanent, today it is only a very occasional survivor. Consequently, Murray's account is of incalculable worth to researchers of weather phenomena. But that was not all Murray had to say; remarkably, this was her second visit to the locality in the space of several years. Indeed, it is likely that her journey to 'the snow house' was inspired by her first visit, where, a few years previously, she noted:

the house [Rothiemurchus] is within sight of Cairngouram [Cairngorm] mountains, whose hollow cliffs are filled with never-melting snow. The cap of winter upon the crown of luxuriant smiling summer below, was a contrast I had never before beheld ... The sun, however, was sufficiently high to gild the mountains and the lovely scenes around Rothamurchus; and for many a mile my eyes were feasted by the white patched hollow sides of Cairngouram.[6]

The view she describes, near to where the current town of Aviemore is situated, is on an early summer's day almost unsurpassable for beauty from an urban environment in the UK. High native fir trees stud the landscape, and Braeriach's northern corries' cliffs grab the viewer's attention especially. They are normally flecked with snow until well into summer and are perfectly described by Murray. It is little wonder she chose to return a few years thereafter.

Another traveller of this period was a redoubtable and eccentric character by the name of Colonel Thomas Thornton. In an extraordinary essay from 1804, *A sporting tour through the northern parts of England and great part of the Highlands of Scotland*, which chronicles him, frankly, rampaging around the Highlands on a prolonged shooting frenzy, Thornton depicts his travels from previous years. One such trip recounted an unconventional journey up on to Sgòr Gaoith, a high hill in the west of the Cairngorms. The book itself is laced with self-importance and bravado (and a *lot* of shooting), as befits an eighteenth-century former army officer. As a result of this bravado, one must exercise an element of circumspection when picking out what is factual. The account in question, though, would appear to be broadly reliable, based on the landscape that he describes. In 1786, Thornton and his party travelled on 6 August from Ruthven near Kingussie to cross the River Feshie and climb to the ridge of Sgòr Gaoith above Loch Einich. In the morning they agreed to make 'one of the gullies of snow' (i.e. there was more than one gully with snow in it) their guide. This would have been a snow

---

6   Ibid.

patch on the Glen Feshie side of the ridge. The party proceeded on a very hot day, when Thornton recorded 29° Celsius in the glen, and at twelve o'clock they arrived at the first snow. On reaching the high ridge, they were:

> depositing our champaign, lime, shrub, porter, etc. in one of the large snow-drifts, beneath an arch, from which ran a charming spring.[7]

This is doubtless Fuaran Diotach or the 'dinner well' south of Sgòr Gaoith. Again, it is revealing that he wrote 'one of' the large snowdrifts, plural. He described the higher mountain above them (Braeriach) as 'chequered with drifts of snow'.

Accounts like Thornton's and Murray's are interesting because of the *by the way* observations they make about snow. To visit these old drifts is not their objective in the wider context of their tours, so it may be assumed that there is no compulsion for them to embellish the story. In August nowadays, the location where Thornton and his party deposited their champagne would have long before lost its snow. To see these patches, as he did, on the Glen Feshie (i.e. the west) side of Sgòr Gaoith is unheard of in this era. It is yet another piece of, albeit anecdotal, evidence that points towards the current epoch not getting as much snow in winter as during the Little Ice Age, and our climate being quite different. Another entry from Thornton's journal from 9 September 1786 is eye-opening:

> Began to shoot at eleven, but found it wonderfully cold, when, looking over towards the Cairngorms, I discovered a very sufficient reason for the chilliness I experienced; for the wind was not only north, but the heights were completely covered with hillocks of snow, which appeared, as well as I could judge from their distance, to be at least four feet thick: we had some sprinkling of snow before; but this, I confess, astonished me.[8]

---

7   Thornton, Thomas, *A sporting tour through the northern parts of England, and great part of the Highlands of Scotland* (1804), London: Vernor and Hood [etc.].

8   Ibid.

In living memory such a depth of snow falling in early September has happened only once, in 1976. It may well be the case that Thornton's experience was similarly rare, but it seems unlikely.

* * *

Some of the other snow-related anecdotes from the period of 1750–1900 also provide fascinating insights into the cultural practices of the day. Parish minister the Reverend Donald McGillivray wrote in 1835 about Ben Nevis in the highly informative *New Statistical Account*:

> The deep clefts on the north-east side of Ben Nevis are never without snow. For two seasons when ice failed, the snow gathered and condensed into ice in these clefts and was of great service to the salmon-curers. The country peasants with their small hardy horses carried it down in panniers on horseback.[9]

This very much understated comment is worthy of elucidation. For fishermen to even *consider* going to the cliffs of Ben Nevis to gather snow shows how important a part it played in the workings of the local economy. Anyone who has ever been to the snows on the north-east face of Ben Nevis knows what is involved. They are situated above testing terrain, characterised by greasy and unstable rock which is liable to give way at any time. On foot the snows are hard to reach, even today where a good footpath has been constructed between sea level and the refuge at 2,200 feet. In the early nineteenth century, taking ponies up rough ground to above 3,000 feet must have been a significant and risky undertaking, necessitating the best part of the day to get up there and back. As an aside, it would be fascinating to know just how many ponies they took up, and how much snow they brought back in their panniers. Clearly it was a trip worth doing, so one would imagine it was quite a few ponies.

One of my very favourite historical accounts of a colder Scottish climate was written by 'that Intelligent Knight Sir George Mackenzy' in 1675:

---

9   *The New Statistical Account of Scotland* (1834–1845), Committee.

There is another little Lake in Straglash at Glencannich on lands belonging to one Chissolm; the Lake lies in a bottom 'twixt the tops of a very high hill, so that the bottom itself is very high. This Lake never wants [i.e. lacks] Ice on it in the middle, even in the hottest summer, though it thaws near the edges; And this Ice is found on it, though the sun by reason of the reflexion from the hills in that country, is very hot, and Lakes lying as high in the neighbourhood have no such Phenomenon. 'Tis observable also, that about the borders of this Lake the Grass keeps a continual verdure, as if it were in a constant spring and feeds and fattens beasts more in a week than any other grass doth in a fortnight. The matter of fact I have fully examined in both these; but to hit the cause, requires a better Philosopher than I am.[10]

The 'little Lake' being referred to is almost certainly Loch Uaine ('green loch') between the two highest hills in the north-west Highlands, Càrn Eighe and Màm Sodhail. Sitting at an altitude of just under 3,000 feet, it is one of the highest bodies of water in that area. It held broken ice at the start of June during the 1900s, but only after very cold snowy winters and springs, notably in 1951 and 1977. No one currently alive has seen ice persist there till July.

As for the 'feeds and fattens beasts' comment, this adds independent weight to the credibility of his observations about ice. What he came across was the fact, now well-known from studies on arctic tundra and alpine land, that vegetation which has recently emerged from snow or ice quickly shows a flush of spring growth, even though the time may be late summer or autumn. On the Cairn Gorm plateau, for instance, studies have found that grazing sheep strongly favoured grass near the edges of long-lying snowbeds. Within three metres of the edge, soil was frozen or waterlogged, and plants dormant, whereas favoured plants at three to ten metres had started to grow. Plants further out than ten metres had grown for a few weeks and had passed the first growth flush, and sheep

---

10 Macfarlane, Walter (1907), *Geographical collections relating to Scotland made by Walter Macfarlane*, edited by A. Mitchell & J.T. Clark, Edinburgh: Scottish History Society.

avoided them. So, as snow receded, grazing sheep ate newly growing plants, rather like a long spring season. Sir George's 'constant spring' was an apt observant phrase more than three centuries ago.

* * *

Of the quirkier tales of historical demands that recur in Scotland, none is more interesting – or charming – than the monarch insisting on snow at midsummer. Indeed, in a tradition similar to the Farquharsons of Invercauld – who, it may be recalled, owned their land on the proviso that the Laird's Tablecloth was set – a number of other families' landed interests apparently depended on snow. The tradition appears to be extremely deeply rooted in Highland folklore, because it occurs widely. For instance, the Comyn (or Cumming) family of Lochaber was given the Lordships of Lochaber and Badenoch about 1229 and their 'right' to the Lochaber lands was as long as there was 'Snow on Ben Nevis, heather on Druim Fada, and ebb and flow of Loch Eil'.[11] This tradition endured for hundreds of years and was noted again in the early nineteenth century: 'It is said that Cameron of Glen Nevis holds his lands by the tenure of an unfailing snowball when demanded.'[12] The story was fleshed out in a book from 1986 on the history of Ben Nevis. It stated that:

> an old legend, shared with some other Scottish mountains, has it that should all the snow vanish from Ben Nevis then the land shall be forfeit. To prevent this, Cameron of Locheil, in a summer when it looked as if the last of the snow might melt away, is said to have sent locals up the mountain with straw, to protect the remaining snow patches.[13]

The Munros owned the east side of Ben Wyvis, near Inverness, and an account states that for centuries they held their land on condition that

---

11   MacCulloch, Donald B. (1971), *Derivation of the Name Lochaber*, Clan Cameron Archives.

12   MacCulloch, John (1824), *The Highlands and Western Isles of Scotland*, Edinburgh: Longman [etc.].

13   Crocket, Ken (1986), *Ben Nevis*, Glasgow: The Scottish Mountaineering Trust.

they 'supplied a bucket of snow at the Palace of Holyroodhouse on Midsummer Day to cool the King's wine'.[14] (How the Munros managed to get the snow from Inverness-shire to Edinburgh before it melted is not explained.) In the same book it was said that the Macintyres of Glen Noe in Argyll paid their rent, on Midsummer's Day, of a 'white-fatted calf, and a bucket of snow' taken from a high corrie of Ben Cruachan. Lastly, a book from 1951 gives a further three examples: 'Certain Highland families traditionally hold their land on the condition that they are able to supply the King with a bucket of snow whenever he should pass that way. The Camerons of Glen Nevis, the Grants of Rothiemurchus, and the Munros of Foulis are all supposed to hold their lands on this condition.'[15]

Though in the main anecdotal, these stories – for their independence of authorship – are unlikely all to be confected. Some may contain, to a greater or lesser degree, an element of exaggeration, but it seems a strand of truth weaves through the stitching of these legends. The stories are consistent with what seems to be an unarguable fact: that our winters are far less snowy than they used to be. Or, to summarise *Cool Britannia*, the 2010 book I co-authored with Adam Watson:

> Although much of the above historical evidence is anecdotal and qualitative, it appeared independently with many observers in different years and regions. Because of this, we infer that observations were unlikely to be atypical, such as being concentrated by chance in an unusually snowy year or area. Comparisons involving quantitative observations from recent decades add confidence to this inference. To sum up, it is reasonable to conclude that more snow lay on the hills of Britain during the 1700s to early 1900s than in decades since 1930. Accounts from the lowlands of Scotland and England fit with this.[16]

* * *

---

14   Gordon, Seton (1971), *Highland Summer*, Abingdon: Littlehampton Book Services Ltd.
15   Gordon, Seton (1951), *Highlands of Scotland*, London: R. Hale.
16   Watson, Adam, and Cameron, Iain (2010), *Cool Britannia*, Bath: Paragon Publishing.

As interesting as these older historical references are, and as much as I admire the fact that they bothered to chronicle them at all, later commentators compiled their records rather more assiduously. These are the people who were the true pioneers of the study that I and others are now engaged in. They were diligent and thorough, keeping superb records. The work that I, and people like me, do tries to follow very much in their footsteps.

Chief amongst these pioneers are two remarkable men, Seton Gordon and Gordon Manley. The former is well known in Scottish outdoor circles. Born in Aberdeen in 1886, he was a superb naturalist and writer. His inherent enthusiasm for wildlife and writing was no doubt improved upon by the private education he enjoyed, and his time at Oxford University. Seton Gordon knew many of the landed families in Scotland and moved in their circles. He had a wide variety of passions, one of them being long-lying snow patches. Luckily, he left a great store of highly valuable information, wonderfully written. He was active at a time when comparatively few people took to the hills, so his observations are of great worth.

Gordon Manley is perhaps less well known than Seton Gordon. However, his body of work on the study of long-lying snow is almost without equal. A native of the Isle of Man, Manley was born in 1902 and grew up in Blackburn, Lancashire. He read Geography at Cambridge University and set up many meteorological stations across the country. He was a founder member of the Association for the Study of Snow and Ice which established the National Snow Survey. His 'The Snowline in Britain' paper, which appeared in a Swedish journal in 1949, is a *tour de force* of its genre.[17] In it he not only discusses snow patches on Ben Nevis and elsewhere, but also how much of a drop in temperature would be required for glaciers to reform on some British hills. From his many books and scientific publications, as well as his field observations, I feel sure that Manley and I would have got on like the proverbial house on fire.

Seton Gordon seems to have been a more complicated character.

17 Manley, Gordon (1949), 'The Snowline in Britain', *Geografiska Annaler* Volume 31, 179–193.

Adam Watson told me much about him, as they met and corresponded frequently from the 1930s right up until Gordon's death in 1977. Despite his complexities, Gordon's writings on the general outdoors – and on snow – were in a league of their own. Nan Shepherd is often lauded as the best writer on the Cairngorms, but for me Gordon's prose captures the essence of these hills more satisfactorily. A snippet of his work, taken from a night-time sortie to a remote part of the Cairngorms, shows this well:

There was silence in the Lairig, save for the croaking of some ptarmigan and the rush of the March Burn. Exactly at midnight we reached the watershed and the Pools of Dee. For some time, the sky southward had foretold of the rising of the moon, and now, at midnight, the pale orb climbed above the shoulder of Ben Mac Dhui and shone into the dark pools that lay as though asleep ...

In early summer the snows of the Garrachory are as yet almost unbroken, and where the moon shone the snows were faintly tinged with gold ...

We reached the head of the corrie, where snow lies from one year's end to another, as the first pale flush of dawn appeared in the sky to the north-east. Behind the black cliffs to the west the cirrus clouds were tinged with lemon where the invisible moon shone, but as we climbed we at last saw the moon herself (dimming her lamp against the dawn) and sat among the confused waste of immense granite boulders to await daylight.

All around us in the half-light were great snowfields, pale grey and mysterious. Up the rocky slopes of Sgor an Lochain Uaine a herd of phantom deer seemed to strive upward, yet ever to remain in the same place. To the unaided eye it appeared that they swayed this way and that in confusion, but the glass showed the black army to be rocks projecting from the snowy surface. From the western sky two stars shone palely, their rays becoming momentarily more dim as the daylight increased.[18]

18 Gordon, Seton, 'The Garrachory by Night', *The Scotsman*, 3 July 1926.

After the 1950s, the number of people interested in observing long-lying snows increased. In part this may have been to do with increased leisure time and opportunity of travel, and no doubt greater disposable income some had during the post-war construction efforts. Because of this, and as the people engaged in it generally did so with a bit more precision, the corresponding records are more valuable. Some observations are better than others, however.

One well-known spotter of the era was a gentleman by the name P.C. Spink. He wrote a paper annually from 1965 to 1980 for the Royal Meteorological Society, chronicling the 'Scottish Snowbeds in Summer'.[19] I have a great deal of respect for Spink and his obvious enthusiasm. Unfortunately, his methods and – therefore – his conclusions, were rather dubious. He lived in Lincolnshire and came up to Scotland only infrequently. Sometimes his pronouncements of which patches of snow had survived the year were based on observations made in July or August. These are months too early to be considered reliable. He often over-estimated or simply missed patches because he wasn't observing from the right place. For example, in August 1964 from Ben Macdui he saw 'no signs of snow survivals on Ben Nevis' but most of the Ben Nevis patches that last till winter are invisible from Ben Macdui. Therefore, despite the relatively large body of work Spink produced, it would be sensible only for information on those sites that he visited at close range or in good visibility on the stated dates to be considered as authoritative.

Two observers from recent decades, John Pottie and David Duncan, were local to the high hills which hold the longest snows. Their homes afforded good views towards these hills, and because they were able to go to them frequently and at short notice, the quality of their records is generally superb. They both took meticulous notes and photographs, especially of the Cairngorms. Our knowledge of survivals from year to year is greatly enhanced by their diligence. For many years both were co-authors of the annual snow-patch paper that I now write. Pottie camped in the snow at 4,000 feet when he was in his seventies, just so he could

---

19 Spink, P.C. (annually from 1965 to 1980) 'Scottish Snowbeds in Summer', *Weather*.

get a sighting of an old patch before it was buried for the winter season. Such dedication is rare even in people less than half his age.

One man, though, has done more for snow-patch observations than any other. Though he has been mentioned already in the first chapter as the person I contacted all those years ago, Dr Adam Watson's work deserves a whole book on its own. I devote an entire chapter to some of it later in the book, especially his observations on snow. He was a visionary and all future work on this subject will continue along the tracks that he laid almost single-handedly.

* * *

That so many historical writers even bothered to mention seeing snow on the hills suggests that the British interest in it is deep. During the writing of our 2010 book, Adam and I came across scores and scores of historical references to snow when travellers were sojourning around the Highlands. Often the references were in passing and of little use as factual observations. Others, as we have seen, were superb.

It is perhaps our latitude and proximity to the Atlantic that fuelled our interest in snow during the past and continues to do so to this day. In many countries, such as Norway or Sweden or Canada, snow is more or less a banker. It comes regularly, so is expected. High hills in those countries carry glaciers, which mean that individual patches that persist are unlikely to attract much attention. (I remember when I went to Egypt in 2001, I woke up on the first day and said to the guide, 'What a lovely day.' He simply shrugged and said that every day was like that and no one talked about it.)

In the UK, however, snow is an occasional visitor, rare enough to be a surprise. It is also hard to predict. It might arrive at the end of October, even at relatively low levels, but it also might not be seen at all until late March, if even then. There is simply no rhyme nor reason to when it shows up. Maybe because of this we find it intriguing, and that is why it has attracted a level of interest from writers over the years out of proportion to its scale.

The final factor that ought to be considered is the slightly eccentric nature of Brits and British hobbies. There is definitely a touch of Noël Coward's 'Mad Dogs and Englishmen' about this pastime. Yes, I am Scottish, and so are many of the people who indulge in this hobby. But I recognise something of that mad-dog tendency that encompasses not just Englishmen but Scots, too. A story I am fond of retelling, which highlights this odd national trait, happened in 2001. In November of that year, twelve British tourists were arrested by Greek authorities after being accused of spying near an airport in Kalamata. The reason given to the Greek authorities by the tourists and the British Embassy was that they were on a plane-spotting holiday. As a pastime, plane-spotting is unheard of in Greece. Their excuse was met with contempt. The police were simply unwilling, or unable, to believe that a group of twelve people would fly from Britain to Greece just to stand at the end of a runway with binoculars and notepads. Surely nobody would be so, frankly, weird. The whole thing was put down to a massive cultural misunderstanding.

I am unlikely ever to be incarcerated for my research, but I do so empathise with my plane-loving countrymen who were jailed for their passion.

four

Braeriach:
place of the eternal snows

The bulk of Braeriach is immense. Apart from Ben Nevis, its only possible rival, the Cairngorms' second-highest hill has no British equal in terms of complexity or majesty. Its name, translated into English as 'the brindled upland', understates its character significantly. There may be other hills in Scotland that are, arguably, more photogenic upon first glance: An Teallach, Liathach, Suilven, Ladhar Bheinn, Bidean nam Bian. All these peaks, with their airy ridges and pointed summits which tear holes in the sky, take the breath away. From many angles Braeriach lacks the steep upwards projection that draws so many photographers to the West Coast in search of the perfect calendar shot. It has a languid, torpid majesty which contrasts starkly to those farther west, ones that seem to huddle close to their neighbours as though fearing their own solitude. Braeriach's various minor tops lean back and stretch, rather than stand upright like a phalanx of granite sentinels.

From no matter which angle one gazes upon the slopes of Braeriach, the eye can never settle. It is drawn here and there by a cliff or gully, lochan or snowbed. Its summit, some 4,251 feet above the sea, lies just a few yards from the shattered granite cliffs of Coire Bhrochain, which themselves fall away for over 1,000 feet. Depending on how one counts them, Braeriach has another twelve corries to go with Bhrochain.

No single British hill, not even the huge primeval edifice of the nearby Beinn a' Bhuird, can compare.

Each one of these thirteen glacial gouges in the hill possesses its own character. Though two may sit adjacent to each other, separated only by a wall of granite, they will be utterly different. Take, by way of example, the magnificent Coire an Lochain. Walk in April by the loch that sits in its bosom and you will be transported to within the Arctic Circle in early summer. In this corrie, winter does not give up easily. As spring's warmth heats the water of the loch, the ice sheet – which covers it for at least five months of the year – starts to reduce in mass. Large ice floes form as the sheet breaks up, and it takes only a little mental embellishment to imagine a hungry polar bear standing atop them, licking its lips as it waits for an unsuspecting seal to surface for air. The cliffs above the loch hold snow late into the year, easily visible from the town of Aviemore, miles to the north.

Coire an Lochain has two neighbours to the east: Coire Ruadh and Coire Beanaidh. Though both, maybe, lack the grandeur and beauty of Coire an Lochain, the streams that issue from them combine to form a watercourse, the Beannaidh Beag, that can be as fierce and beautiful a barrier to a snow-patch observer as any high cliff. In full spate, just as it enters the larger Am Beannaidh river, the ford across it is insuperable. I recall some years ago having to adjust my plans of a sortie to the long-lying snows of Braeriach on account of the foaming torrent being a one-way ticket to a certain early grave. Granite boulders weighing a third of a tonne had been gouged from the riverbed by the storm and tossed to the side. Had I gone there just a few days earlier I'd have been able to ford the river without bothering to remove my boots. Such are the vagaries of the Cairngorms' weather systems.

High above the Beannaidh Beag's ford, Braeriach's huge and open plateau stretches for miles. It is not so much a typical Scottish summit but, rather, an Arctic-like desert where only the hardiest plants can get a toehold and the doughtiest animals inhabit. Loose gravel, broken down by many thousands of years of weathering, lies haphazardly among the larger rocks that nature has not yet crushed. Skin-flailing winds are commonplace. But, for all its bare harshness, the plateau has its own beauty.

On a March day a few years ago, I zigzagged on skis 2,000 feet up a gully from Glen Eanaich on to the plateau, only to be met by thick and heavy cloud. Snow cover on the plateau ran complete, but flat light made the going difficult and hard to read. Care was required near the eastern rim, where multiple corries yawned and bit deeply into the hill. On terrain such as this many people have paid for misnavigation with their lives. When one cannot differentiate between ground and sky, walking over the edge of a cliff is not an unheard-of occurrence. Suddenly, and without warning, a strong wind blew and catapulted the thick mist clear, revealing that I was nearer the edge than I reckoned, though far enough away for it not to concern me. Lying to the south-east, across a vast opening and expanse of cliffs and cloud, loomed the unforgettable vision of Sgòr an Lochain Uaine and Cairn Toul, two 4,000-foot hills that adjoin the Braeriach massif. Spindrift whipped upwards from their summits like spiralling ghosts, twisting to the music of the gale. Smoke-like wispy cloud was thrown over the edges of cliffs at high speed. A memorable ski tour around the three high hills followed, with no other soul in sight. The ski descent back down to the glen floor, on deep and uniform spring snow, was the best I had experienced in my short skiing career.

Nearby, and lying buried under deep snow that same day, were the plateau's best-known curiosities: the Wells of Dee. These springs are unique in Britain. From an unknowable source, deep within the rock, various small spouts emerge within a few hundred yards of each other, then conjoin and begin their journey to the sea at Aberdeen, many miles distant. The immediate landscape around each individual spring is a sea of colour as the water issues from the ground. The primitive mosses and liverworts that nestle down in the small hollows depend on this unfailing supply for nourishment. Few other plants could survive at such an altitude and in such difficult conditions. It is the highest source of any watercourse in the British Isles, well above 4,000 feet.

After these individual strands of the infant Dee have fused together, the young – but already impressively sized – river then must find an exit from the plateau, which it does, suddenly, by tumbling as a waterfall over cliffs and into Garbh Choire Dhè – the rough corrie of the Dee.[1] This beautiful cataract, known as the Falls of Dee, is invisible in winter, buried under untold amounts of snow. Not until late spring does the water show itself, punching a hole through the sugary, receding April snow.

After it regains its form and exits Garbh Choire Dhè, the nascent Dee flows into the very heart of Braeriach's greatest corrie: An Garbh Choire – Seton Gordon's *Garrachory*. Its meaning in Gaelic is simple: 'the rough corrie'. Many such corries across the Highlands are called Garbh Choire, but only this one gets the article – *An* Garbh Choire. It is *the* rough corrie, not just *any* rough corrie.

The sheer scale of this place is difficult to describe satisfactorily. Even when a walker is in its proximity in clear weather – a relatively rare occurrence – the cliffs guarding each smaller, subsidiary corrie that run off it constrain the full panorama. The summit of Ben Macdui, on the other side of the huge glacial pass of the Lairig Ghru, allows the walker to grasp the full enormity of the Garrachory.[2]

At its south-western extremity lies the highest and most distant subordinate, Garbh Choire Mòr.[3] For every snow-patch devotee, this is our Mecca. It is the place in Scotland – and indeed the whole of Britain – where snow virtually always lies longest.[4] Every trip here in summer and autumn is a mini-adventure. One could almost say it is like a pilgrimage. It is the most isolated place in the Cairngorms, and because of this it is rarely visited, even today. It is fitting, then, that the most difficult place to get to in the Cairngorms is the one that harbours the longest-lasting snows. Only those who harbour real desire will reach it. This is the place where, as a teenager, I saw Muriel Gray on TV standing on a rocky spur looking down upon the snow.

Over the last few years the semi-perennial snow patches at Garbh

1 Garbh Choire Dhè is pronounced as 'garra-chor YAY'. The Ordnance Survey incorrectly styles it as *Garbh Choire Dhaidh*.
2 Lairig Ghru is pronounced as 'lahrig GROO'.
3 Pronounced as 'garra-chor MORE', with the emphasis on the third word.
4 Only in 2018 did any snow last longer than here. This was at Aonach Beag, near Ben Nevis.

Choire Mòr have gained a level of exposure they have not previously seen. But it would be wrong to say that they had hitherto been ignored or even unknown. Indeed, from the 1700s until the early 1900s game-keepers and stalkers of the area passed word down, quietly, from parent to child that the last vestige of snow was never known to disappear at Garbh Choire Mòr.

Other pioneering walkers mentioned it as well. In 1829, the writer John Hill Burton (biographer of David Hume, the well-known Scottish philosopher) walked on the summit of Ben Macdui in summer and observed the snow from across the Lairig Ghru. Describing the scene towards Garbh Choire Mòr he saw:

a long wall of precipice, extending several miles along the valley of the Dee. Even in the sunniest weather it is black as midnight, but in a few inequalities on its smooth surface, the snow lies perpetually.

This last word, 'perpetually', occurs sometimes in obscure outdoors literature when used to describe Garbh Choire Mòr's snows. Prior to 1933 the epithet 'perpetually' would have been justified. No man or woman for many generations would have seen this location (and by extension, Scotland) entirely bereft of snow. But this situation wasn't to last forever. The Little Ice Age, that long, cold period in Europe when snow fell harder and for longer, had breathed its last several decades back. So, one warm, sunny September afternoon at Garbh Choire Mòr in 1933, the final snow patch in Scotland could withstand the heat no longer and expired. This hardened piece of post-glacial history, this deformed ice, hundreds of years old, dissolved in the warm air of a Cairn-gorms summer with no one present to give a eulogy or toast its passing.

Nevertheless, it is pleasing, at least to me, to know that despite its quiet, unwitnessed departure, the earnest members of the Scottish Mountaineering Club did not let the snow's disappearance go un-heralded. After they had written a note to various estimable publications, the story gained national interest and was reproduced around Britain. Even the *Belfast Telegraph* of 2 January 1934 ran a piece:

In Scotland there has not been so hot a July since 1903, and a feature of this intensity of the 1933 summer heat was recorded in the 'Meteorological Magazine' which stated that the heat last summer caused all the Highland snowbeds to succumb to its thermal power. This is regarded as something unique, never having occurred in living memory before … [T]he Garbh Choire snowbed under Braeriach – the 'rock where the snowflake reposes' … vanished for the first time in memory, so that by October 1 Scotland was entirely without snow.[5]

So unusual an occurrence was this melted snow that none of the mountaineers or people who frequented the higher Scottish summits expected to see such a thing happen again. And for twenty-six years after 1933 they were to be correct. Normal service resumed, and the snowbeds took on their customary position. That is, up until 1959. By then, more people were in the hills and more eyes were watching. None were sharper than Sandy Tewnion's. By day he earned his living as head biology teacher at Dollar Academy, but during holidays and weekends he was a hill man. On one such day of free time, Tewnion walked across heather and rock to Garbh Choire Mòr to climb one of the corrie's many cliffs. Climbs all over the Cairngorms were being pioneered by an increasing number of people who had more leisure time and an inclination to explore. Tewnion was one such figure. Even at this early date in 1959, 22 July, he worried that the 'perpetual snows' were in trouble once more. Pictures taken by him on this date show his worry was well-founded. Just four patches remained, none of which were large. With another complete disappearance seemingly inevitable, a return visit was necessary just to confirm this. Therefore, on 13 September, Tewnion returned. Expecting the worst, he was not to be surprised. He walked to the lip of the hollow that held the last patch, only to find it empty. Quite probably the first person in hundreds of years to do so, he stood on the bare ground which had been exposed by the second melting this century. No record exists

---

5   *Belfast Telegraph*, 2 January 1934.

of anyone having done so in 1933. Perhaps he was the first person *ever* to do so. It is quite a thought.

As befits his scientific background, Tewnion took various photographs of what he found, in colour. These offer us a unique insight into what was contained in the hollows that had only been exposed to the elements for a matter of weeks in the last several hundred years at least. The spectacle was extraordinary. Where the very last patch had lain, a kind of primordial black residue coated the ground and rocks. To the untrained eye it looked as though an oil tanker had shed some of its load, such was the darkness of the substance. We shall never know what it contained, though it was likely to be many years' worth of decayed grass, mosses and ptarmigan dung that had found its way on to the snow and had become trapped. The melting snow had liberated it on to the boulders that had been covered for so long.

Notable too was the lack of any form of life that surrounded the snow-patch site. The cliff faces here retained their virgin-pink-granite hue. Not even the most antediluvian life form such as lichen or moss could or can establish itself here, so short is the growing season. As a result, the rocks of the cliff face remain blemish-free, as though they've been jet-washed.

Few people noticed all the snow melting this time round. Though it was noted, no letters were despatched to the newspapers. A few people jotted down in their diaries that it had gone, and quietly went about their business otherwise.

The new winter snow that filled the hollow in which Tewnion stood in September 1959 arrived just a few weeks later, in early October. There it lay, undisturbed, for the next thirty-seven years. Normal service had once again been resumed. During this period the two longest-lasting patches at Garbh Choire Mòr acquired names. Not official names as such, but ones bestowed on them by people who were apt to visit these places. It wasn't due to a desire to anthropomorphise the snow, because such things would have been an anathema to the sober members of the Scottish Mountaineering Club. It was a way, rather, of differentiating between things that would otherwise necessitate a convoluted identi-fication system. The most durable patch of all was given the title of

'Sphinx', after the rock climb directly above it. First completed in 1924, this route gained its name years later, in 1942, by W. Thomson Hendry. Apparently, some of the rocks resembled that most ancient of sandstone carvings in Egypt. Today the Sphinx is the best-known snow patch in Britain.

The second longest-lasting, just fifty metres away from the Sphinx, took its name from the gully that starts directly above it. Tall pinnacles of rock soar from the gully's northern wall, so it seemed only fitting to nod to this prominent feature. By extension, the patch took the same name: 'Pinnacles'. Although separated from its near neighbour by no more than a stone's throw, Pinnacles survives much less often. The fact that two patches of snow located in virtually the same place can have such different outcomes is an oddity.

The Sphinx and Pinnacles patches' longevity is very much predicated on the direction of the prevailing winter winds. Moisture-laden Atlantic storms can blow massive quantities of snow into Garbh Choire Mòr, and Pinnacles Gully acts as a conduit, funnelling the snow down through cliffs until it reaches the hollows that the two patches occupy. We can only speculate as to the depth of the snow that can accumulate here in exceptional years. 'Mind-boggling' would cover it, though sixty to seventy feet if one prefers more precise measurements. In such snowy winters the survival of the Sphinx and Pinnacles patches through the entire year is almost assured, even from a very early date in spring.

From 1959, when all the snow melted for the second time, the run of winters thereafter was generally snowy. 1963, 1967, 1975, 1977, 1979, 1983, 1986 and 1994 produced lots of it. Others such as 1972, 1987, 1990, 1991 and 1993 had only slightly less.

However, in 1996 a seemingly seminal event occurred. The preceding winter of 1995–96 was not vintage for snowfall at all. Much of what tumbled from the sky did so on southeasterly winds. This type of weather system may benefit some ski areas, but it is much less useful at filling up the east- and north-east-facing hollows that hold the most durable snow. As a result of the lack of southwesterly storms, the relatively shallow snow patches melted, and quickly. On 1 October 1996 the Sphinx and

Pinnacles hollows were visited, and the latter had gone, with the former critically small. It vanished utterly a week or so later. The reason this event appears to be seminal is because of what came next.

It would have been reasonable in mid-October 1996 to conclude that, given it had occurred only three times since the 1700s, another complete vanishing of all snow in Scotland would not happen for many years to come. After all, the most recent disappearance before 1996 was thirty-seven years previously, and before that twenty-six years. But this was to be quite wrong. Thirty-seven years had elapsed between 1959 and 1996, but we had to wait a mere seven years for the next one. Not only that, 2003 saw the Sphinx expire by the almost unimaginably early date of 23 August. To be fair, the summer of that year was as good as any I can remember. Months of warm weather and bright sunshine dealt a death blow to the meagre depths that had accumulated in a poor and mild winter. Just seven years now separated the last two vanishings. Could it get worse?

The next two years after 2003 were touch and go, too. In 2004 only the Sphinx survived the year in Scotland, and in 2005 just it and Pinnacles. The run of extremely poor years continued into 2006 when, yet again, all snow vanished. This run was unprecedented.

Alarm bells by now were starting to ring out. The once 'perpetual' snows of Garbh Choire Mòr had succumbed to the weather three times in just eleven years (1996, 2003 and 2006), having done so only twice in the previous 300 (1933 and 1959). The media were running more and more stories on what at the time was called 'global warming', although we now refer to this as 'climate change'. Climatologists suggested that worldwide temperatures were on the up, and that glaciers were in retreat. The same phenomenon was being put forward for Scotland's vanishing snows. Ski centres were in financial difficulty, and there seemed to be a correlation between each poor trading year and how much snow survived to winter. Both were in retreat.

But then, a reprieve. Garbh Choire Mòr's Sphinx and Pinnacles both rallied in 2007 and were big when visited in late October of that year. Several other patches of snow survived in Scotland for the first time in a few years. The story was repeated in 2008, with even larger remnants in place by the time they were buried by winter's new falls. Indeed, good quantities of snow persisted at Garbh Choire Mòr for the next few years, culminating in the outrageous volumes which could be seen in summer 2015. In that year, it was more to do with the fact that spring seems to have been entirely bypassed, rather than any exceptional snowfall during winter. In April and May there were no appreciable thaws, and winter refused to go. During the first week in June the high Cairngorms looked as though they were still in deepest midwinter, and Garbh Choire Mòr was a massive, snow-filled cauldron.

I recall walking there on 13 September 2015, scarcely able to take in what I was seeing. The cliffs of the corrie headwall had at their foot a massive wreath of snow that stretched unbroken for hundreds of metres. By this stage of the year the individual patches along the cliffs would normally have formed, but on that day they were still conjoined, and showed no inclination in splitting any time soon. The quantity of snow that survived that year was more than any other since 1994 and deserved to stand among the very best years of the last hundred. Proof that you write off the Scottish winter only at your peril.

Or do you?

The dark days of the mid-1990s and 2000s had passed, and a new spirit of optimism had emerged on the back of a series of decent winters. Maybe this period had been an aberration. A mere blip. Unfortunately, it wasn't, because the years 2017 and 2018 would be unprecedented.

All told, I went to Garbh Choire Mòr five times in 2017, and saw it numerous times from afar. Each time I visited I could see the Sphinx and Pinnacles reduce. I hoped that heavy snow would come to their rescue and bury them, ensuring their persistence for at least another year. But, alas, no snow came. On my last trip there that year, on 30 September, the snow was so small and light I could lift it up. It felt odd to be holding this lunch-platter-sized piece of old, dirty snow that had fallen from the sky late in 2006, eleven years previously, the oldest relic of its type in Britain. For some divine amusement, the weather gods laughed at us as we stood by the dying patch, sending a few derisory flakes downward. For a minute it got quite heavy, but then it stopped just as it looked as though it might do something meaningful. So it was, then. Another year, another complete disappearance. The tally now stood at four since 1996.

The next year, 2018, brought a dose of *déjà vu* with it. Worryingly little snow fell in winter across the Cairngorms, and at the start of August the Sphinx and Pinnacles were looking sickly and feeble, as though diseased with some form of illness. Numerous visits were undertaken, and I managed to rouse enough interest in others so that the burden of getting there was shared. But it made no difference how many people went there, as the result remained the same. Each set of photographs showed the Sphinx and Pinnacles reducing in size. By mid-September I had written them off. Curiosity got the better of me a couple of weeks later, and almost a year to the day since I went in 2017, I ventured back on 29 September 2018.

Clambering up the stony ground I knew it wouldn't – couldn't – be there, but always there is a modicum of optimistic hope that, somehow, it has clung on. On that September day, as I peered over the hollow where the snow usually lies, I saw nothing. Not a lick of anything white. Here I was, almost sixty years after Sandy Tewnion, standing in exactly the same spot and seeing exactly the same thing – black matter deposited on bare rock. It then occurred to me that I had been one of only a very select few who'd stood on this hallowed piece of ground. It felt like a little comfort, but my overriding feeling was sadness. For the first time in recorded history the Sphinx had disappeared two years in succession. When I brought news of this disappearance to the attention of the wider scientific community, it attracted – just as in 1933 – the attention of news agencies not just in the UK but around the world. 'Scotland's glacier melts' was a common headline. Preposterous, of course, but the level of interest in the story had been quite remarkable. I spoke to radio stations in Canada and New Zealand, where both countries have enormous reserves of glacial ice. It seemed incredible to me. Why the interest? Probably it had to do with the heightened awareness of climate-change issues, where parochial bellwether incidents like these attracted attention out of all proportion to their scale. For a time in 2018, the Sphinx was the most famous patch of snow in the world.

* * *

Garbh Choire Mòr is a unique environment. In large part this is to do with its remote location. It is the most isolated corrie in the Cairngorms, and just getting to it, never mind the snows, is an undertaking in itself. Any ambitions to see it need to be worked around a calendar, with the best time of the year to visit being early September. The days have sufficient light in them to allow a fairly steady walk there and back in a day from wherever one wants to start. The weather in September can often be agreeable, but as ever with the Cairngorms a large dose of salt needs to be taken with any such recommendation.

I have, to date, been to the corrie in excess of twenty-five times, and by every conceivable route. I have tested all approaches, each slope, many gullies, looking always for the easiest way in and out. I settled long ago on there being no such thing as 'easy' when talking about getting to and from Garbh Choire Mòr. From every angle it is a long, hard and demanding place to attain. If venturing there from the plateau above, with the intention of dropping in, there is only one chink in its armour. It is an unnamed gully with enough of a slope as to permit the walker downwards access. If this gully were a degree or two steeper it would be impossible to descend. As it is, loose stone and moss abound, with each footstep a lottery. A forward fall would be catastrophic, ending almost certainly in a serious injury or, God forbid, worse. When in pairs or threes, one walker must go ahead and then stand well to the side while the others come down. The whole gully is a ticking time bomb of loose rock, ready to be dislodged by one misplaced boot. Even a medium-sized boulder set off just ten yards up could knock a lower walker flying. And this route is my preferred one! As with most gullies and steep slopes, though, exiting is a far easier and more pleasant experience than dropping in.

The safest way of getting into the heart of the corrie is via the Lairig Ghru. Only from that magnificent glacial trough is there an absence of any steep ground. Once off the indistinct track that runs through the Lairig Ghru, the walker will turn to their right if approaching from the north, into the huge opening of An Garbh Choire – the Garrachory.

At this point, Garbh Choire Mòr is visible in the distance, at the terminus. It appears laughably distant.

The walker is then obliged to negotiate ankle-breaking peat hags, cunningly disguised by thigh-height heather. Stout footwear is essential here to stop one's feet getting wet from the interminable bog that besets every person who ventures here. After two miles of this terrain are disposed of, there comes the not insignificant matter of fording the embryonic River Dee. In spate, crossing it is next to impossible. Sticking to the north bank is the safest option, but by then most walkers are so beguiled by the terrain, and so tempted by the sight of the An Garbh Choire refuge on the opposite bank, that crossing the stream becomes an imperative.

When standing at the refuge, with perhaps a flask of tea consumed and breath regained, one is right in the heart of the corrie. It is a sight of sights. Coire Bhrochain and Garbh Choire Dhè are prominent, with ancient granite cliffs seemingly encircling the walker. Garbh Choire Mòr seems no longer to be unobtainable, being only about an hour or so farther on.

When one enters Garbh Choire Mòr – the Garrachory's innermost sanctum, and the home of the snows, something *feels* different. Coming in via this route has an almost religious quality to it. Because all the while one has been walking towards the snow along a walled corridor of granite on each side, it has the feeling of walking through a massive cathedral, towards the altar. In this instance the altar is the uppermost cliff of Garbh Choire Mòr. The innermost holy shrine is where the snow is contained. The pilgrims come to it, and the process of getting there feels like a purification. The final push up to the foot of this cliff is the final penance before absolution.

Arriving at the snow has been described by some as anticlimactic. Because such effort has been expended in reaching the Sphinx or Pinnacles, the sight of a twenty-foot-long piece of dirty snow can apparently be underwhelming. I must confess that I have never felt that way. I savour each trip to this location, as there is always so much to see and hear, to experience. The snow is never quite the same, and a visit is often accompanied by the cruck-crucking of a raven above, or the comb-across-paper rasp of the ptarmigan. The mellifluous warbling of the snow bunting is frequently heard in the misty gloom of the summit cliffs, invisible to the flat-footed, transient visitor to their domain. Carcasses of unknown animals are dotted around, often killed either by falling from above or being caught in an avalanche. To some people this place can feel almost malevolent: a giant cul-de-sac of intimidating proportions, a cold dead end.

No visit to An Garbh Choire is underwhelming. It is a place that sears itself on the memory. Because of its isolation it will never be a place of popular attention but, because of this, it will remain forever the most special of places. No single location in Scotland that I have been to carries so much beauty within it, nor as much magic or shades of intrigue. It may no longer be the 'place of the perpetual snows', but it is a place which no one who has been to can forget as long as their memory holds:

> ... the Ridge of Braeriach, said, by an eminent calculator of altitudes, to have 2,000 feet of sheer precipice; that 2,000 feet of precipice is the object which it now almost aches your eyes to look upon – a flat black mass, streaked with snow, and sometimes intruded on by a cloud which divides the upper regions from the lower. It is probable that now, in mid-day, a hot sun gilds its black front, and mocks its streaks of snow, while a dead, unearthly silence pervades the mass.[6]

6   *North British Daily Mail*, 2 October 1848.

five

The back of beyond

Over the years, I have made innumerable journeys into the hills to chronicle patches of snow. I am fortunate to have seen a good deal of the country and ventured to many places that others would have little reason to go to. Many of these trips have been uneventful, save for the snow itself. Quite a few, though, have been a combination of nerve-wracking, funny, terrifying, and sometimes all these things. Some of the most eventful and memorable are described below.

* * *

29 August 2017, Garbh Choire Mòr

'Are you sure he understands what's involved in this?' I asked.

'Oh yes, he's a keen hillwalker and is used to big days out,' replied the editor of *The Great Outdoors* magazine.

'You told him about the rough terrain, and the long bike haul just to get to the start of the walk?' I added.

'Definitely. We're hiring a bike for him tomorrow and he'll meet you at Whitewell car park at 9.30 a.m. as agreed. He's looking forward to it.'

I hung up, satisfied that my companion the next day was sufficiently well briefed on the trip he wanted to go on.

It was about a week previously that I'd received an email from the editor of *The Great Outdoors* magazine asking if the well-known Irish comedian Ed Byrne could accompany me on a trip to see the by now famous snow patches of Garbh Choire Mòr. He found it fascinating that people had such a passion, rather than, say, collecting football shirts. He wrote a monthly column for the magazine and was always on the lookout for unusual outings to file copy on. He had most certainly come to the right place.

We met the next morning as planned at Whitewell car park, near Aviemore. The weather was overcast but dry. (Any day that starts dry is a blessing. It can be demoralising to start a long journey in full waterproofs.) After some pleasantries we set off on the first five and a half miles of the day, which were to be done on two wheels. We started up the well-constructed track towards Glen Einich. All was well as the first couple of miles went by, but then the road started to steepen. It was at this point that I had my first suspicions that we could be in for quite a testing day. We were obliged to dismount from the bikes and push them for several sections of the journey in. Conversation was necessarily brief and only then when information needed to be relayed, such was the effort expended. But, at last, after about an hour of huffing and puffing into the wind, we arrived at the spot where the bikes could be set aside for the journey back. I assured Ed that the return downhill was well worth the effort of the tiring inward cycle.

If he thought the next section would offer any respite, this was quickly dispelled by the sight of what lay ahead. To attain the summit plateau of Braeriach we would need to follow the long line of the Allt Buidheneach, a small stream that gallops down from about 3,500 feet. In winter and spring, the gully that this stream empties is known simply as 'the escalator' to ski tourers, as it offers reliable and straightforward access upwards at a constant (steep) angle. At times of the year when snow is not present, like that day, it's an unremittingly testing upwards trudge through high heather and slippery grass. It is, though, the quickest and most direct route up. I thought that Ed, being an experienced hillwalker, would make short work of it. Perhaps I was just used to the terrain, or had lots of experience of this particular gully, but in the end the upwards traverse proved to be rather trickier for Ed than I had envisaged. In fact, during the walk up he said, (half) jokingly, 'You know, Iain, I'm starting to go off you!'

At the top of the gully we paused for a bite to eat and caught our breath. It had been a lung-bursting ascent after all. Once refuelled and refreshed we made our way across the plateau. With easy ground in front of us we enjoyed decent progress and had time to engage in some chit-chat. The cool late-August day kept the temperature pleasant. Things were starting to look up. That is, until two miles later when we reached the drop-in point to Garbh Choire Mòr.

As described in the previous chapter, the gully that affords the quickest and safest ('safe' is a relative term when talking about Cairngorms gullies) access is a steep defile where what little solid ground that does exist is never more than a misjudged footstep away from disintegrating into a welter of exploding gravel. We negotiated the gully with great care, taking over twice as long as I usually would. Such care and time are absolutely necessary when more than one person is going up or down a gully. On the way down, the Sphinx and Pinnacles snow patches came into view. Despite my assurances that they weren't as far away as they looked, Ed affirmed to me afterwards that they were.

With the gully safely negotiated, all that was left to do was to walk to the snow and do the necessary: measure and photograph it for posterity. When we arrived at the two white remnants I could sense that Ed found the whole thing bordering on the absurd. Here we were, about four hours away from relative civilisation, down a stony chute at the foot of crumbling cliffs. Beside us were two melting patches of snow that were barely thirty feet long. Though there was a grudging admiration from him for the dedication that people like me displayed, it was just too esoteric for him to appreciate fully.

All that said, he was genuinely interested in the Sphinx and Pinnacles patches. That such things could exist in late August, at the tail end of a long summer, seemed to intrigue him. I got him to pose beside the Sphinx so that we could take some photographs for his piece.

'What way are we going out?' Ed asked me when we were set to leave.

'Back the way we came in,' I replied.

'Isn't there another way out?'

There are several ways in and out of Garbh Choire Mòr, but all of them involve varying degrees of risk. When Ed asked me the question, all these routes flashed through my head. I discounted Great Gully immediately. Though it was in close proximity to us, two people going up there would be asking for trouble. It is a ticking time bomb of loose rock which is held together – if at all – by moss and gravel. We would need to be glued together going upwards so as to avoid one of us sending a rock hurtling down and taking the other with it. Great Gully is too narrow for close-quarter climbing in places, so that was out. The other major gully in the corrie, between Sgòr an Lochain Uaine and Garbh Choire Mòr, I discounted for the same reasons. I'd been in and out of them both before, but I vowed the last time never to do so again. The third option, leaving via the Lairig Ghru, would necessitate such a diversion that it would mean us getting back in the dark, even if it were easily the safest way out across uncomplicated ground.

'The only real option,' I said at last, 'is Pinnacles Gully.' I pointed directly upwards to the improbable gouge in the cliffs above. 'I don't really recommend it, though.'

'Have you done it before?' Ed asked.

'Several times. It's doable, but as I said I don't recommend it as it's a bit scrambly.'

'Let's go for it,' said Ed, and so off we went. He seemed confident enough to attempt this exit, so I assumed he was no stranger to a bit of sport.

The first section of the gully was relatively straightforward. The terrain, steep but not overly so, offered only a little difficulty. There were good handholds and no real tricky sections. Before long, however, the gully kinked to the left and the going became sportier. We encountered a couple of cruxes, which with some jiggery-pokery we managed to negotiate. All the while, though, I had to be vigilant with loose rock. With me taking the lead, I needed to ensure that each upwards movement was on as good a standing as possible, lest I send one of Britain's best-known comedians hurtling to a sticky end hundreds of feet below.

The 'good standing' I searched for at each turn was not always easy to find. The ground in this soaking defile is a constantly changing melange of splintered rock, moss, grit, football-sized boulders and streaming water. Booby traps lie in constant wait, requiring only one misplaced hand or foot to rend a boulder asunder.

But, by good fortune and basic hill competence, we were nearing the top. The end was definitely in sight. However, there was one final chockstone to negotiate, and it was – to use the words of an acquaintance of mine who coined it when we did the same gully – 'a beezer'. It blocked the entire gully and had a slight overhang. Remembering it from the previous times I had ascended, I recalled what move to employ to circumvent it. Once I was up, Ed tried to do the same. Over and over he contrived to pull himself up, but each time the rock defeated him. He prodded and probed with both hands and feet, trying desperately to get some traction, but nothing worked.

'How on earth did you do that?' he asked, looking up at me from his precarious position.

'I'm not sure. I just sort of manufactured a handhold and used my legs to crawl up,' was my unhelpful response. Once again he tried, and once again he was unsuccessful.

There was no possible way of going back down, for to do so would be akin to signing our own death warrants. Going up a steep gully is one thing, but descending it is another entirely. Upwards momentum is relatively easy to control because gravity constantly checks it. Downwards momentum has no such check. If one were to lose any sort of footing or handhold, and in this place there were many only too happy to give way, the outcome was likely to be grave. It would probably involve moving downwards at great speed until halted by a large rock or some other stony feature. I need give no explanation as to the associated medical repercussions of such an event.

It was at that point I thought what an embarrassment it would be to have to call Mountain Rescue. Braemar Mountain Rescue follow me and Ed on Twitter. Having to explain to them why we came up this preposterous route would be the outdoors' equivalent of a trip to the headteacher's office. It was unthinkable. So, there was nothing else for it. I had to do the honourable thing.

'Grab my hand. I'll haul you up,' I said, thrusting my arm downwards. In order to secure myself I wedged my right foot into a good crack in the granite, and my left against the chockstone blocking Ed's escape. My knees and one of my elbows had a decent placement on the rock, allowing for good purchase. It must be said at this juncture that the method of extrication I am about to describe is absolutely discouraged. Hauling someone up by their hand, when a slip would mean the direst of outcomes, will be found in no guidebooks and nor will it be taught in mountain leader courses. It was born of absolute necessity.

'Are you sure?' replied Ed, looking slightly worried. I assured him that my stance was sound, and that it was either this or calling 999. Resigned, and perhaps anticipating unfavourable headlines, he grabbed my hand using the 'monkey grip' (that lock where fingers are curled round the fingers of the other hand, the safest method of ensuring no slippage). With an almighty pull I heaved him up until he was able to wedge a foot into a wide crack in the rock. This, in turn, allowed him to obtain sufficient purchase on another rock so as to permit upwards momentum. At last! We had negotiated the chockstone and we exited the gully. In the interests of decency, I shall omit the industrial language that both of us used when we emerged on to the plateau once again. It was a huge relief to finally get back on to some flat ground. Alas, for Ed the trial wasn't yet over. The downwards 'walk' was done in knee-high heather which often concealed boulders that were apt to twist the ankle. Many, many curses later we were back at the track and bikes.

The return to the cars on two wheels was, as it always is, a joy. A tailwind pushed us back down the hill, with very limited effort required. It took barely twenty-five minutes, and we parted after loading the bikes on to our respective cars.

I read Ed's subsequent write-up in *The Great Outdoors* magazine a month or so later. It was clear from his piece that he is unlikely ever to join the ranks of us who venture into the high hills to look for snow, but there was a generous testimonial. He also said that this was one of

.

the hardest days he'd ever had on the hill. Secretly I was rather pleased at this admission.

* * *

25 October 2015, Beinn Bhrotain

I had long lost count of the trips I'd made into the hills that year. Twenty? Thirty? Who knows?

Spring 2015's obdurate refusal to pass on the seasonal baton had left the high peaks and passes of the Cairngorms looking unusually snowy, and it was now early summer. The frightful cliffs that rise vertiginously from many of Braeriach's corries had upon them a full winter jacket, yet it was now June. At the same time, the pregnant-like summit bump of Ben Macdui rose white and unblemished above everything else within sight.

Summer passed, though. Much of the snow melted, of course, but relics of winter and spring remained across Am Monadh Ruadh – the Gaelic name of the Cairngorms – into September and even October. It was no exaggeration to say that the scene was a facsimile of one from many autumns past. John Taylor, King James VI's 'Water Poet', who journeyed to Braemar in 1610, would have recognised it:

> There I saw Mount Ben Avon, with a furred mist upon his snowy head instead of a nightcap: (for you must understand, that the oldest man alive never saw but the snow was on the top of [many] of those hills, both in summer, as well as in winter.)[1]

In a normal year – so far as any year in these parishes can be thus described – counting the number of patches of snow that endure to the tenth month would be an exercise able to be conducted over the course of a weekend. This, however, was not such a year.

For some time leading up to the start of October, my Saturdays and

---

1   Taylor, John (1618), *The pennyles pilgrimage, or the money-lesse perambulation, of John Taylor, alias the Kings Majesties water-poet*, London: Printed by Edw. Allde, at the charges of the author.

Sundays, as well as a good chunk of my annual leave entitlement from work, were given over to what sometimes felt like a never-ending circle of repeat visits to the inaccessible nooks and crannies of Scotland's highest tops, gathering data for the annual snow-patch paper.

24 October saw one such visit. That day's target was the immense bulk of Beinn Bhrotain, 'the hill of the mastiff'. Like so many of its Cairngorms brethren it has a whale-like appearance from distance. Traverse around its northern flank, however, and the rolling countenance is brutally sliced open by the shattered and splintered granite cliffs above Glen Geusachan. For me, no lover of heights or steep cliffs, my path was to be more benign.

Cycling from the spate-engorged Linn o' Dee at daybreak, I made for White Bridge. Though progress on two wheels would have been easier on the east side of the adolescent River Dee, its fording would have been impossible, given the quantity of rain that was now trying solemnly to get back to the sea. Wiser counsel suggested the western approach, crossing White Bridge over the Geldie – which swells the Dee to double its size – and on towards the foot of the mastiff hill.

Arriving at White Bridge I paused briefly and marked the clouds that were lifting. The mature orange and browns of the now dormant autumn vegetation were in stark contrast to the gleaming white of Ben Macdui and Braeriach, whose top 500 feet were resplendent in a castor-sugar covering of fresh snow. It was then, also, that I caught my first glimpse of Bhrotain's snowy white spot. It was exactly as reported: sitting in the upper reaches of Coire an t-Sneachda, 'the corrie of the snow'. The Gaels were, apparently, noting long-lasting snow locations hundreds of years ago.

Onwards.

But, alas, not for long. The normally placid Allt Iarnaidh, which drains but a small area of the southern slopes of Beinn Bhrotain, was a seething, foaming torrent of angry water. Luckily, just upstream, its course was constricted by a narrowing of the gully, and a simple hop over with the bike was sufficient to overcome what would have otherwise been an impassable barrier.

Ten minutes or so later, I was at the hardly discernible start of the path which led up the course of the Allt Garbh. This handsome brook reached upwards right into the heart of the hill, emanating directly from the snowy corrie that I was aiming at. For the next three miles or so it would be my noisy but unwavering companion.

The terrain was in no hurry to lend me height. A slow and steady upwards march through thick, tussocky grass and heather necessitated close alignment to the chatty burn that cascaded over the virgin granite outcrops, stripping anything unlucky enough to grow within its cold reach.

Eventually, some two hours after parking the bike, I reached the corrie. Last year's patch of snow sat in a large hollow just below the horizon's edge. Now, on easier terrain, I made for it, noticing a large inverted 'V' carved in its southern edge. I knew immediately what this meant: a tunnel. But this was surely no ordinary tunnel. It was one that had been months in the making. Water and wind had carved it out during the short summer and autumn. I hurried towards it across good ground. Two ptarmigan took to the air, their winter coats of brilliant white not yet developed.

Upon reaching the opening, I peered in excitedly. Snow tunnels are one of the great sights of the Highlands, and I had been at enough of them to know how wondrous they could be. Even so, seldom had I seen anything like this one in Scotland. I crouched motionless, then dumbfounded. A cold wind, far cooler than the ambient air temperature, passed down through the tunnel and across my face. A coldness that had been laid down some eleven months previously was blowing away with the wind.

Mighty pillars of white snow on each side of the small burn supported the most stunning of arches, which ran the full length of the seventy-five-foot-long patch. Translucent blues and whites, caused by thinning snow being pierced by the daylight, lent an otherworldly quality to what would normally be just one of a million anonymous watercourses. The tiny stream's babbling noise offered a lyrical soundtrack to the atmosphere.

In addition to the colours of the snow, mosses of luminous green hugged the waterside. Large, pink granite blocks also contributed to the kaleidoscope of colours, which were almost overwhelming. Today, the small stream was not anonymous: it was the most beautiful place in the whole of the Cairngorms, and probably the entire country. At least, to me it was.

I thanked whichever of the responsible gods that, for this brief window of perhaps a few days, I was fortunate enough to be there and see the snow before it collapsed on to the stream. Its blocks would dissolve and end up feeding the River Dee, as all precipitation on this hill must ultimately do. In my head I drew parallels between this scene and when blossom appears on the trees in Japan: an ephemeral window of staggering beauty which disappears as quickly as it comes into being. I drank it all in, unsure if I would ever come to this location again and witness such an exhibition of colour and beauty.

* * *

19–20 July 2019, Ben Nevis

All told, there were eight of us. Initially there had been seven, but a last-minute call went out on social media to recruit one more Sherpa. The amount of kit we'd need over the course of the next day and a half was significant. Luckily, the internet did its thing and a local guide, suitably remunerated, was roped in for the expedition. We all met up at Ben Nevis's north face car park, near Fort William, and introduced ourselves. It was good to finally match faces and names.

About two months previously I had been contacted by a film crew in Wales with a view to doing a piece for TV on the north face of 'the Ben'. The producer had seen some photographs online of the snow tunnels on Ben Nevis and thought that a visit there would make for superb and unusual visuals. She aimed to wrap these visuals around some commentary on my snow-patch work, and how that fed into the wider climate-change debate. The piece was to be broadcast on national TV.

The plan for the trip, previously hatched, was to make for the Charles Inglis Clark (CIC) memorial hut that is situated at the foot of Ben Nevis's towering cliffs. It is the only alpine hut in the UK and serves as a superb base in winter for climbers not wishing to make the six-mile round trip on top of an already long and exhausting day.

As we chatted in the car park on that Friday afternoon the weather was mild and cloudy, with a hint of moisture in the air. There was also very little wind. A mere breeze, in fact. Readers familiar with the climate of Scotland's west coast will be unsurprised to learn what came next: midges. And they were on a scale that I had seldom seen, even in the insect paradise that is Lochaber. Thick black clouds of them hung in the air so profusely that they appeared to darken the very light. No one was to be spared. They attacked us indiscriminately and remorselessly. We decided almost immediately that any chit-chat ought to be confined to the walk up to the hut, lest we be bitten half to death by these tiny, though indomitable, bloodsuckers.

Our group consisted of me, the presenter, a producer, a cameraman, a sound man, a Sherpa and two guides. On the walk up I started a conversation with the presenter, Lucy. She did weather forecasting on local and sometimes national TV. A serious though not overly restrictive disability meant that she had the use of only one hand. I did think prior to meeting up that this might be an issue, given where we were going and the nature of the terrain. But my fears were to be absolutely unfounded.

The path that led from the car park to the CIC hut was relatively recently constructed and in excellent repair. Good progress was made and by teatime we were safely ensconced within the solidly built granite walls of the hut. To prove to the group that I was more than just a one-dimensional snow enthusiast, I served all eight of us up a homemade vegetarian chilli that I had made the previous evening. (I surmised that because everyone woke up the following morning without complaining of poor health then it was a reasonable meal.)

There was a party of three people already at the CIC hut that Friday evening. A middle-aged couple were accompanied by the woman's mother, who was in her mid-eighties. To our astonishment they informed us that all three of them intended to climb Tower Ridge the next day. This ridge is the longest sustained climb in Britain and is a significant undertaking even for a fit climber. We wished them well and at about 11 p.m. we all hit the sack for as good a night's sleep as eleven people in a dormitory can have. Other than walking boots and waterproofs, earplugs are the single most important piece of kit that I pack in an overnight bag.

By the time we had awoken the next morning, the three climbers had already departed. The weather was not looking too promising, with low cloud and rain threatening. As we all left the hut after a hearty breakfast, we bumped into a local climber who had already walked up from the car park with the intention of also doing Tower Ridge with a group he was leading. We told him about the elderly lady who was with her daughter and son-in-law, and said that they might catch them up on the climb. We said our goodbyes and started on the day's walk.

The idea for this trip originated with the TV producer. She'd seen photographs that were taken by *Guardian* photographer Murdo Mac-Leod after he and I visited the same place back in 2016. *The Guardian* put the best one of that day in the centrefold of their national edition. The photo showed me standing underneath a huge cavern of snow that had been sculpted out by the wind. The shape of this cavern resembled the dome of a cathedral, and it made for a striking image. If, somehow, we could replicate that today, then the producer would be over the moon. I had warned her, though, that because of winter's poor snowfall the chances of getting anything so impressive were small.

Our destination was Ben Nevis's biggest, longest and most notorious gully: Observatory Gully. Though its entrance is only thirty minutes or so's walk from the CIC hut, it is not a place for the faint of heart. Like so many Scottish gullies it is a welter of loose rock and rough terrain. Ankle-breaking boulders and holes are everywhere, and one must be vigilant when walking into the upper reaches. It is also very long. Its jaws open at an altitude of about 2,800 feet and it runs all the way to the summit, 1,600 feet above. To cap it all, the gully is in a constant state of decay. Some of Scotland's highest cliffs lie on each side of the gully and are prone to shedding their skin from time to time. A few times I have been in there and heard invisible rocks scuttling downwards through the mist. There is little doubt that if one of these rocks were to collide with a person then it would be curtains for them. With eight of us up there the chances of being hit were increased dramatically.

Of all the places I visit regularly, Observatory Gully is the one where I have never felt entirely comfortable. The reasons listed above are part of it, but it goes deeper. Whether my impression of it is shaped by events that have occurred there in the last few years I couldn't say for sure, but I think there is at least some truth to that. The gully is full of ghosts, created by tragic events. Many people have in the past met their ends there, often in grizzly ways. Though admittedly the vast majority happen in winter, and to winter climbers, the place resonates with sadness. Even at the gully's entrance, before setting foot in it, one is reminded of one recent incident.

In February 2016, a young couple entered the ravine at its narrow opening point. (Though the gully is massive and wide for much of its length, the exact point of entry is squeezed like an hourglass's middle.) They were likely going up part of the gully to attain the start point of Tower Ridge. Before they got there, though, they were hit by a massive avalanche and swept hundreds of feet out of the gully. Their bodies were found some five weeks later.

The gully is also full of the hill's detritus. In summer, when the snow melts, one can often find climbing gear such as ropes, ice axes, ice screws, gloves, hats, you name it. Each time I pick one of these up, particularly an axe or screw, I wonder if its owner had to be carried off the hill. They are solemn reminders of what this place is capable of visiting on un-suspecting or unfortunate climbers.

Anyway, despite having two guides with us (one of whom, Heather, was Scotland's best known and respected), I was still a tad apprehensive about how individuals would cope once we got into the bowels of Observatory Gully.

At first it was all fine. We kept the pace steady, with me leading through the boulders. I knew well the faint path that wove upwards, having used it dozens of times over the years. On the way up I kept an eye out for our three roommates from the CIC hut but, despite looking at their most likely route, I saw nothing. The rest of our group were in fine fettle, all things considered. Unfortunately, the weather gods were not smiling on us, with cloud coming down and reducing the visibility. This had the added effect of making the gully feel quite claustrophobic. It was at this point I first sensed some uneasiness within the less experienced members of the group.

I watched with admiration as Lucy clambered up the gully. A native of Tottenham in London, she'd never done anything like this before (I didn't know this previously) but, although I could see that she was nervous, she displayed a mountaineer's attitude. Always smiling, she did exactly as instructed by the guides: no mean feat considering she only had the use of her left hand. The sound man and cameraman both plodded up as well, displaying an admirable stoicism in the face of the testing conditions. Of most concern to me was the producer. She was starting to struggle with the terrain as we ventured higher up the gully. The pace had slowed dramatically, and I began to wonder if she was going to make it up to the snows, which were still some 1,000 feet above us. However, she was in good hands with Heather, and I thought that a slow and steady approach would see us get there in the end.

Up we went again after a short break. The ground upon which we walked was unrelenting in its steepness and difficulty. Though in theory the amount of ascent we were doing wasn't too bad, it was inordinately tiring. For each three steps one takes up the gully, one needs to add another for all the backwards sliding that occurs on the loose gravel. Then, about ten minutes after the previous break, we heard a loud cracking noise from above. I knew exactly what this noise meant and so did Heather. It was a rockfall. Admittedly it sounded minor, but with eight people present no chances could be taken. Heather shouted to everyone to get close to a small cliff face that offered excellent protection from anything falling from above. In the gloom we couldn't see where this rock came from, nor where it was going. The noise of it crashing down the gully slowly dissipated. We looked at each other in total silence until we were happy it was safe to continue. The sense of nervousness amongst the group was now palpable.

It became obvious at that point the producer was at the end of her journey upwards. The rockfall was the last straw. Heather agreed to take her back down, as it had – understandably – all become a bit too difficult.

She remained in good spirits for all that and said, smiling through some tears, 'I can't believe you do this for a hobby, Iain.'

Shortly after the producer departed in Heather's good care, we reached our destination. As feared, the lower of the two patches that I thought might give our stunning visuals was in a sorry state. Mild weather over the last two weeks had decimated it to the point where it would be dangerous to enter one of the tunnels. Lumps of snow lay around the entrance to the main passageway, indicating that it was unstable and prone to collapse. Plan B was enacted. This involved a short climb to the longest-lasting patch of snow on Ben Nevis, at the head of the gully. It was in much better shape than the last one, and we got some footage. Not quite the stunning shots of previous years, but enough to justify the extra effort in getting there.

The crew were thankful to now be on the way down. For every step we took, I could almost feel the relief grow amongst them. This was magnified when we descended below the mist, and the views opened up. Thick mist has a peculiar way of distorting scale and distance. If a prolonged period of time is spent in it, then it can demoralise the unprepared walker. This is especially true if the walker is unfamiliar with the terrain. By the time we exited the gully proper, the weather was much improved, as were the spirits of the party.

We arrived back at the CIC hut to pack our belongings and make our way off the hill. Fortuitously we again bumped into the local climber who had been up Tower Ridge. He told us that his group had not encountered the three climbers who had been in the hut the previous evening. I thought this most odd, with only three explanations. Firstly, the eighty-something-year-old climber and her daughter and son-in-law had gone at such a pace that the climber hadn't caught them. The second was that they had looked at the weather and decided to do something else. The third reason was that they had become lost. As it transpired, it was the third reason. We heard the following day that they missed the turn-off for Tower Ridge and ended up going up some other route, got lost, and had to be airlifted off the hill. Our adventure was much tamer by comparison.

On the way down, I reflected on the previous twenty-four hours and vowed never again to take anyone up on to that part of Ben Nevis unless they were experienced hillwalkers. Though the crew had been admirably stoic in the face of some difficult conditions, the north-east face of the Ben is no place for the inexperienced.

* * *

25 November 2018, Aonach Beag

The following section is in note form, taken in part from a diary entry of a visit I made. It describes the thought process of the day: a series of situations and decisions taken in real time.

Glen Nevis. Stop car quickly. 9.30 a.m. Four shafts of sunlight splitting the clouds on the shoulder of Ben Nevis. Beautiful. Take photograph. Stand for five minutes. Three stags spring upwards on the crags to the east. Cloud regroups. Shafts of sunlight shrink and disappear. Back to gloom. Drive to Steall car park at head of glen. Get out of car. Cold. Pack extra fleece. Winter or summer boots? Hmm. Snow sitting on Mamores above 3,000 feet. Winter boots. They're heavy and cumbersome. Don't really want to wear them but OK then. Tell myself I'll regret it if I don't wear them. Don't forget tape measure. Lock car. Time to go. 10 a.m.

Take path and go around to Steall ruins, or straight up pathless and steep hill to Bealach Cumhann? Short-term pain for two-mile saving? Yes. I go up. Thick grass. Slippy. Urgh. Too steep. Why did I come this way?

Two golden eagles overhead. Rewarded. Stand motionless. Eagles loop above, calling to each other. Effortless soaring on northwest wind. Must get on but can't stop watching. They sense my hurry and disappear over Meall Cumhann. Must keep going. Many miles left.

Reach Bealach Cumhann, 'the narrow pass'. Apt name. Aonach Beag appears ahead. Wears a skirt of white. Not thick. Castor sugar on a cake. Its ancient gaze sees me and laughs across the corrie of the cattle. So far to go. 11 a.m.

Drop into Coire Giubhsachan, 'corry GOOSachan'. Another one. They're everywhere. All over the Highlands. Boggy. Very boggy. Stick to the side of the stream. Grassy and dry. Just jump over the tributaries. Better than walking on bog. Two miles till drier land. Feels longer. Sun comes out briefly. Cold. Think it's snowing on Aonach Beag high above me. Can't tell 'cos of cloud. Winter boots heavy.

Ground steepening. Far less boggy. Sun comes out briefly. Phenomenal scenery. Ben Nevis towering above me on left. Aonach Beag on right. An amphitheatre. The gods are in their seats. Feel like I'm being used as entertainment. Take many pictures. Nobody about. Why aren't more people here?

At last, 830 metres. Final push on to Aonach Mòr. Three-hundred-metre ascent in less than one kilometre. That means 1:3 gradient? Deep breath. Winter boots still heavy. Should've left them. Never mind. Up. Up. Up. Hit snowline. Thicker than it looked from below. Catch breath. Turn around to admire view. Very steep ground. Exposure kicks in again. Hold grass to steady myself. Views across to Carn Mòr Dearg exceptional. Those colours. Snow devil whips across the hill above me. Still no one about. Making good progress. Up.

Reach plateau of Aonach Mòr at last. Cloud lifts. Thick snow. Glad of winter boots. Take it all back. Two walkers on plateau half a mile away. Come from mountain gondola to the north. Cheats. Wait, maybe they walked. Perhaps not cheats. Walk to col of Aonach Beag. Will snow patch be there? Been a while since it's been seen. If so is last one in Scotland. Everything else melted ages ago. How has it lasted? So strange. Please be there.

Reach col. Look down to see if patch is there. Think it is. Can't be sure. Fresh snow obscuring it. Please be there. If gone will be second year in a row. Please be there.

Drop into An Cuil Choire. Steep. Very steep. Loose rock underfoot. Must be careful. No phone signal. No rescue possible if slip. Please be there. Cloud rising. Stags watching me across the corrie. 'What is he doing?' they say in deer language. It's there. I see it. It's big, I think. It's covered in new snow. It's survived. The only old snow in Scotland. Here it is. Tape measure out. Twenty-five metres long. Amazing. So glad it's there. What's the date again? 25 November. Oh yeah. Snow can't melt now. It's too late in year for it to melt.

Stop to have sandwiches. One-thousand-foot cliffs of Aonach Beag above. Utter silence. Unlimited beauty everywhere. Snow patch sits quietly. Sees the same view every day. Sit and work out how it managed to survive. Cliffs? Avalanches? Spindrift? All of those, probably. Never mind. Time to go.

Back the way I came. Hard trudge back on to plateau. Down off Aonach Mòr precarious in snow. Should've gone another way. Rewards worth it. Allt Coire Giubhsachan shining like fire with sun on it. A golden snake slithering for miles. Another eagle overhead. Circles me. Same one as earlier? Has to be. Flies off quickly.

Arrive back at Bealach Cumhann. Feet now wet. Winter boots heavier than ever. Descend. Catch glimpse of car. Not far. View up and down Glen Nevis staggering. Late-afternoon light illuminating silver birch wood opposite. Solitary rowan tree grows out of rock. Steall waterfall behind. Mesmeric. More photos. Too many photos. Tired now. Downwards. Only fifteen minutes to go.

Arrive at car. Last one here. Getting dark. Satisfied at day's work. Last snow in Scotland from last year measured and photographed. Glad.

Winter boots still heavy. Take them off. Ahh.

* * *

26 October 2019, Garbh Choire Mòr

On Saturday 26 October I set out from the Sugarbowl car park at the head of Glenmore, near the Cairngorm ski centre. Heavily laden with a twenty-kilogram pack, my intention was to make for the recently refurbished Garbh Coire refuge, in the shadow of Braeriach. After overnighting there I would, weather permitting, head for Garbh Choire Mòr to see if any old patches of snow had made it from last year. The fresh thick snow that was in evidence on certain aspects meant that if any of the patches from 2018 had persisted to this weekend then it was likely they'd be 'safe' for another year.

Not being in a particular hurry, and wanting to see the England versus New Zealand rugby semi-final, I delayed departure until early afternoon, walking at moderate pace towards the Lairig Ghru via the Chalamain Gap.

The weather was generally fair, but a few squally hail and sleet showers made the going unpleasant at times. Not until I was in the bosom of the Lairig did I have the wind at my heels, which made for an easy enough saunter.

On the way up the Lairig I fortuitously bumped into a couple who had just come from the Garbh Coire refuge, the overnight stop where I was headed. They told me that four Czech walkers were an hour ahead and were likely to get there well before me. Bugger. Given its location, and that it was a cold October day, I'd been sure I'd have it to myself. Not that there's anything wrong with sharing a refuge, of course, but this particular one is very small, and four would be more than cosy. A fifth occupant would have been too crowded for comfort, so I changed to Plan B and instead made for Corrour bothy. This would add a few miles to my journey, but never mind.

It is said that diligence is the mother of good fortune, and so it proved that evening. Upon entering Corrour bothy, some three and a half hours after leaving the car, I came across the Mountain Bothy Association's Neil Reid sitting in front of a roaring fire. He and fellow bothy-fixer-upper Andy had come up to do some maintenance, and we had a good feed and a few drams. A couple of Edinburgh walkers turned up as well, with one of them eschewing his windblown tent for a roll mat on the bothy floor. Wise man.

After a decent sleep and a hearty breakfast, I said my goodbyes and set off back up the path I had traversed the previous evening. About three miles or so battling into a nasty headwind, with freezing rain and sleet thrown in for good measure, I turned off the main path and headed straight for the yawning opening of the mighty Garbh Coire: the Garra-chory. The wade across bog and heather, and leaping the nascent River Dee, was not what you could reasonably call enjoyable. As much as I relish being in the wilds, trudging through boggy ground with a heavy pack is few people's idea of fun. As I drew near to the Garbh Coire refuge, which lay on my path, I could see coloured rucksack covers outside. The snow came on thick now, and the wind howled. The four Czech walkers clearly weren't in any hurry to leave, and I passed the refuge without stopping. Had I knocked on the door I suspect they'd have jumped out of their skins.

The walk up from the refuge to Garbh Choire Mòr, the innermost recess of the Cairngorms, was absolutely foul. Heavy squalls bit into my face, necessitating the use of ski goggles. In addition, the ground was slippery with freshly fallen wet snow, and the goat track that's normally visible in summer was indiscernible. I hopped across bogs and stones, tussocks and holes. Despite decent boots, my feet were getting wet. It was at this point I gave serious consideration to turning back. Looking up to where I was going all I could see was thick mist being blown about by squally snow showers. 'I must be mad,' I said to myself, more than once. But I had come too far to turn back now. I thought about the times I'd been in blizzards so fierce that I could barely even see my own feet. This was nowhere near as bad. Plus, I knew many people were waiting with bated breath to see if any snow patches had survived. Being an avowed mindless optimist, the brief clear(er) spell that arrived gave me hope that the weather would improve. Onwards.

After another half an hour or so's walking, the terrain unyielding in its difficulty, I came to the foot of the Garbh Choire Mòr cliffs. I was greeted upon my arrival by yet another squall. Heavy spindrift danced off the summit plateau and funnelled down the various gullies, giving the corrie an altogether Arctic edge. Despite it being the twenty-fifth time I had visited this location, I had never done so in what was basically full winter condition. It was transformed from its usual benign (if achingly beautiful) state into a place that had no particular liking for *Homo sapiens*.

After telling myself that I was here with a job to do – and an important one at that – I set about it. Depositing my rucksack behind a large stone, I took my small spade to the location that holds the UK's most permanent patch of snow: the Sphinx. Pristine white with freshly fallen snow, there was no discernible old firn, and I feared it was gone.[2] The mild weather of the last two weeks had surely finished it. I stood on the virgin snow and sunk my spade into where I thought any old *névé* would be located.

---

2   *Firn* is from the German for 'of last year', referring to snow that has survived from the previous year.

An odd clunk sounded as the metal hit something about six inches below the surface. I knew instantly that this couldn't be terra firma. I hurriedly cleared away the fluffy white new stuff, and with immense satisfaction saw the dirty remnants of 2018's Sphinx slumbering beneath its new winter duvet. With the weather worsening, I took a few pictures and made a crude measurement attempt (nine metres square). I shovelled as much of the cleared snow as I could back on to the Sphinx to leave it as I found it and returned to my rucksack. Job done. Now for the journey out.

The idea of retracing my steps was not filling me with joy, so I took the decision to go up on to the Braeriach plateau and down to the car via Sròn na Lairig. Though involving more climbing and a risky exit through a steep gully, I figured this would be preferable to the swamp-fest I had ascended by. This seemed like a good idea at first, especially as the snow wasn't very deep on the grassy slope I was going up. When the going got steeper I donned my crampons, giving extra purchase. This was fine at the start, but then the grass gave way to icy snow about thirty metres short of the exit. Looking above I could see quite a cornice had formed. With a twenty-kilogram pack on my back I started to question the wisdom of my exit strategy. I thought about descending, but this would have been riskier than going on. Kicking steps in the increasingly icy and steep gully, I inched up nervously. A slip would have had grave repercussions, and the prospect of being rescued in this seldom-trodden corrie seemed distant. But, eventually, calm and methodical kicking and stepping saw me emerge on to the plateau. My heart was beating quickly from the exertion and the adrenaline. But now there was another fresh challenge.

As I edged on to the plateau I was greeted by a howling gale and bitterly cold temperatures. Allied to that there was virtually no visibility. This wasn't a problem in itself, as I know the Braeriach plateau fairly well (you just follow the rim of the corrie, using it as a handrail). Walking into the wind was energy-sapping, particularly across a seemingly never-ending boulder field. No matter how many times I thought, 'The summit's just over the next rise', there was inevitably another one. When the summit eventually did come into view, I was no more than ten metres away from it, such was the poor visibility. From there it was a long, long descent down Sròn na Lairig and back through the Chalamain Gap and to the car. Fifteen hard miles.

With all that said, I drove back home with an incredible sense of satisfaction and achievement, knowing that at least one of Scotland's tiny relics from last winter had lived on to fight another year.

* * *

26 September 2015, Sgùrr na Lapaich

I was awoken at 5.30 a.m. by my alarm. It was going to be a long day, so an early start was essential. Rousing myself from the warmth of the sleeping bag, I peered out of the tent. It was dry, thankfully, and the clouds sufficiently few as to allow the first metallic blue streaks of the approaching sunrise to announce themselves. Two stags already roared on the hillside above me. It was, after all, the eve of the rutting season. I took a quick breakfast but decided not to pack the tent because of the amount of dew on the outer sheet. The forecast was dry so it could wait until I came back later in the day.

I set off with heavy legs along the northern shore of Loch Mullardoch. The sun was taking its time to rise, so I had to be careful picking my way through the bracken and stones. This was not a place to have a mishap. It is the very wildest of country, far from help and phone reception. The more sensible walkers, when picking off the four Munros on the north side of the loch, opt to hire the boatman who ferries people to the far end of Mullardoch. I had no such luxury. The walk to the base of An Riabhachan took over two and a half hours, and by the end had turned into a penance. I was happy, therefore, to leave the unceasing rock and bracken behind as I started to gain height.

This was the first time I had set foot in these fine hills. Loch Mullardoch and its northern neighbour Loch Monar occupy some of the most isolated countryside the Highlands have to offer. Long had I promised myself I would visit them, but always found a reason not to. This autumn there really was no option. So much snow still lay on the hills from the previous winter that a visit there was essential to gauge how much was likely to survive until the following winter. Such was the remoteness and roughness of the terrain that I could not rely on waiting for someone else to do it.

The walk up An Riabhachan's south-east shoulder, Sròn na Frìthe, was a steep but enjoyable one. The ground was dry and the view across the loch to the high hills opposite was glorious in the now first proper light of the day. Coming this way, where so few had ventured before, meant that there were no paths to make the ascent easier. This hardly mattered to me. My approach to climbing hills is, and has always been, 'If in doubt, keep going up.' (If one adopts this principle in general, the only logical end point is the summit cairn.)

Before long, the summit ridge had been attained and from there it was an easy saunter to the summit of this majestic hill. On the cliffs of its north-east corrie, Creagan Toll an Lochain, lay a huge wreath of snow. In excess of 100 metres long, it was certain to survive until next winter – for the first time in twenty years. I took some photographs and had a sandwich. The next hill to be climbed was Sgùrr na Lapaich. There were rumours of large patches still on this hill. I was keen to see them.

Duly climbed, I dropped off the summit of Sgùrr na Lapaich. It is the highest point on the long range of hills which rise above Loch Mullardoch's northern flank, and the most handsome. Its shapely ridges glide down gracefully and evenly on three sides. As I eased down the middle one, which runs due east, I became aware of a huge scar on the head-wall of the hill's unnamed south-east corrie. The farther down the shoulder towards the Bealach na Cloiche Duibhe – 'the mountain pass of the black stones' – I descended, the more of the corrie I could see.

'That can't be snow. No way,' I said to myself. 'God almighty! It is.' I stood rooted to the spot, my brain quite literally unable to process the vision that was now before me. I was at this point about 300 feet above the corrie floor, still high on the ridge above. Down on that corrie floor was a massive bank of what appeared at first glance to be snow. But it wasn't ordinary snow. It was almost entirely covered in a brown, muddy mixture of earth, rock and water. A small stream had gouged a channel all the way through it, though this itself was partially hidden by the sheer volume of material surrounding it. I had to get closer. Down I traversed through steep grass and rock for the next ten minutes, until at last I was as close to the snow as I dared get.

It was a scene of utter devastation. Nowhere was there any sign of ground level or vegetation. The floor's expanse in front of me looked like the beginnings of a large construction site. Mud and large rocks lay everywhere. Some of these rocks were huge, many tonnes in weight. Even more strange-looking than this, however, was what this debris sat on. At the edges of the rubble and wreckage, deep snow could quite clearly be seen. It was completely discoloured by the mud that was leaching into it. It looked, to all intents and purposes, like the snout of a glacier. It looked old. So, what had caused this ruin? It could be only one thing: a massive avalanche.

As mentioned previously in this chapter, 2015 was an unusually snowy year. Huge accumulations of snow had built up on some aspects all the way through April and even May. As a result, there were colossal, drooping cornices across the Highlands in June just waiting to fall. And fall they did, nowhere bigger than here. I was to find out later in September that, amazingly, there was a witness to this avalanche.

On 20 June 2015 a walker was out rambling on Sgùrr na Lapaich in thick mist. Soon after summiting she heard 'an almighty rumbling and crashing as the snow slope on the eastern side of the coire detached. We were above it but saw the debris later.' It is incredibly fortunate to get an exact fix on this date, not least because it shows that big avalanches can occur in Scotland even at midsummer: literally. Eight days later another walker passed this way and took photographs. What they show is remarkable. Thousands upon thousands of tonnes of snow, rock and earth had slid right from the top of the hill and smashed aside everything in their way.

As I stood that day on 26 September, three months after the event, the layers were still very much in evidence. The profile of the snow was fascinating. It was like a birthday cake with two very distinct layers. On the top, perhaps two feet or so thick, was a disfigured and mangled layer, interspersed with rock and mud. This was clearly the avalanche layer which had slid down. It was lying on top of the winter snow layer, which was far more uniform and complete. It was also far deeper, maybe five or six feet in places. Capping both these layers was the mud, earth and rock. Oddly, despite the chaotic scene, the top layer of earth and mud was giving the snow beneath a protective layer, hence preserving it. In other words, had the avalanche not happened then all snow here would long ago have melted. As it happens, this particular snow survived the year, where it had not done for many before.

The scene remains the most memorable I have ever encountered in Scotland, and the most unusual. To see so much material dislodged was slightly disconcerting, as was the volume of snow that it preserved for the year. It was yet another experience that, had it not been for my hobby, I would never have had.

Top The view from the author's parents' house across to Ben Lomond, the hill which was the catalyst for the author's passion.

Above Shafts of sunlight illuminate a glen near Aonach Beag.

Top Ecologist, mountaineer, polymath. The late Adam Watson conducting the annual snow-patch survey from Meall Odhar near Glenshee ski centre in July 2009. © *Allan Cameron.*

Above Britain's snowiest place, Garbh Choire Mòr on Braeriach, taken during the snowiest winter of the twentieth century, 1951. Exceptional depths here, perhaps up to seventy-five feet deep. © *Adam Watson.*

Top The place of the eternal snows: Braeriach's Garbh Choire Mòr in typical late summer condition.

Above The author at Britain's longest-lasting patch, the Sphinx, at Garbh Choire Mòr. The snow was to survive that year, but by the time it was buried it measured the size of a small dining table. © *Murdo MacLeod.*

The towering cliffs of Garbh Choire Mòr on 11 October 2019. The patch in the foreground, Pinnacles, is named after the granite tors above and melted completely just a few days later. It is Scotland's second longest-lasting patch of snow historically.

Top Britain's longest-lasting snow patch, the Sphinx, a day or so away from melting on 30 September 2017. The snow in the author's hands fell in late 2006 and was almost eleven years old.

Above The author at the site of the Sphinx. The ground here had been exposed to the elements for perhaps only a few months in the last 300 years when visited by the author in late September 2018. This was the second year in succession that it had melted.

A walker is dwarfed by the huge remains of winter snow still present in the Cairngorms in early September 2015.

Top Only certain colours of the light spectrum can penetrate snow, amply shown here in a magnificent snow tunnel near Drumochter in the Cairngorms National Park.

Above The author sits under a remarkable snow tunnel at Beinn Bhrotain in the Cairngorms. The range of colours displayed both on the snow and the ground were exceptional.

six

Dr Adam Watson

Exploring for yourself by your own free will, without formal courses,
training or detailed planning, is the best joy the hills can give.
– Dr Adam Watson[1]

With the honourable exception of my parents, no one has had a bigger influence on my thinking, and nobody at all with regard to my approach to the outdoors, than Adam Watson. In the period I knew him, from 2005 to his death in 2019, I learnt more about research, writing, snow and the general outdoors than I did in the previous thirty-odd years.

It is almost impossible to overstate the importance of Dr Adam Watson to the understanding of Scotland's upland ecology. Knowing where to begin when listing his achievements, awards, testimonials, publications, etc. is almost a futile task, and would need its own chapter.

Adam was known universally in Scottish scientific and outdoors circles. Whether it was birds, place names, flora, fauna, it didn't matter: if he said something, it automatically commanded respect, with people drawn to him because of his unparalleled knowledge. Widely acknowledged in

---

1   Watson, Adam (2011), *It's a Fine Day for the Hill*, Bath: Paragon Publishing.

his time, even amongst chiselled mountaineers, as *the* pre-eminent expert on the Cairngorms, he walked, skied and climbed on that range for the best part of seventy-five years. His wide array of studies and associated published output, not just on the Cairngorms but across Scotland and as far as Arctic Canada, almost beggar belief.

As if all this weren't enough, he has, single-handedly, done more on the systematic study of snow and snow patches in Scotland over the last fifty years than virtually all other researchers combined. Moreover, to be frank, were it not for him the chances of this book being written would be small. Remote, even.

There is room only in this chapter to touch on some of his output outside of snow research. It would be remiss, though, not to recall just a few of his general writings and the backstory of this most remarkable of men.

* * *

Adam Watson FRSE, FRSB, FINA, FRMS, FCEH, was born on 13 April 1930, in Turriff, Aberdeenshire.[2] He grew up in the town, where his father was a solicitor and had his own practice. Despite a serious childhood illness which kept him from school for some time, Adam excelled academically. He was awarded the *dux* prize in secondary school and went on to attend Aberdeen University, where he studied zoology and graduated with a first-class honours degree in 1952.[3]

As a young boy, Adam developed a fascination for a multitude of subjects. A keen birdwatcher from an early age, he would frequently note nesting sites and behaviours in his diary. So precise were his notes that when a respected local birdwatcher turned up at his father's practice to speak to an 'Adam Watson' he was amazed when the teenage Adam (not the forty-eight-year-old father, also called Adam) appeared and started chatting to him.

---

2   Fellow of the Royal Society of Edinburgh, Fellow of the Royal Society of Biology, Fellow of the Arctic Institute of North America, Fellow of the Royal Meteorological Society, Fellow of the Centre for Ecology and Hydrology.

3   The *dux* prize is a Scottish academic award, usually given to the highest-performing pupil in a school in a given year.

Adam's epiphany on snow, the subject which it is fair to say fascinated him more than any other in life, happened in 1937 when he was just seven years old. He later wrote:

> I looked from a window as pale columns of snow crossed a dark sky and big flakes tumbled. For hours I watched flakes melt on a slate roof and then gradually persist until the snow hid all the slates. In the garden I waded in the deep snow, feeling it cold in my fingers until it turned grey and melted. A foot of white became an inch of grey as I stood on it. Each snowflake had a variety, and each of the crystals comprising a flake. The airy mantle on branch, pavement and road absorbed sound. A strange hush pervaded the normally noisy town.[4]

This was his spark. From being fascinated by snow in general, his attention soon turned to observing snow patches in summer. By the age of nine he had already drawn the shapes of some of them in his diary and notepads. Mirroring my own interest as a nine-year-old, he was curious as to why they lasted so long.

It was also at this age when Adam picked up a book. A small, blue book. On a wet July day in 1939, during a family holiday on Deeside, he happened across a copy of Seton Gordon's seminal *The Cairngorm Hills of Scotland*. It literally changed the young Adam's life:

> Books are one of the pinnacles of human culture and achievement. Few things can have a more revolutionary effect on the attitudes and beliefs of young minds in a receptive mood. That was so for me, with Seton Gordon's books. Only perhaps once or twice in a lifetime may a brief event, such as a casual glance at a book, or a sudden union of two like minds, become a clear turning point which transforms the rest of one's life.[5]

---

4   Watson, Adam (2011), *It's a Fine Day for the Hill*, Bath: Paragon Publishing.
5   Gordon, Seton (1979), *The Immortal Isles*, Lewiston: Melven Press (from the foreword by Adam Watson).

Contained in Gordon's book were notes and photographs on a host of topics, and all of them struck a chord with the nine-year-old Adam. He studied each and every page with wide-eyed interest. One topic, however, caught the young boy's eye more any other: snow. Adam was excited, as I was years later, to learn that other people were looking at patches of snow on remote Scottish hills and recording them. A kindred spirit, he thought!

There was nothing else for it: he decided after reading the book that he *must* write to Seton Gordon to share some information with him. To use his own words, he had a 'burning desire to write to the man who changed my life'. So, on 10 October 1939 he did just that. Expecting no response, Adam was beyond excited when Gordon wrote back a week later. Thus commenced a correspondence that would last for the next thirty-eight years.

Along with snow, another of Adam's great passions was the ptarmigan. This wonderfully adapted member of the grouse family can, in the UK, only be found in the Highlands of Scotland, and even then generally above 3,000 feet. It is the only bird in the UK whose plumage changes in winter to brilliant white. As well as being the topic of Adam's degree dissertation, the ptarmigan was the focus of his PhD study and thesis, which was achieved in 1956. All through his life he would return to the ptarmigan, and he wrote on it extensively – becoming one of the world's foremost authorities. Though I never asked him about why he found the ptarmigan – specifically – so interesting, I suspect that part of his great love for the bird was that he envied it. Often he spoke of how we, humans, were but a fleeting visitor to the upland environment. Though we may enjoy a day on the highest tops, we are always obliged to return to our homes either by hunger, a need for shelter or to escape harsh weather. The ptarmigan need not return to low ground for any of these things. To sate its hunger, it eats plants which grow on the rough ground that covers the highest hills. For shelter, and to escape the worst of the weather, it can fashion a snow hole in a matter of seconds. From there it sees out the worst that the weather can throw at it, in relative comfort. The bird also, Adam stated more than once, taught him humility.

A diary entry from 10 April 1944, when he was just thirteen years old, recalls his first encounter with the bird. The quality of the writing and the eye for detail is astonishing for such a young boy:

> At the top of the shoulder I had lunch above a big snowdrift, and passed more drifts further up. The summit has many boulders and a cairn at 2,860 feet. Mr Seton Gordon told me that if I climbed Morven I might see ptarmigan at the top. Not having seen any, I determined to try and find one before stopping to look at the view. A moment after, when I had gone ten yards down the north side, a beautiful cock ptarmigan with a fine crimson comb suddenly rose from my feet and ran down the hill after a short distance. When I followed, it ran further, and then flew to alight at 2,500 feet. Returning to the cairn, I looked at the view, simply magnificent. Loch Muick lay in a hollow, and I saw Lochnagar ... sharply set against a slightly hazy sky ... A spotless snow blanket covered the rounded top of Ben Macdui, while through the telescope I saw the big black cairn standing out sharply against the surrounding white. Ben Avon looked as though whitewash had been emptied over it.[6]

His precociousness and supreme self-confidence were in evidence all through his adolescence and into early adulthood. Never, though, was this self-assurance worn with any hint of arrogance. It came out of a profound and deep knowledge of the upland environment and was attained and maintained by operating extensively within it from a very early age. His confidence manifested itself in some of the extraordinary adventures he undertook. None exemplify this better than his trip to Iceland in 1949, not long after his nineteenth birthday. Armed with only ten pounds in his pocket, he and two university friends 'hitched' on an Icelandic trawler that had put ashore at Aberdeen for a brief time. Gathering their gear after a botany exam at Aberdeen University, they

6   Watson, Adam (2012), *Some Days from a Hill Diary: Scotland, Iceland, Norway, 1943–50*, Bath: Paragon Publishing.

rushed to the harbour and boarded, not really knowing what to expect when they arrived. All that mattered was the adventure that awaited them. When they reached their destination, those adventures began, and in earnest. Skiing and climbing among the snowy peaks were the order of the day, with nature-spotting thrown in during the more relaxing periods. Being short of cash, they had to improvise with foodstuffs. The country was very expensive due to inflation, so they'd often flush wildfowl off the nest and fry or boil the eggs that the birds sat on. The three young men would have salted cod during most meals, which they'd bought a great store of for not very much money, due to its great abundance in Icelandic waters. Reading about the escapades of skiing off great snowy peaks one day, then climbing cliff faces the next, it is obvious the trio enjoyed a memorable time in the country. Adam enjoyed it so much he stayed on an extra week himself.

Before leaving Iceland, after a month's adventuring, Adam was put up in a police cell whilst waiting for a trawler to take him back to Scotland. There was no crime in the area, so the prison was empty. Adam was very grateful to the local policemen for the bed in this most unusual of digs.

The exploits of the three friends are chronicled in several of Adam's books, and they leave readers shaking their heads at the audaciousness of it all. Trying to imagine three nineteen-year-olds undertaking the same thing today is very difficult. The red tape alone would be enough to deter the most resolute and intrepid students.

* * *

By the time Adam was nineteen and had been to Iceland he was already a good friend of Tom Weir, that most well-known of Scottish outdoors' characters. They met near Braemar, at the home of Cairngorms stalwart Bob Scott, in December 1947. Tommy Weir, as he was known to his friends at the time, introduced Adam to ski mountaineering: a skill that would transform his enjoyment of the hills and his ability to get around them. Weir influenced Adam's attitudes to hillwalking and mountaineering more than any other person he'd ever known, Seton Gordon included. Often in Adam's writings he references the role Weir played in his understanding of what was possible when on the hill. He found Weir's enthusiasm most infectious. A fine example that he wrote about was from March 1949, when Adam and Weir were skiing in 'very poor conditions' on Braeriach, but with the hope that the weather would improve. Sure enough, it did:

> As we reached the summit of Braeriach, we suddenly saw blue sky and immediately the sun began to turn firm in the frosty air. Just as we began to see extraordinary views of snowy slopes and cliffs leaping out of clouds, ... Tommy shouted 'Great stuff!' and on hearing this I felt like leaping into the air with excitement.[7]

Tom Weir and Adam enjoyed many days out on the hill together. Adam's chapter on Weir in his book *It's a Fine Day for the Hill* goes into some of them in detail. It shows the enormous respect Adam had for Weir, as well as some of the latter's less attractive sides. As is often the case with great people, their lack of ability sometimes to compromise or see the other's point of view can be to their detriment. That certainly seems to be the case with Weir.

But of all the foreign expeditions Adam made in his teens and twenties, I got the distinct impression from him, in person and via his writings, that the one he held most dear was the trip to Baffin Island, Canada, in 1953. He spent four months there in summer as part of an international expedition by the Arctic Institute of North America. His role in the group was to study birds and mammals. It is little exaggeration to say that I lost count of the amount of times that Adam mentioned this trip. The accompanying book that he wrote about the journey, *A Zoologist on Baffin Island, 1953,* is, to my mind, one of the finest things that he wrote. None of the writings are more touching, or better evoke the sense of joy that the trip gave him, than the last paragraph of the journal:

---

7   Watson, Adam (2011), *It's a Fine Day for the Hill*, Bath: Paragon Publishing.

On the day I left … [I was gifted a] copy of the Swiss song-book *Poly Liederbuch*, which contained most of the songs that we sang on the expedition. By a nice coincidence it was published in 1953. Later I wrote on it 'From Hans Röthlisberger 1953'. During 1981, when Fritz called at my house on his way to give expert advice at a Public Inquiry, he and I sang *Aprite le porte*. Then he added his signature 'Franz Hans Schwarzenbach 1981'. I treasure my wee green book. Now as I open the pages fifty-eight years later and see the favourite songs of our Swiss team, my eyes fill with tears.[8]

When one reads the book that contains the accounts of the Baffin trip it is obvious to see the seriousness with which the team took the expedition. Belonging to an era where, academically, rigour was paramount, Adam believed passionately in testing his theories in the real world to see if they stood up. If they didn't, he would discard them. As he saw it, scientists who were uncritical in dismissing the possibility of an alternative hypothesis were doing no one any favours. He very much disliked it when academics were so precious about their work that they recoiled from proper scrutiny. For Adam, the truth was far more important than hurt feelings. How we need such thinking today.

Time and time again he spoke with passion on the value of fieldwork. He lived and breathed for getting out on the hill and gathering evidence. So much so, in fact, that by his own admission it was sometimes difficult to tell where work stopped and free time began. 'I just liked working,' he would say.

There are simply too many other stories and people for this book to go into when discussing Adam's past. Luckily he left a large body of work for us to enjoy and peruse. It is seldom far from reach on my bookshelves.

* * *

Adam's work on the subject of snow and snow patches is astonishing. From the age of eight he would note and record individual patches.

---

8   Watson, Adam (2011), *A Zoologist on Baffin Island, 1953*, Bath: Paragon Publishing.

He continued this throughout adolescence and adulthood, right up until his very last days. In the many thousands of emails, telephone calls and physical meetings I had with him from 2005 to 2019, he never tired of talking about them. He would often say to me on the phone that he had been out on a drive on Deeside with his friend Derek Pyper, and that they had pulled into some lay-by or other to observe the patches on Lochnagar, or Beinn a' Bhuird, or Brown Cow Hill, and so on. Every trip in good weather, to count hare, grouse, ptarmigan or whatever, he would also mark the patches in his little notepad. Not only did he do it in the name of research, but for the sheer love and joy of it.

The annual 1 July survey from Meall Odhar near Glenshee ski centre, which is discussed at length in the next chapter, was and is, for me, one of the highlights of the year. So important did I find doing it with Adam that in 2010 it got me into trouble.

On 25 June 2010, I married my fiancée in a church in Essex. I lived in Chelmsford – Essex's county town – at the time, and my fiancée was from that part of the world. During the preparations for the wedding, the subject of the honeymoon raised its head. In truth I was not looking forward to this conversation, even though I knew it was something I could not avoid. It is established tradition that newlyweds honeymoon very soon after a wedding, perhaps a couple of days thereafter. This tradition did not suit the timing of the 1 July survey, however. I did not want to miss it, so I decided that attack was the best form of defence. I boldly announced one evening that any honeymoon would be impossible before at least 3 July that year. The annual snow survey *must* take priority. The *only* way around this would be for us to honeymoon in Scotland. (I knew there was little to no chance of this happening, so it was a calculated risk with very long odds.) My fiancée looked surprised when I announced this, but I was resolute, and she accepted this short delay. In the intervening years I have told this story to various people, to varying levels of astonishment. I am normally a very easy-going person and don't rock boats any more than I need or have to. In this, though, she could see I was resolved, and she did not object in the end, seeing how much it meant to me. In the event we departed for Morocco well after survey day, much to Adam's amusement. (The story about how I absconded from the honeymoon for a day to go on a trip up Jbel Toubkal, North Africa's highest mountain, to look at Moroccan snow patches, is for another day.)

The 1 July survey was also the first time I met Adam, in 2008. I remember it well. We were to meet at Glenshee ski centre, whereupon the then centre manager, Graham McCabe, would drive us up in his Land Rover to the summit of Meall Odhar. It was a lovely morning as my car pulled into the car park that day. The air was clear and the cloud base high: perfect for snow-spotting. Soon after I arrived, a small blue/silver vehicle pulled into the car park and I instantly recognised Adam by his long, flowing white beard. I introduced myself to him and was met with a large, beaming smile. On that day I spent many hours in his company, learning much even in that time. I did not miss a subsequent July survey in all the time Adam was alive. In fact, in 2018, when he was unwell, I was charged with undertaking it myself. Determined, I went on to the Cairngorms immediately after doing the Glenshee survey, thereby completing an additional ten-mile round-trip to conclude the survey. In his reply to me the next day he called it a 'tour de force'. I was extremely pleased and proud at this.

The memories of each 1 July survey are precious to me. Though in reality we endured a lot of different weather over the years as we did it, in my mind all I can see is us sitting on the warm grass in brilliant sunshine, each with binoculars, counting every patch together. Latterly, either because he wasn't sure himself or he trusted my judgment an increasing amount, he would ask me how big I thought a particular patch was.

'Is that a four or a five, would you say, Iain?'

'Looks like a five to me, Adam.'

'Aye, I think that's right enough.'

The 'four' or 'five' he talked about was a numbering system he had devised for measuring patches from great distance. It is very difficult to judge whether something is ten metres or fifteen metres long when one is several miles away. It is much easier, though, to estimate if its size is between five and ten metres, or ten and twenty metres. Adam had a system for everything. I still use it now.

The 2018 survey was the first one that Adam had missed since he started doing it in 1974. It was a worry. For a number of years, we had seen that he was becoming increasingly frail. He walked with a stick constantly by then and was unsteady on his feet. Despite this physical frailty, however, his mind remained razor-sharp. At a gathering of friends some time before the 2018 survey he regaled the assembled crowd with stories on place names, folk tales and snow patches. He never missed a beat with his timing.

At his eighty-sixth birthday party, in Braemar, many of Adam's friends and acquaintances gathered to celebrate. It was quite a large group and some of the attendees did not know each other, or at least had never met. Adam did a round-the-table, introducing each person and giving a brief synopsis of who they were. There were some highly impressive people present, with some renowned authorities in their fields, and as some of them were announced I suddenly felt quite inadequate. With so many academics and experts, I lowered my head when he got to me. In his characteristically generous fashion, Adam introduced me as 'probably *the* expert on snow patches in Scotland'. I was taken aback. Immediately I replied, 'Oh, I don't think so!' or words similar. Without pausing he replied, 'Oh, I think so, Iain. Nobody has put more work in than you have.' Such generosity and sincerity one does not easily forget. What, I think, he was doing when he said this was figuratively passing on the baton. He was happy that I had earned my stripes and was ready to take on the task that he'd been doing for almost eighty years.

Though there are many people fascinated by snow and – more recently – snow patches, the amount of folk who are willing to put their boots on and go on to the hill to measure, record and photograph them is still relatively small. But there are innumerably more people now than when Adam started doing it in the 1940s. This is what makes his archive of information so impressive. Even today, with modern technology and the availability of information, our records cannot compete with his. He listed literally thousands of patches across the Highlands and was able to monitor a lot of these when out doing other fieldwork. I am still working through an archive of his files and am consistently amazed at the depth and sheer amount of information that is contained within. It seems bottomless.

* * *

On the morning of 23 January 2019 my phone rang. It was Adam's daughter, Jenny. I knew that Adam had been ill and was in hospital, as I had spoken with Jenny just a few days before. She had been concerned as he wasn't eating or drinking much. Immediately I knew what the news was going to be. I answered the call with an already shaking hand.

'Oh, Iain. Dad died this morning,' said Jenny, her voice cracking. My heart sank. I confess that at that point my mind went blank and I remember little of the conversation. It was a gut-wrenching piece of news to hear. I think I tried to say something comforting and consoling, but I probably didn't. Her grief was all too evident down the line.

I do not profess to have known Adam nearly as well as many others. To them his loss will be more keenly felt than for me. But, on that morning in January, it felt like a little piece of the soul of Scotland had slipped away. Hyperbole is all too easy when talking about great people, but I think in Adam's case every bit of it is deserved. His work, completed over seventy-five years of being among the birds, hills, lochs, rivers, people, trees and snow, has left an indelible imprint on the collective understanding of the country that he loved so dearly.

Adam taught me the value of discipline, rigour, patience and diligence when it came to record-keeping and the general gathering of information. As befitted his academic background and training, he also cautioned against overexaggeration – something that I must admit I have a natural preponderance towards. I can see his face in my mind if ever now I think about plumping up a story.

Adam's childlike wonder and awe for the works of Mother Nature were attributes that never left him. Whether he was watching snow clouds form over the North Sea, or inspecting a grouse walking across a mountain path to ward off humans from its chicks, he always had that glint in his eye. His deep, deep passion for the natural environment – especially his beloved Cairngorms – was infectious. Though my own passion for what I do is significant, it has been enhanced by knowing Adam. I have never met anyone like him, and I don't suppose I shall ever again. He was, truly, a one-off.

The very simplest and best writing I can think of to sum Adam up is that which is contained in Seton Gordon's first reply to Adam's letter to him in October 1939. Never has there been a finer, more prophetic or more apt summation of such an important man written in so few words:

> It is a fine thing for you to have a love of the hills, because on the hills you find yourself near grand and beautiful things. And, as you grow older, you will love them more and more.

I last saw Adam in December 2018, just a month before his passing. By then he was extremely frail and had been diagnosed with a terminal condition, but he managed a scone and a small bowl of soup at his favourite café, near his house on Deeside. With us was another of his friends, Mick Marquiss – himself a renowned and highly respected bird-watcher. We chatted for a couple of hours about digitising Adam's snow-patch observations. He was full of ideas and chat. He also had the same beaming smile and glint in his eye as when I met him all those years before at Glenshee. As I left that day I wondered if I would see him again. I am glad that our last meeting was a happy one.

* * *

In April 2019 I was asked to write an obituary for Adam. It appeared in the Mountaineering Scotland magazine shortly afterwards. It is tran-scribed in its entirety below. Reading it for the first time in a while, I smile as I remember some of the memories:

> A number of years ago, while we were conducting the annual July 1st snow survey at Glenshee ski centre, I asked Adam what his favourite bird call was. With his almost unparalleled experience of fauna in upland Scotland and similar arctic environments, I expected something a bit more exotic than the answer he gave. Perhaps the

snowy owl, I thought, or the metallic drum of the snipe? Maybe even the evocative call of the peewit as it swoops over the Aberdeenshire farmland?

'I like the ptarmigan,' was his reply.

No one who knew Adam will be surprised by this answer. I shouldn't have been. But it so was typical of the man. Understated and considered, his reply was –in many ways–a mirror image of his character.

I first contacted Adam in 2005, after having read a piece he had written about snow patches in Glen Coe. His email address was at the bottom of the text, and I thought he might be interested to learn of my diary entry for 1993, which ran contrary to the information he had presented. I was, of course, wary of doing so. He was a well-known Scottish scientist of immense standing and respect. Here was I, a nobody, who had left school at sixteen and became an electrician, without an academic bone in my body. However, as I came to understand, our interest in patches of snow that clung on to the shady recesses of Scotland's highest hills was something that united us in fascination, bemusement and wonder. It transpired that he, like me, became hooked by this esoteric area of study as a child. Anyway, I needn't have worried. In his effusive reply he thanked me for my note and asked if I had any observations for that year, and any others that I might have. I was thrilled! He was apparently completely unconcerned by my lack of academic background. 'Enthusiasm is the most important characteristic a person can have in the field of study, Iain,' he told me. It was a word I heard him use many times.

Over the course of a normal person's life you might meet a few people who you really rate. People, for whatever reason, whose knowledge seems almost limitless, or whose grasp of a subject, or subjects, is boundless. Years – decades – of study and immersion have generally been spent in pursuit of this knowledge. But that will only be part of it. When they speak to you about their subjects they generally do so with an accompanying passion and elucidation that few others can.

Folk like these are rare. Adam was one. In the field of outdoor studies on the high Scottish landscape he was virtually peerless, especially in his beloved Cairngorms. It is doubtful that any one person has known as much as he did on the flora, fauna, people, place names, and ecology of this range, and nor is there ever likely to be one in the future. He spent his formative years in the Cairngorms, and the list of people he met, befriended and spent time with there during the course of his life, whether walking, climbing or skiing, reads like a *Who's Who* of the great and the good of the Scottish outdoor world. Folk like Tom Weir, Bob Scott, Tom Patey, Mac Smith, Seton Gordon. The list goes on.

As much as spending time in Adam's company was informative, it was also good fun. He had a childlike quality of mischief in some of his dealings. Well do I remember sitting having lunch with him and his beloved wife, Jenny (who died in 2016), and being regaled with possibly the best story I had ever heard. (The tale involved a dispute between a member of the royal family and a local, with the latter telling the former to remove himself from his land, using very Anglo-Saxon language. But that's a story best not committed to print.) Another episode was during the winter of 2010 when he and I went to Arbuthnott kirk near Stonehaven to see the grave of Lewis Grassic Gibbon, author of the classic novel *Sunset Song*. The graveyard was like an ice rink. Literally. Water had poured down an adjacent embankment on to the grass and had frozen solid. We entered the church grounds and had to hold each other up, laughing as we slid over to where Grassic Gibbon was buried. Adam commented that it would be ironic if we were to slip and meet our ends in a graveyard.

One of Adam's biggest assets, though, was his generosity of spirit and his patience. This generosity was very evident in 2011, when he asked me to become lead author of the annual snow-patch paper that he had written every year since 1995, and which appeared in the Royal Meteorological Society's *Weather* journal. Perhaps aware that he wasn't going to be able to do it forever, he suggested I had a bash. I recall my first attempt. It was returned with so much red pen that I thought maybe I wasn't cut out for it. Adam wouldn't hear any of this and encouraged me with what words to use (and, crucially, what words I had used too often, or incorrectly). It makes me immensely sad now to think that I shall never get another opportunity to send Adam the draft for his comment.

Adam Watson was an inspiration. He will be mourned by his family, and by the Scottish scientific community. All lovers of the Scottish outdoors are in his debt. I shall miss him.

seven

Scottish snow and climate change
(with a eulogy to winter)

It is no exaggeration to say that the very longest-lasting patches of snow in Scotland exist right on the edges of being able to survive from year to year. Whereas in previous decades it could be said with confidence that year-round snow was the norm, it is increasingly the case that this mindset is an artefact of history. How dated the outlook of the Scottish Mountaineering Club members who wrote to *The Times* in the 1930s looks to us now, when they pronounced that none of them expected snow ever to vanish again from Garbh Choire Mòr. In the modern era any notion that our snow could be permanent is misplaced; the patches at Garbh Choire Mòr are walking a tightrope without a pole.

Even the language we use to describe these snows is at risk of becoming obsolete. Once upon a time the adjective that best described them was *perpetual*; later it became *semi-permanent*, and then *semi-perennial*. But this phrase, too, will not be able to withstand the pressure put on it by the snow's continuing disappearance. Does something that vanished entirely in 2017 and 2018, and only just made it through the year in 2019, deserve the epithet of *semi-perennial*? This may seem a semantic consideration of little consequence in the real world but, on a planet where epochal changes have – in the past – been measured in millennia, the relatively rapid realignment of descriptive wording should be of concern to us. Small, upward deviations in temperature will put further pressure on our, for now, *semi-perennial* patches of snow.

Because I am not a climatologist, nor even an academic, I have always resisted being drawn on whether the apparent increasing scarcity of snow and associated disappearance – and continuing disappearances in the future – of Scottish snow patches is or will be a direct result of so-called anthropogenic climate change. This is not because I don't have an opinion on it, but, rather, that I lack the qualifications to discuss it. I am primarily interested in whether the snow patches survive or not. Visiting these remnants and measuring them requires no academic aptitude. All an observer needs, apart from a general level of fitness, are willingness and the ability to use a tape measure and camera. The only real skill needed is when compiling the results of a season's observations, where compressing them into 2,500 words for the annual paper is often tricky. Accordingly, this chapter will deal with what has been published and what we *know* about our winters and snowfall over the last forty or so years.

* * *

Climate change

There is very little uncertainty that our climate is changing. Though the twenty-four-hour news cycle and increased visibility of world events has undoubtedly distorted our impressions, recent events from around the globe seem to confirm that the trend is towards a warming planet. Semi-apocalyptic reports are commonplace, with every new day seemingly bringing a fresh story to the news websites and print media: the Greenland ice sheet shedding uncountable millions of its tonnes of ice, Kilimanjaro's glaciers massively reduced since the 1960s, the Arctic ice cap recording its lowest levels on record. So it goes on. Story after story, headline after headline. Hardcore denialists shrug their shoulders and say we can't solely blame human activity for these worrying stories as we lack the ability and know-how to adequately interrogate the data. At the other end of the scale, an uncompromising lobby of activists blame such things as Australian forest fires and severe weather events on climate change, without rigorous scientific analysis to back their theories up. Where does the truth lie? As is often the case, it is likely to be somewhere in the middle. What I think undeniable, though, is that the climate *is* changing, and rather quickly. The academic consensus is also overwhelming.[1] According to NASA:

> 2016 and 2020 are tied for the warmest year since 1880, continuing a long-term trend of rising global temperatures. The ten warmest years in the 141-year record have occurred since 2005, with the seven most recent years being the warmest.[2]

It must also be said, though, by way of a counterargument, that climate has always changed, and sometimes rapidly. During the Medieval Warm Period from about AD 800 to 1300, Norse colonists grew crops and kept cattle in Greenland, grapes were much cultivated in southern England, and farmers in England and Scotland extended the altitudinal limit of ploughed cultivation on to large areas of exposed moorland. Next there came a cold climate, as we have discussed, termed the Little Ice Age, between 1300 and 1850 across Britain, the rest of Europe and most of the globe. This was followed by warming, especially since the early 1900s. It is perhaps the scale and speed of the recent increase in global temperatures that has caused alarm for many, notwithstanding the historical fluctuations. There are many examples of the upwards trend in temperature to pick from, but some notable ones are:

» In 2020, the average temperature across global land and ocean surfaces was 0.98° Celsius above the twentieth-century average of 13.9° Celsius, making it the second-warmest year on record.
» The seven warmest years between 1880 and 2020 have all occurred since 2014, and the ten warmest years have occurred since 2005.

---

1   *Quantifying the consensus on anthropogenic global warming in the scientific literature.* Available at: https://iopscience.iop.org/article/10.1088/1748-9326/8/2/024024/pdf [accessed 13 May 2021].
2   Climate Change: Vital Signs of the Planet (2021), 'Scientific Consensus: Earth's Climate is Warming'. Available at: https://climate.nasa.gov/scientific-consensus [accessed 13 May 2021].

» From 1900 to 1980 a new temperature record was set on average every 13.5 years. From 1981 to 2019, it increased to every three years.[3]

These figures are very persuasive. And, although it is extremely difficult – not to mention potentially unwise – to extrapolate one set of data and link it to local variations elsewhere on the planet, as a general trend it is relevant to introduce evidence of a warming planet into any discussions about a decrease in snow cover generally.

* * *

The snow's demise

Since 1974, an annual Scottish snow survey has been done on 1 July. This survey is conducted each year from the same place: Meall Odhar, a hill overlooking the Glenshee ski area in the North-East Highlands. Sitting just above 3,000 feet, its summit is one of the finest vantage points in Scotland for the number of hills one can see that still carry snow. The object of the study is to count *every single* extant patch of snow lying on hills east of the A9 road. It is an ambitious thing to undertake every year. Not only because of the sheer scale of the survey area, but also because of the weather. Often 1 July is a washout, so a day or two either side needs to be considered if the forecast is poor. Likewise, if the number of snow patches visible on this date is huge – as it so often has been – then extra eyes, and accompanying legs, are drafted in to assist. Each patch is recorded visually because this is the only practicable way of counting (measuring each patch would be impossible unless we could draw on literally hundreds of volunteers). When seen from the Meall Odhar viewpoint, the largest snow wreath is almost always on Ben Macdui, and is often as much as a kilometre long. Others can be less than five metres. If one splits the survey totals since 1974 into two sixteen-year chunks (i.e. the first sixteen years, 1974–1989, and the last sixteen years, 2004–2019) then one can see the results. They are sobering.

---

3   www.climate.gov/news-features/understanding-climate/climate-change-global-temperature [accessed 13 May 2021].

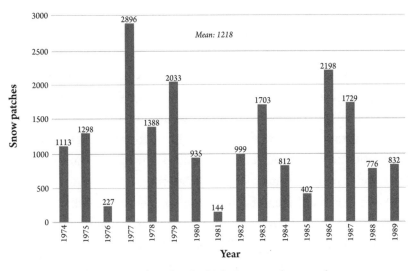

Figure 1: Total number of individual snow patches counted
in the North-East Highlands on 1 July (1974–1989)

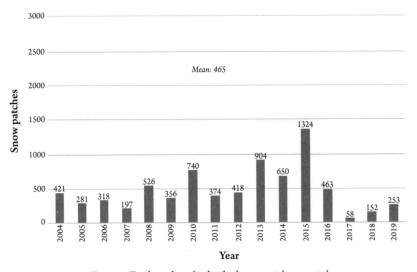

Figure 2: Total number of individual snow patches counted
in the North-East Highlands on 1 July (2004–2019)

The mean (average) total of patches has more than halved from 1,218 in the period 1974–1989 to 465 in 2004–2019. This is a quite remarkable reduction in a relatively short space of time. During the first sixteen-year range, totals of 1,000 or more were commonplace, and in 1977 getting on for triple that amount. In the most recent period only once did it exceed 1,000 (in 2015). If one plots a trendline on a graph of these totals then it is obvious to see which way it is going: down. It also seems to suggest a correlation with the NASA figures, which show the ten warmest years on record have been some of the most recent.

As we may recall from chapter 4, the number of times that all snow in Scotland vanished has increased markedly since 1996 (four times between 1996 and 2019, compared to just twice in the previous 200 years). This on its own ought to be a cause for concern. If we look slightly beyond this, though, and examine the times that all snow *nearly* disappeared between 1996 and 2019 then the picture becomes even more stark.

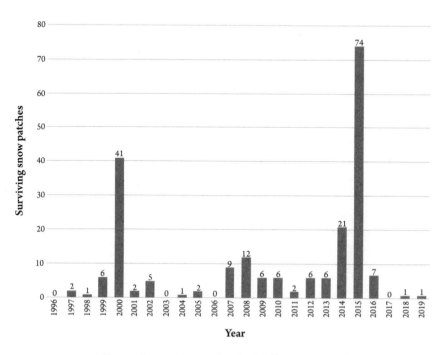

Figure 3: Snow patch survivals in Scotland between 1996 and 2019.
Data from annual Weather journal of the Royal Meteorological Society.

Figure 3 shows that, along with the four years when no patches survived (1996, 2003, 2006 and 2017), there were another eight years when only one or two did. Or, to put it another way, in half of the twenty-four years since 1996, only two or fewer patches persisted from one year to the next. The period of 2017–2019 saw the lowest combined three-yearly total (two patches) since records began, narrowly eclipsing the previous lows of 2003–2005 and 2004–2006 (three patches in each). Never since these snows have been studied have we seen such low numbers.

Interspersed throughout Figure 1 it will be seen that occasionally there is a large spike. Most obvious of all is 2015. In that year seventy-four drifts of old snow made it through the year to the following winter, which is undeniably a remarkable total. Many places that had not seen snow surviving for decades did so again. Perhaps years like 2015 are not such a surprise when we consider that climatologists tell us severe weather may become more common in the future, and that although the general trend may be a downwards one in terms of the amount of snow that falls, big-spike years will endure.

One recent study done on the subject of snow in the Cairngorms predicts that cover will decrease dramatically in the coming years. It states, rather ominously:

Our modelled future estimates indicate a potential for snow cover in the next decade to continue at a similar quantity to the recent past, with large inter-annual variability. However, from c.2030–2040 there is likely to be a substantial decline in the number of days of snow cover. By c.2050 the trend seen in the past may have continued to the extent that the number of days of snow cover is about half of the long-term observed average. However, variation from year to year, both observed and modelled, suggests the potential that snow cover in some future years may be comparable with past records. The long-term trend is towards greatly reduced snow cover with the possibility of some years of very little to no snow by 2080.[4]

4   Rivington, Mike et al. (2019), 'Snow Cover and Climate Change in the Cairngorms National Park: Summary Assessment', www.climatexchange.org.uk [accessed 13 May 2021].

'Very little to no snow by 2080' is a worrying consideration. Scarcely believable, even. How I wonder what those who walked on our hills in the eighteenth and nineteenth centuries would have thought, seeing huge amounts of it even in August, if they'd been told all of it would soon be consigned to history. For my part, the idea of a Scotland without long-lying patches is one which sees the hills diminished. Wreaths of snow in summer give the highest peaks a grace and splendour which elevates them above their modest heights. If their ermine is stripped away by the cruel heat of the sun and the rain, then it will see their beauty lessened. To witness the north face of Ben Nevis from Carn Mòr Dearg in July, with its multitude of gullies filled for hundreds of feet with white, pristine drifts of snow, is to see one of the greatest visual feasts that these islands can offer. How sad it will be for us all to be deprived of such scenes.

Stories from continental Europe, where glaciers are being covered by blankets of foil in order to reduce their melting, have surfaced in the last year or two. I have also heard many people in the UK suggest such a thing for some of our own long-lying patches. My view on this is unequivocal: there's no point. It smacks of desperation. Human nature being what it is, the tendency is to want to *do* something: 'The glaciers are shrinking, we must protect them.' And many have sympathised and been inspired enough to do it. But it is a howl in the wind. All the blankets in the world can only delay the inevitable. This sort of action is reminiscent of the parable of King Canute, where he exhorts the tide to go back out again. It is a futile endeavour. I also dislike the idea of covering a Scottish snow patch up just to artificially preserve it from one year to the next. It feels a little like cheating to me. I have heard people riposte this position with comments like, 'Well, it would only be to counteract the damage mankind has done in the first place.' Well, possibly. That veers too close to the anthropogenic climate-change argument that I try to avoid whenever I can.

If the apparent trend continues on the trajectory it has been on over the last twenty years, it is very difficult to see any other outcome than where patches of snow become the exception rather than the rule. Whether or not that comes to pass remains to be seen, but this scenario is one which no right-thinking person can or should look forward to.

* * *

'Winter'

On the highest Scottish hills, the dates of meteorological winter are extremely difficult to define. In theoretical terms, of course, they fall on the same dates as a lowland winter (December, January and February). The reality rarely agrees with the theory, though. In practice, *proper* winter weather can be encountered on Ben Nevis or the highest of the Cairngorms in any one of ten months of the year.

Now, it is often said that snow can fall on *any* day of the year on Scotland's hills. This is, however, inaccurate. Certainly, very *poor* weather can be experienced on any given day of the year. But that is different from *winter* weather. According to unpublished records, since the 1940s there have been nineteen days in August and twelve days in July when it has never been known to snow on the Scottish hills.[5] August, therefore, is the least snowy month of the year in Scotland, followed by July. In all other months snow can be expected to fall reasonably regularly on the highest peaks. But *not*, as is often claimed, on any day of the year.

If snow does fall during the meteorological summer then it tends, though not always, to be ephemeral. Here today, gone tomorrow falls are still relatively commonplace. All it takes is an incursion of Arctic air and some convection across the North Sea to promote showers. After all, Scotland isn't so very far away from the Arctic when one looks on a map. On the rare occasion when fresh snowdrifts *do* persist for more than a few days, then it is usually on only the very highest hills. A good, relatively recent example is 2011. On 28 August that year, a heavy fall, driven on a strong northwesterly wind, caused severe drifting on the Cairngorms. Wreaths from this storm were still visible in mid-September. Fresh August snow has not been known to endure so long as this since at least 1945.

A truly unique occurrence of an early fall of snow on Scottish hills is worth mentioning at length. It is entirely without precedent in the reckoning of people whose memories of such things extend back to the earlier parts of the twentieth century, and it has not happened since.

---

5   Adam Watson kept meticulous records of snowfall in the Cairngorms.

Given the evidence laid out earlier in this chapter, it is debatable if such a thing is ever likely to happen again in the lifetimes of those reading this book.

The event itself happened in 1976. Those old enough to remember that year often wax lyrical about the long days of unbroken sunshine and high daytime temperatures, especially during August. Those of us who weren't old enough to remember it, or had not yet been born, have read about it. It was a summer of very high temperatures, sustained for weeks at a time. Droughts were experienced all over the UK, with water being rationed as reservoirs dried up. Near Aviemore, temperatures of above 20° Celsius were recorded for nineteen days during August.[6] By the end of that month, only two snow patches remained in the whole of Scotland (at the Sphinx and Pinnacles locations in Garbh Choire Mòr, as usual). Both were very small. It was thought that if the hot weather continued then both would not last more than a couple of weeks.

However, the next day, 1 September, fresh snow fell on the high Cairngorms at around 10 p.m. There was, according to one observer, 'a fair covering the next morning at altitudes above 1,080 metres.'[7] Late on the 6th there was more fresh snow, and heavy falls all day on the 7th were accompanied by a north gale. Ten centimetres of wet snow lay at the Lecht and the road summit of the Cairnwell pass beside Glenshee ski centre. During the evening of Wednesday 8th, the weather turned cold again, and fresh snow fell on the high hills. Next day, the 9th, fifty millimetres of rain fell in lower Deeside, which would have equated to snow high up. It has been calculated by I.C. Hudson (as per the previous footnotes) that this would have been equivalent to about sixty centimetres of snowfall. Hudson also reported that fresh snow on 9 September was deposited down to 490 metres on lee slopes due to drifting. In a note sent to me, Adam Watson described the astonishing circumstances and scenes:

---

6   Hudson, I.C. (1977a), 'Cairngorm snowfield report 1976', *Journal of Meteorology* Volume 2.

7   Hudson, I.C. (1977b), 'Record early cold spell in the Scottish Highlands, 9–16 September 1976', *Journal of Meteorology* Volume 2.

On 10 September, deep snow covered the high Cairngorms but warmer air had arrived on a northeast wind and the snow had started to thaw on the highest tops. My father went to Cairn Gorm that day and reported snow showers and a bitter wind. He walked from the top chairlift round to the hollow of Ciste Mhearad to the east, and wrote in his diary that it was well filled with new drifted snow. His diary also holds his comment that his Norwegian wooden cross-country skis would have been useful. He took some remarkable photographs showing exceptional snow cover. My father told me that he found fresh drifts up to ten feet deep in hollows on the north and east and south sides of Cairn Gorm, and reported deep snow lying all the way down to Loch Avon. A party of Belgian tourists, equipped for summer conditions, narrowly escaped with their lives on 9 September after a two-day visit to Loch Avon, and 'had to be rescued. They were suffering from exposure. If the party had not chanced to meet two experienced mountaineers, at least three would have died.'

Heavy snow showers returned on 12 September, and – all told – it has been surmised that nearly a metre of snow fell during this unprecedented period, with drifts in excess of three metres observed. Adam's note continued:

On 17 September, there were still deep extensive drifts in many places despite warm weather with sunshine. The whole of the southern slope of Cairn Gorm lay under a deep blanket of con-solidated packed drifted snow, which had blown in the gale off the north side. I was on Cairn Gorm plateau on Sunday 19th with my father, a day with bright sunshine and strong glare off the fresh snow. Deep snow drifts lay on the plateau and on the corries of Braeriach, completely covering the old snow in Garbh Choire Mòr. I wrote that the heavy melting and hot sun made it feel like early June. The south side of Cairn Gorm still had a complete cover of deep drifted snow up to three metres in depth, and likewise some hollows on

Cairn Gorm such as Ciste Mhearad. I wished that we had brought our touring skis, for the snow had a firm uniform surface and skiing would have been much easier than walking.[8]

So, what happened to the Sphinx and Pinnacles snow patches that were very small at the end of August? In a trip to Garbh Choire Mòr on 2 October, John Pottie found that much of the early September falls had melted. However, there was still a fair amount visible near where the old patches were located. Upon closer inspection he found that the Sphinx and Pinnacles patches were still lying buried under the remnants of the new falls. The photograph he took shows a birthday cake effect, with the newer snow lying pristine on the top, and the old snow (which by October 1976 was nearly eighteen years old) very much darker and dirtier. Had these new snows not fallen in early September, then it is almost certain the old ones would have disappeared. I have always thought it ironic the best summer of the twentieth century failed to melt all Scottish snow on account of an early blast of winter riding to its rescue.

Whilst September 1976 was an example of a premature taste of winter, another year worthy of note for its unseasonal weather was far more recent: 2015. Though March and April of that year were broadly average in terms of temperature and precipitation, the subsequent three months were anything but. Temperatures in May were 2.5° Celsius below normal, with 175 per cent of the usual precipitation. This translated into very significant amounts of snow falling on the hills in a month that normally sees the snow pack reduce. The cold weather continued into June, which itself had eight days of falling snow over the hills: the most for decades. Pictures of the Ben Nevis range in early June showed the hills more like March, with a thick covering of snow running down their shoulders to well below 2,000 feet. It is quite likely that the deepest snow all year on Ben Nevis was to be found in the first week of June. Spring hardly got started.

---

8   Letter from Adam Watson to author.

The annual August bank holiday survey that I began in 2008 turned into an exercise in damage limitation. For the years before 2015, the total amount of snow patches across the whole of Scotland could be counted in August by about ten people, covering perhaps half a dozen areas. Thirty to forty patches were considered an 'average' total since the survey's inception. But because July 2015 was also cool, a huge amount of snow persisted at the start of August. Anticipating a massive total on survey day, I started desperately trying to recruit volunteers to go to places that I never expected I would ever need to worry about. All told, about thirty people contributed that year, counting a colossal 678 patches across the country, of which 74 made it through to winter. This was the highest total since 1994, and one unlikely to be bettered any time soon. During that year I did over fifty trips to the hills, going through several pairs of boots in the process. As much as I'd like to see a repeat of 2015, I can happily live without another like it for a few years yet.

In conclusion, therefore, it is entirely fair to say that winter is never too far away from Scotland's highest hills, even during these warming years.

* * *

A eulogy to winter
On Friday 10 February several years ago, I received a text message from Alistair, a mountaineer friend.

'A bit last minute, Iain, but with the great forecast and a full moon, I'm heading up to the far north-west tomorrow afternoon, if you fancy a wee snow plod and an amazing-views adventure?'

I checked in my head if I had anything on that weekend. Fortuitously, and unusually, I didn't.

'I'm in,' I replied. 'I'll bring my sleeping bag and will see you at yours.'

Now, the 'north-west' is a huge area, so we needed to nail down the venue. Where exactly would be best to see this snowy landscape lit up, hopefully, by a full moon? Skye was discussed. The view to the Cuillin from Sgùrr na Strì was likely to be unsurpassable. But given its increasing popularity as a viewpoint and the superb cloudless forecast, we thought it might be a bit crowded up there. Places to pitch a tent are scarce, so if we got up there after someone else then it could be a bit of an uncomfortable sleep. The decision made itself in the end: Torridon. And there was only one hill that was going to tick all the boxes.

Liathach is a supermodel. It is a hill of almost unearthly beauty and dimensions. It rises to over 3,000 feet in little under a mile from the road that runs at its foot in Glen Torridon. An air of impregnability surrounds it when viewed from below. A castellated ridge runs along Liathach for much of its length, giving some superb walking and scrambling if one feels so inclined. In winter the ridge is a highly challenging mountaineering undertaking. The seriousness of it increases under snow, but so too does its beauty. Snow embellishes height with most hills, but with Liathach the sensation is magnified even more dramatically. Almost to the point, even, that it can be mistaken for an Alpine summit. Photographs of the hill adorn virtually every calendar that features the Scottish outdoors, and its name pulls on the memory of folk who've been on its many and varied pinnacles. Some hills pass from recollection quickly, but Liathach lingers long. With the Aonach Eagach and An Teallach, Liathach makes up the most revered triumvirate of ridges on the British mainland.[9]

When we eventually pulled up in the car on the afternoon of 11 February it looked glorious. Though meteorologically we were into the heart of winter, only the top half of the hill carried snow. The weak and insipid February sun had little in the way of apricity to trouble Liathach's white veil. It was an inviting view, and soon after 1 p.m. Alistair and I were packed and on our way.

We each carried a sleeping bag and plenty of warm clothing in our rucksacks. The overnight temperature on the summit was forecast to be well below freezing, accompanied by strong gusts. The wind chill was going to be correspondingly fierce. The cold would be exacerbated by having to camp on snow as opposed to solid ground. On this trip, body heat was the currency that was needed to pay for the visual returns.

Despite Alistair's mobility being hampered as a result of a nasty cycling

---

9   Respect is due to Snowdon's Crib Goch and Helvellyn's Striding Edge, but for me they fall just short when compared to this trio.

accident some months previously, we made decent progress up the steep south flank of Coire Liath Mhòr. A well-worn path afforded solid and quick access upwards, even if the going was a little tiring. We bumped into a few people who were descending. They looked somewhat puzzled to see us going upwards at this time of the day, especially with a tent strapped to my rucksack.

It wasn't too long before we reached the crest of the ridge itself. By now we were into thick snow and looking for a suitable camping spot. The sun was still quite well up in the sky and the views were grand. If we could get camped and set for the evening then we would have plenty of time to witness what we hoped would be a memorable sunset.

In order to pitch our tent, we had to clear an area of snow from the only flat(ish) piece of ground visible for what seemed like miles. Specialist pegs were needed to lash the tent into the thick and loose snow. The wind was bitter. It sought out every chink in our clothes. The forecast had been spot on. At a guess it was about -17° Celsius and that was with the sun still in the sky. Eventually, though, we were set. The tent was pitched, and the sleeping bags unfurled; the camping stove was lit, and hot drinks were taken. All we had to do now was to wait about thirty minutes for the sun to set. The moon was already climbing above the horizon near to Beinn Alligin. Everything was set up perfectly. What happened over the next hour I shall never forget as long as my memory holds.

As the sun dipped behind the westernmost peak of the ridge, Mullach an Rathain, the sky began to reflect its dying light. A palette of reds and yellows and blues and purples coated the clouds. The nearby peak of Beinn Eighe was set ablaze by the low-trajectory shafts of sunlight. One of its summits, Sail Mhòr, was cut in half by the shadow of Liathach. On the bottom part, semi-darkness; on the top, an iridescent pyramid of sandstone. The snow that lay on the summit cones of the surrounding hills sparkled with a golden light as though layered with gilt. Despite having seen snow in many guises over the years, I had seldom known it to turn such a colour. Further afield, in this ragged edge of the country, the shattered, ancient sandstone edifices of Torridon were coated in a blood-red hue. The magnificence of the colour was exacerbated by the rich white blanket of pristine snow that sat atop it. In over one billion years, the age of this landscape, surely it had never looked so good.

Three hours before, the weak February sun that shone overhead had struggled to add any definition to the prehistoric peaks. Its azimuth was too high. Now, as it sank beyond our vision, it concentrated all its magic on the thousand-million-year-old hill, serving up a visual treat the like of which can surely only be witnessed on a handful of days every few years.

Finally, the sun did set. The purple and blue hues that illuminated the clouds faded. On the horizon, though, colours changed from deep blue to a glowing orange as the light refracted off the stratosphere. Crystal-clear air revealed the Cuillin of Skye in the distance. So sharp was it that individual peaks could be discerned even with the naked eye. Eventually the orange faded as well and was replaced by dark blue, then black. The light show put on by the sun had ended. We sat on the snow, dressed head to foot in every fleece and coat we had brought up with us. We watched the very last of the colours empty from the horizon. It was now punishingly cold with no sun left to draw any warmth from. Phoebus had now gone for his long winter slumber.

We stood up and walked around a little to ward off the bitter gusts that by now brought with them wind-chill temperatures in the minus twenties. Although the sun had gone and the sky was dark, we had superb visibility. This was due to the full moon overhead lighting up the snow as though it were floodlit. In the distance, Beinn Alligin's snow-capped horns shone almost as brightly as they would at sunrise. Liathach's burn-ished peaks stretched skywards, the snow picking out each rock and cliff. Moon-shadows abounded. I had in the past walked on snow that had been illuminated by a full moon, but never in air as clear as this, nor with such views. The experience was almost transcendental. We took pictures to try and capture the moment, but they were a pale imitation of the real thing.

In due course we were beaten by the cold and had to call it a day. We scurried into the tent and settled down for the long night ahead. Never have I been so glad of a sleeping bag. During the night, as I lay waiting for sleep to arrive, I thought of what our experience would have been like had our trip been done without snow on the hills. The clouds may have been just as pretty, but the hills would have been diminished. It may have been physically more pleasant to be there in summer, but for the memories it bestowed on us winter was unbeatable.

* * *

Climate change may yet claim Scotland's winter. Many people will say that the loss of snow from our hills isn't nearly as pressing an issue as the havoc that may be wrought across the world by its effects. That may be true enough, and I do not wish to make light of the troubles that might accompany these potential changes. But a loss of snow on our hills is not without consequences either. A privation of biodiversity and civic amenity can be anticipated, and that's just for starters. Wildlife is sure to suffer just as businesses are certain to fail.

Many walkers live for the day when they open their curtains and see the ground and trees coated in thick drifts. Snow not only transforms the landscape, it transforms us. It has the power of atavism, to turn normally sensible adults into playful children, drawing out the dormant magic that many have lying deep within their souls. One only needs to look outside on a cold, wintry Sunday morning to see and hear the smiles and laughter of parents sledging down hills with their sons and daughters. Online, many folk become unaccountably excited even at the prospect of snow *falling*. No other regular weather event causes such fervour and derangement.

One can only hope that the predictions of climate-change doom do not come to pass, for all our sakes.

eight

The grand tour

We are fortunate in this relatively small country of ours to possess so many different and distinctive regions. From the Ochil Hills near Stirling through the Trossachs and down to the Southern Uplands, the snow-spotter's lot is a varied and happy one. In this chapter, I list some of the finest places in Scotland to visit old relics of winter snow. By adding a map reference of the snow positions, I am not necessarily seeking to encourage others to go there. Rather, this allows the reader to see the spot on a map should they so desire. Any literature that encourages someone either to explore for themselves or to open a map can only be a good thing.

* * *

Aonach Mòr (snow location: NN193736)

Aonach Mòr sits at Scotland's top table. It is a hill bettered in height only seven times across the country, topping out at 4,006 feet. But it is not just for its height that it deserves to be in such exalted company. Uniquely amongst West Coast hills, one can walk for two miles in a straight line and not once dip below an altitude of 3,000 feet. This may seem like a trifle, but to those of us who enjoy a bit of respite from the West's constant up and down ridges, it is bordering on blissful. Not even Ben Nevis can boast such constant terrain. This is but an aside, though. Its credentials for being at the top table are many and varied.

Speaking of terrain, in this regard it is a peak of contradictions. Whereas the summit of virtually every Scottish hill in excess of 3,500 feet is a jumble of stony chaos, Aonach Mòr's is a study in green serenity. Thick jade and olive-coloured carpets of grass and rare mosses spread across the plateau to the extent that the walker would be forgiven for thinking they were striding across Devon pasture, rather than on some of the highest land in the UK. Camping on this verdure is a little treat that I give myself every now and again, especially if it's on a still summer's evening, high above where even the most intrepid midges venture. The thick moss that pockets some parts of the summit plateau is so soft that no camping mat is necessary to facilitate a good night's sleep.

Another contradiction is its name. Aonach Mòr in Gaelic means 'the big ridge' or 'the big ridged hill'. However, it is not as high as its immediate neighbour, Aonach Beag, which means – conversely – 'the small ridged hill'. The Gaels who named it were patently referring to the shape of the hill rather than absolute height. (Either that or they had a rather wry sense of humour.) Its shape is pleasingly uniform from the glen below when viewed from the east. The summit ridge is, as mentioned, relatively flat for quite a distance.

Guarding the plateau on both the east and west sides are a series of insurmountable cliffs and gullies. The only realistic access the walker enjoys is from either the north or the south. Come from the former, though, and one is obliged to negotiate the ski centre paraphernalia that reaches to the very edge of the plateau. Come from the latter and one is compelled to ascend via a rather precarious and indistinct ridge which, though doable without too much technical difficulty, provides a stern test for those who suffer with a fear of heights (this includes the author).

As part of writing this chapter I tried to count how often I have been to the summit of this hill. I gave up quickly but reckon it must be close to fifty. The reason for this multitude of visits is that Aonach Mòr is home to one of Scotland's longest-lasting snow-patch sites. Moreover, in its great north-east corrie, Coire an Lochain, lies one of the country's most interesting and unusual locations.

If one were asked to draw a picture of the stereotypical Highland corrie, Coire an Lochain ('the corrie of the small loch') would be the outcome. Tall cliffs fall away from the summit plateau for a couple of hundred feet before giving way to steep, stony ground that angles towards the floor of the glacial trough. At its base lies the eponymous small loch. All told, from summit to lochan, 1,200 feet separate top from bottom.

Because of its great height, in Scottish terms, Aonach Mòr sees a lot of snowfall. Being the largest of the corries, Coire an Lochain gets a fair proportion of that snow. Consequently, the bravest skiers and snow-boarders flock here in spring to put their skills to the test against the challenging terrain of Coire Dubh, Easy Gully (a misnomer if ever there were one) and Summit Gully. Hidden from view for much of winter, spring and even sometimes in early summer, there lies in Coire an Lochain Scotland's most impressive (in my opinion) periglacial feature.[1] Namely, a protalus rampart. This grandiose title can be explained simply as:

> a simple landform. It requires the existence of long-lived snow banks below rock cliffs. Frost weathering leads to rockfall from the cliff and the blocks roll down the snow slope to accumulate at its base.[2]

Like so many of the long-lying snow locations that are described in this book, it is a demanding place to get to and hardly ever visited in summer. As a result, it retains a real sense of isolation and wilderness, despite its reasonably close proximity to civilisation. In large part this isolated feeling is to do with what the walker finds when he or she gets there. It is a beguiling spot. When one visits during a 'normal' summer in, say, July then it is possible to peer into a glacial past. Stand by the rampart, similar to that shown in the diagram overleaf, and one is standing amidst an evolving landscape, shaped by the elements to this day.

---

1 Of or relating to an area or feature close to an ice sheet or long-lying snow, which influences the climate or geological processes of that locality.

2 Further details about protalus ramparts may be found at https://gh.copernicus.org/articles/70/135/2015/gh-70-135-2015.pdf [accessed 13 May 2021].

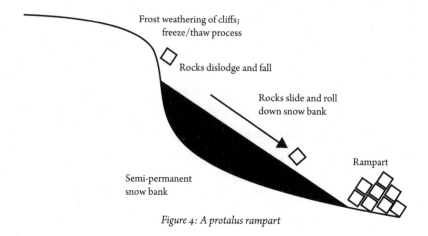

Figure 4: A protalus rampart

On the cliffs above the long-lying snow patch, water percolates into the fissures and cracks of the granite, either because of winter rain or snow-melt. Once this water freezes it expands in the joints and renders the granite unstable. Eventually, by the slow hand of erosive time, a piece of this cliff will surrender its grip and fall. Waiting below, the large patch of long-lying snow acts as a chute, deflecting the flying boulder to the bottom. Over many years the pile of stones gets larger and larger, becoming a rampart many metres high. In fact, it perpetuates its own success. The larger the rampart gets, the deeper the hollow that it protects becomes. More snow is added the following year and the cycle is repeated. On and on it goes. Each year, slowly but very definitely surely, the rampart's defences are added to by the presence of the snow it inhibits.

This landscape is not static. It is in a constant state of change. Imperceptible differences accumulate over hundreds and thousands of years. Our own fleeting time on the planet allows us but a brief glimpse of this relentless process which will, ultimately, turn the hill back into the dust it was fashioned with.

* * *

Geal-Chàrn (snow location: NN480748)

Though it sits far closer to Ben Nevis than Ben Macdui, Geal-Chàrn's appearance – like that of its immediate neighbour Ben Alder – is definitely more akin to the Cairngorms than to Lochaber. Both of these magnificent hills have high plateaux which rise gradually from the west, only to drop vertiginously into east- and north-east-facing cliffs. Of the two hills, Ben Alder takes most of the awards. In height it has an extra sixty feet. In visual splendour it surpasses virtually every hill in Scotland, let alone Geal-Chàrn. It also contains one of Scotland's highest named bodies of water: Lochan a' Garbh Choire – at just shy of 3,700 feet. Geal-Chàrn, however, has Aisre Cham.

Aisre Cham – pronounced, improbably, as 'ashra CHOWM' – means in Badenoch Gaelic 'crooked pass' or 'crooked ridge'. Why it has this name is a mystery. No ridge or pass exists here, so its true meaning will probably stay lost in the haze of time. The Aisre Cham is, in effect, a hollow that sits just below the eastern edge of the massive plateau of Geal-Chàrn. Like so many other long-lying snow locations it benefits from windblown snow piling into it over winter and spring. It is my view, though I stress this is just a hunch and based on nothing in the way of science, that this site is one that can rival even Ben Nevis for sheer depth. In the snowiest years I have seen accumulations there that bury virtually every rock and crag. When one ventures there in late summer or early autumn one realises just how much snow it takes to do that. We are talking in the scores of feet deep here. And what an expedition it is to get there!

The 'Ben Alder round' in Scottish mountaineering circles is considered a classic. Before the closure of Culra bothy on health and safety grounds (it has an asbestos roof), most weekend warriors would walk there on a Friday evening from Dalwhinnie before attempting the six Munros the next day. Because the walk back was ten miles, they might even decide to stay another night and saunter home at leisure on the Sunday morning. With that option no longer available, walkers must either take their own tents, sleeping bags and camping equipment in; or, as is becoming increasingly common nowadays, do the whole thing in a day, travelling from and back to Dalwhinnie by bike. I first did this round in 2012, and I used a bike. It is a huge day, not recommended unless one is very fit. The three other times I have journeyed back and forward to this range

I have also used this method of transport, but eschewed doing the six Munros. Whenever I venture there nowadays it is to the Aisre Cham I go. The snow patch there is one of the few that has been known to survive from year to year outside the Cairngorms or Ben Nevis range. For that reason alone, it is worth going. I also find the location grand and wild in a way that only the most remote of hills can be.

During a memorable visit of 12 October 2014, I went to the Aisre Cham via a ridge called the 'Lancet Edge', which connects Geal-Chàrn with its lower neighbour Sgòr Iutharn. The clue to the memorability of the visit is tied up in the ridge's name. One section of the razor-like ridge almost got the better of me. Not for its technical complexity – it was a simple step – but, rather, its exposure. Not even on the famous Aonach Eagach ridge did I feel so much trepidation as I did there. My old nemesis of being afraid of heights made my head swirl at the prospect of stepping over this minor gap in the ridge. Perhaps the fog of time has made it seem more difficult, but my memory of it lingers so vividly that I will not attempt it again. At the time I remember making a pact with myself that this was the first and last time I would attempt it. The reward of seeing two ptarmigan on the rocks just as I'd crossed the gap went some way to making up for it.

As to the location of the Aisre Cham itself, it is the classic long-lying snow hollow. Vegetation is in short supply, with the usual collection of primitive liverworts and mosses huddled in little clumps wherever possible. The ground is perpetually wet due to snowmelt, even during the very driest summers. It is tricky to access and, when one does, the moving, waterlogged ground means one has to tread with caution.

But it is worth the effort. How well I remember walking off the summit of Càrn Dearg on 1 October some years ago, in wonderful autumn sunshine. As I descended on soft, dry grass I could see Geal-Chàrn ahead of me. Two large patches of snow glowed in the early afternoon sunshine at Aisre Cham. I plodded at a leisurely pace toward them. As the ground steepened along the twisting ridge of Aisre Ghobhainn it felt like I was hardly expending any energy, my feet moving almost of their own accord. The heat of the sun was on my back and time didn't count for anything.

Reaching the snows, I sat beside them and looked out across the barren expanse of the Ben Alder deer forest. It was little short of magnificent.

* * *

Glen Almond (snow locations: NN771370 and NN786371)

In an isolated part of southern Perthshire sits one of Southern Scotland's finest glens. That said, if popularity were a measure of how fine something is, you wouldn't know it if you were to walk along the Land Rover track that runs through it. Even on a weekend you might only come across a few mountain bikers or the occasional walker. In point of fact, I have encountered more people fishing here than any other activity. I know why. Munros. Or, rather, the lack of them.

Glen Almond benefits enormously from having no hills higher than 3,000 feet. Technically Ben Chonzie qualifies as being accessible from the glen, but in reality no Munro-bagger is going to come at it from this direction, for to do so would mean a massive diversion. Munro-baggers generally take the path of least resistance. This lack of high hills means that one can walk there in relative peace and tranquillity, save for the odd estate vehicle (more on this later, alas). It is also a cul-de-sac. Many glens act as conduits through the Highlands and as a result have attracted through-roads. Not Glen Almond. In this respect it is like Glen Lyon – a princess amongst Scottish glens. There are few houses to be seen, and those that are visible often seem to be unoccupied. Auchnafree House, the large estate building, seems to be the population hub of the glen.

When I moved back to Scotland in 2014, after having lived in England for the best part of ten years, Glen Almond was the first place I decided to camp in. Its wildness, relative proximity to my house in Stirling and lack of footfall were the main reasons for this. The other, as you may now have guessed, was to see some snow.

Apart from the peacefulness of the glen, which is certainly one of its main draws, the unusually flat terrain aids the walker or cyclist. In nine miles, from the glen's eastern terminus at Newton Bridge, the ground rises only 300 feet. In so doing the road passes a host of interesting historical features like the Neolithic chambered cairn Clach na Tiompan and its nearby standing stones, the Stuck Chapel and millennium memorial, and – of course – the wonderful crags and cliffs above on either side of the glen.[3]

A host of estate roads peel off from the valley floor to attain the summits of the hills on the northern flank of the glen. These are predominantly to aid the 4x4 vehicles taking shooting parties up to the innumerable grouse butts that exist all over the hills here. The best, and the steepest, of these roads is the one that runs up the shoulder of Sròn an Fhèidh ('the point of the deer') on to Sròn a' Chaonineidh ('the point of mourning'). Any pretensions of staying on one's bike for most of the traverse upwards are laughably curtailed within a minute or two. Even an electric bike, I fancy, would toil for grip on the relentlessly steep 'nose of the deer'.

Sròn a' Chaonineidh is the central hill in a line of three that consists of Creagan na Beinne to the west and Meall nan Eanchainn to the east.[4] All three have pronounced northern corries that, despite their relatively modest height, can gather huge volumes of snow and which often persist far longer than hills of greater height. Sròn a' Chaonineidh's Coire Quaich is the finest of the triumvirate. The best time to visit this corrie is in mid- to late April. If the winter has been snowy, extend that by a couple of weeks. By a distance the most impressive I've ever seen these snows was in June 2015, when I took a tent up and camped on the lush grass that carpets the wide summit ridge. On that visit I was expecting to find some snow, but not a huge amount. How wrong I was. Upon getting there the whole rim of the corrie, for hundreds of metres, had its own glacier, complete with mini crevasses. I climbed down into a couple of them and walked along for quite a distance. This made up for the camping disaster I had that night by forgetting to pack the tent pegs. (To cap it all it started raining. Were it not for military tarpaulin I threw in at the last minute it would have been a thoroughly miserable evening.)

---

3   Technically the glen carries on through the Sma' Glen and past Glenalmond College, but for the purposes of this chapter it is the section west of the A822 we are concerned with.

4   Creagan na Beinne is another name the Ordnance Survey got wrong. It's pronounced locally as 'crehgan a VEHYn'. Meall nan Eanchainn means 'the hill of the brains', according to Dwelly's *Gaelic–English Dictionary*, Glasgow: Akerbeltz Publishing.

Slightly farther east, Meall nan Enchainn's equivalent corrie is Coire nan Gabhar.[5] For some reason, probably that the walls of the corrie are a bit less steep, the build-up of snow here is usually a bit less, but it is still a fine location.

Without wishing to labour the point too much, these hills in spring are – for their isolation, peacefulness and snow – among the very best that the Southern Highlands have to offer. They are also a superb place to see certain wildlife. A lack of walkers in general means some birds abound. Golden plover in particular are common, their shrill cries a pleasure to hear in the cool breeze of the summit ridge. Mountain hares – *Lepus timidus* – are often seen, but more skittish. If one is lucky, eagles may be viewed soaring overhead.

All of that said, recent events on these hills have soured their appeal for me. In 2019, a golden eagle went missing on the estate that these hills lie in. Its satellite tag stopped functioning after it spent an inordinate amount of time on the ground. To compound this the eagle itself was called Adam, so named by member of the Scottish parliament Andy Wightman after he adopted it. The Adam in question was – sadly – Adam Watson. Andy, like so many others, looked up to Adam and had a huge respect for his work. (I know Adam was delighted when Andy named the eagle after him.) The exact location of the disappearance was narrowed down to the southern spur of Sròn a' Chaonineidh, right at the end of an estate road. Like so many eagle disappearances on managed grouse moor, the whole thing is remarkably suspicious. I avoided the area for a while afterwards because of this. It left too bitter a taste in the mouth.

Of course, it's not the hills' fault that this happened. But, humans being the irrational creatures they are, sometimes take things like this a bit too personally. I will be back soon enough to see those hills, and their wonderful snowdrifts.

* * *

---

5   Coire nan Gabhar, 'the corrie of the goats', is pronounced as 'carn nan GO-er'.

Ben Macdui (snow location: NH994014)

The Feith Buidhe ('the yellow bog stream') is a small and short burn that has a big story to tell.[6] Its water is not yellow, as the name might suggest. Rather, it is the pale yellow mat-grass that is to be found adjacent to the stream for much of the year that gives it its name. The source of the burn is the Lochan Buidhe ('the small yellow loch') which, at 3,700 feet, happens to be Britain's highest body of named water.[7] The Feith Buidhe runs for only 1.5 miles and empties into Loch Avon.[8] The journey cannot take the water more than an hour to complete. Seldom has such a short route encountered so much visual beauty or tragedy.

Regarding the latter, much has already been written about the dreadful Cairngorm plateau disaster of November 1971. Its alternative name is the Feith Buidhe disaster. For those not familiar with the awful story, it involved a group of six fifteen-year-old schoolchildren and two leaders. Of the eight people in the party, five of the six children and one of the two leaders were killed when they decided to open-camp near the Feith Buidhe in horrendous winter weather. They had been heading for a hut, but – due to the atrocious conditions and lack of experience – never made it and most perished after the disastrous decision to bivouac in a blizzard. The group was but a few hundred metres from the shelter they were aiming for, which makes the story even more tragic. In summer I often pass the site of their bivouac. In benign conditions it is scarcely believable that such an attractive spot can turn so lethal.

A few hundred metres downstream from this solemn site, the Feith Buidhe tumbles off the plateau and down what locals used to call Creagan na Feith Buidhe ('rock of the yellow bog stream'). Once it has crashed down these granite terraces it joins with several other streams and morphs into the wonderfully named Meur na Banaraich ('the finger stream of the milkmaid'). This short, new river has barely had enough time to settle when it is subsumed by the queen of all Cairngorms lochs: Loch Avon.

---

6   Feith Buidhe is pronounced as 'fay BOOee', with a thin 'B', almost like a 'P'.
7   Lochan a' Garbh Choire on Ben Alder is the other claimant to this title. Both sit at 3,700 feet, but Lochan a' Garbh Choire is more a puddle than a loch and has no outflow.
8   Pronounced as 'loch AAhn'.

The granite terraces the Feith Buidhe tumbles down are no longer called Creagan na Feith Buidhe. They are now styled (the) 'Feith Buidhe slabs'. The old name has all but passed from use. The view down Loch Avon from the top of these slabs is one of the classic visions of the Cairngorms. Personally, I consider it to be bettered only by the one from the summit of Sgòr Gaoith across to Braeriach. The Loch Avon view is enhanced considerably when one ventures there in early or mid-summer. Huge volumes of snow gather on these east-facing slabs in winter and spring. The plateau to the west has no major dips or hollows to contain the snow's eastward journey. When it blows over the edge it gathers on the neatly symmetrical slabs and piles high. Very high. There can be fewer places in the Cairngorms where so much gathers, in fact.

Like many fascinating snow-patch locations in the Cairngorms, the site is missed by almost every walker. Maybe this is down to the snow being invisible from the path that connects the two highest points on this upland shelf: Ben Macdui and Cairn Gorm itself. I doubt that's it, though. Even if it were not invisible, it is unlikely that many would bother to make the detour. Most are intent on 'bagging' the Munros in an already long day's walk. It is their loss, for at the slabs – in the right summer conditions – are some of the greatest snow sculptures that Scotland has to offer. Certainly, they are among the most visually impressive. In this regard the Feith Buidhe slabs themselves are the architects.

Imagine a series of terraces. Think of the stepped pyramid of Djoser in Egypt, or rice paddies on a hill in Vietnam. Now imagine them covered entirely in snow during winter. When spring arrives, the snow starts to melt from the top. A few warm days in May expose the top terrace, and the uncovered ground absorbs the heat produced by the sun. As this ground warms it starts to emit heat in exactly the same way a household radiator does. What happens next is that a gap appears between the top of the terrace and the snow. This gap is now the weak link in the chain. Warmer summer air starts to pour in and eventually reaches the bottom, opening a diagonal gap between the terraces and the snow. To help visualise this, imagine cutting a block of cheese or butter in two. But instead of cutting it straight down the middle, start from the top right edge and slice it diagonally to the bottom edge on the left. Two wedges have been produced. Slide them apart an inch and you have the scene at Feith Buidhe slabs in summer. The left wedge represents the snow, and the right wedge the granite terraces. But this is where it gets *really* interesting.

The space between the snow and the granite terraces is such a rare and little-seen phenomenon in the UK that there are no English words for it. Instead we need to look to the continent for suitable nouns. The Germans call this space 'randkluft'.[9] In June and July this gap is often wide enough to admit the passage of several people at a time. *Randkluft* is rare even in Scotland, but it is nowhere better seen than at the Feith Buidhe slabs. The results can be extraordinary.

The terraces of the Feith Buidhe slabs are tens of metres high. Each shelf is made up of grippy, solid granite. When one walks on to the topmost terrace, then clambers down the rest of them to where the foot of the snow is, it is like nothing else in the country. The first thing one notices is the huge wall of blue ice that one is standing beside. So high is it that it's impossible to gauge its depth. It slopes over one's head and away, angling and stretching away from the walker. The ice's foot is rock hard, as though glacial. It has been compressed by tens of thousands of tonnes weighing down upon it and pressing against the ground. The air is cold there, many degrees below the outside ambient temperature. So much snow and ice are there that the very atmosphere has changed. This is the deep heart of the *randkluft*; the place with no English name. The blue and green hues of the snow and ice's fringe project a science-fiction feel. Shouting at the top of one's voice, a weird and unsettling echo rebounds off the terraces and snow. It is altogether a disconcerting place to be, albeit one of stunning and ethereal beauty.

But it is also dangerous. Though in the deepest bowels of the *randkluft* the risks are minor, at the fringes, where melting is most pronounced, danger lurks. Several years ago, as I traversed the slabs, looking down at a suitable place through which to enter the yawning *randkluft*, I saw some precarious and unsupported snow at the edge of the huge drift.

---

9   Meaning 'marginal cleft or crevasse'.

I declined to go down to it for fear of it collapsing, instead opting for another mode of entry. No sooner was the thought out of my head than the unsupported snow gave way, taking a huge chunk of the whole pack with it. From a distance of twenty metres or so I felt the air pass across my face, as though a great fan had been wafted. Perhaps a hundred tonnes of collapsed snow lay on the slabs, and had I been in the vicinity I would no longer be alive to recount the story. As can be the case in life, beauty is often accompanied by danger.

* * *

Beinn Ime (snow location: NN256084)

This peaceful peak ('the butter hill') is an altogether more sedate affair than most other hills that I visit regularly to ascertain the condition of snow. The journey to get there is short, lying around an hour's drive from Stirling or Glasgow. If one is smart, one will avoid the ascent up the hill from Arrochar and its car parks, which attract hordes of people wishing to walk up The Cobbler – perhaps Scotland's finest hill under 3,000 feet. There is also the matter of the not inconsiderable parking charges that Argyll and Bute Council have now started to levy on the motorist at The Cobbler car park. Being a canny Scot, such taxes on hillwalking are always avoided if I can possibly get away with it.

Mostly, for me, however, it is the hordes that I try to avoid rather than the cost. And for that, one should start on the north-west flank of the hill, close to the junction of Glen Kinglass and the A83 'Rest and Be Thankful' road. From there it is only a 1.5-mile walk to the summit of Beinn Ime – pronounced as 'ben EEm'. This distance, too, is more sedate than some of the mega-walks required to get to other long-lying patches. I say 'sedate', but this – like 'long-lasting' – is a relative term. One must ascend over 2,500 feet in those 1.5 miles from bottom to top. That can quite reasonably be categorised as 'lung-bursting' in most people's reckoning. It is unforgiving, though brief. Do not tarry in summer, mind. It is a place of biblical swarms of flies and – worst of all – clegs: the beasts of the hill. Midges are but an annoyance compared to the bloodthirsty cleg. Even the way that Scots say the word (with a very thick *cl*) is onomatopoeic. It is spat out.

The summit of Beinn Ime is the highest point in the so-called Arrochar Alps, standing at a proud 3,316 feet. Arran can readily be seen to the south-west, as can Ben Nevis to the north. The Paps of Jura light up the western view, and on a *very* clear day with binoculars it is apparently possible to see Sawel Mountain, the highest of inland Northern Ireland's hills – at a distance of 131 miles. I have yet to have such clarity. (On that note, I have spent a little time researching the theoretical longest lines of sight in the UK. The longest I have seen postulated is between Carnedd Llewelyn in Wales and Cairnsmore of Carsphairn in southern Scotland, a distance of a scarcely believable 145.3 miles.)

My annual jaunt to Beinn Ime usually happens in late May or early June. By that time, it generally carries the UK's most southerly patch of snow. That particular baton is handed to it by Ben Lomond, which – in turn – is passed it by one of the Moffat hills in the Southern Uplands. That being the case Beinn Ime is one of the bellwether patches and a close eye is kept on it.

On a blisteringly hot day in late May 2017 I went up Beinn Ime with a Stirling University professor and one of his students. They were conducting some research on vegetation and its relationship to long-lying snow. (At this particular patch site, the vegetation spends anything up to six months of the year buried by snow.) On the way up the heat was oppressive. The car thermometer showed 28° Celsius, and it felt every bit of that. The fair-haired student toiled on the upward walk, unsurprisingly, and even when we reached the summit it was still over 20° Celsius, according to the calibrated thermometer we had with us. The snow that we had come to check was pitifully small, far more so than in previous years. But it was snow nonetheless. The professor and student carried out their experiment and took some data. The heat continued for the downwards walk. When we eventually started the drive home the student suddenly projectile-vomited profusely all over the inside of the professor's car. The professor showed a remarkable calmness despite this. The heat had got to his student, badly. It is the only time I have seen heatstroke when conducting a snow survey. Usually the worry is at the other end of the temperature spectrum.

The location of the snow itself is in a classic north-east-facing basin below the summit cone. It gathers readily and fills to a great extent. In 2015, I went there in July and was astonished to find a good deal of depth still present. It is the only year I have known where it lasted into August: unheard of for such a relatively low hill so far south. For me, the visit to Beinn Ime in late spring signifies the start of the snow-patch-spotting season *proper*. I look forward to my next trip with relish.

* * *

Ben More, Crianlarich (snow location: NN432246)

There are steep hills, steeper hills, and then there is Ben More. Its summit – like Beinn Ime's – is easily accessible from the roadside, a mere 1.5 miles away. Only, pile another 500 feet on top of Beinn Ime, with no added distance. If one goes straight up it from the A85, several deep breaths are advised before setting off. Once one hits its main slope, the angle and effort are unrelenting.

Ben More has the simplest of Gaelic names – 'big hill'. From some angles it looks elegant: conical, smooth, enticing. From others it looks like a bruiser: wide, muscular, solid. With its inseparable partner, Stob Binnein, it dominates the landscape around Crianlarich, the small village where the roads from Glasgow and Edinburgh to the West Highlands meet.

Of these two high peaks (the sixteenth- and eighteenth-tallest in the UK), Ben More is the higher by thirty feet, at 3,852. However, Stob Binnein – 'the anvil peak' – is the shapelier. The former is devoid of the graceful edges and rounded curves that the latter possesses. To see Stob Binnein's crest stretching up to the hill's summit from the top of Ben More is to witness one of the grandest sights in the Southern Highlands, especially when covered in snow. Handsome cornices creep out from the ridge's edge, creating a scalpel-like rim almost to the summit. Discounting Ben Lawers, there are no higher hills farther north until one gets to Ben Nevis. Because of this great height, Ben More – especially of the two – enjoys a lot of snow. All aspects of this hill can receive large amounts because of its conical nature.

In early March 2010, at the end of a very long and cold winter, the aftermath of a huge avalanche was seen on Ben More by a couple of walkers. I recall being informed of it at the time, but because I was living in England I was unable to attend. Fortunately, the walkers who witnessed the debris took many photographs and posted them online for others to view.[10] To my mind it is the largest and longest single avalanche event in Scotland for many years. For the most part the slide was caused by the hill's steepness, allied to the amount of snow that was present on the hills in general that year. The pack would have been poorly bonded (by a lack of freeze/thaw cycles), therefore when the temperatures warmed up slightly the snow was highly unstable. Such was the force of the Ben More avalanche that it 'blasted' its way across the opposite side of the small glen and had started to travel *upwards*. Events like this are rare in Scotland and are reminders that in full winter condition the hills need to be treated with a great deal of respect.

The main place of interest for the chionophile on these two hills is high on Ben More's north-east shoulder: the Cuidhe Chròm.[11] It is unique in Scotland in that it is the only place where a snow patch is named on an Ordnance Survey map. The Cuidhe Chròm ('the bent or curved wreath') is incredibly conspicuous in late spring and early summer for many miles, adopting a pronounced boomerang shape as it crests round a steep slope below the summit cone. It shares its name with two other Cuidhe Chròms in Scotland: one on Lochnagar and one on Cairn Gorm, though the latter is spelled *cuithe* rather than *cuidhe*. The name is evidently hundreds of years old, stretching back to when Gaelic was the spoken language in this area. The appellation was established enough to be included on a map from 1888. Its conspicuousness is almost certainly the reason why it is thus named.

The location itself lies very close to the path that hillwalkers use to attain the summit and it is one of the most accessible long-lying patches in the country because of this. It is, though, an incredibly steep place to visit. A slip here would have grave consequences, almost certainly

---

10   'Crianlarich Avalanche': thepeeler-lawrie.blogspot.com, available at: http://thepeeler-lawrie.blogspot.com/2010/03/crianlarich-avalanche.html?m=1 [accessed 13 May 2021].

11   Pronounced as 'KOOee chrome'.

involving tumbling many hundreds of feet on to rocks below. Another few degrees of steepness would render the location impossible to visit without specialist gear. So steep is it, in fact, that I have used crampons on my boots even when on the grass that surrounds it.

On a glorious, warm May afternoon it is a special place to be. During dry spells the melting snow is a constant source of water. As a result, insects buzz on the fringes of the snow and birds seek it out for food and water. Ptarmigan and eagles are frequent visitors to the vicinity. As well as the wildlife, the views eastward to Loch Tay are sublime, as are those north towards Ben Nevis.

The Cuidhe Chròm is one of the most visible and observed patches in Scotland and as a consequence, excellent melt-date data exists for it. Every summer driver heading into the West Highlands from the south-east of Scotland, on a clear day, sees the Cuidhe Chròm. I wonder how many of them realise its significance.

* * *

Ben Cleuch, Ochil Hills (snow location: NN901005)

Nae silks or satins I'll put on,
Nae flooers shall bloom for me –
But Lady Alva's snawy web
My winding-sheet shall be.[12]

The Ochil Hills are special to me. Not for their height, their cliffs, their wildlife or their drama. They possess none of these attributes in any comparable way to the dramatic peaks of the Highlands. Also, if I'm honest, there is much about the Ochils that I would like to change (there are far, far too many sheep on the hills, for example, and too few trees).

The reason the Ochils mean so much to me is that they are my local hills. I see them every day at close quarters and as such I know their shapes and outline intimately. It is my go-to range when time is short or

12   Crawford, John (1850), 'Katie Glen', *Doric Lays: Being Snatches of Song & Ballad*, Edinburgh: Alloa [accessed via www.electricscotland.com/culture/features/scots/doriclaysoocraw.pdf].

long travel is unappealing. I have been to their summit, Ben Cleuch, over a hundred times in all weathers. I love their grassy, steep slopes. Rain drains from the southern flanks of these hills like no others. No matter how hard the torrent, give it twenty-four hours and it's possible to walk up in training shoes without getting wet feet.

Because they sit close to the edge of the Highlands, as well as some of the major population centres in Scotland, they offer almost unrivalled views of both on a clear day. Arran is visible from Ben Cleuch's summit (just 2,365 feet), as is Ben Macdui – just. The panorama from here also extends south to the Pentlands, which are visible above the three Forth bridges. If one looks closely, Berwick Law and the Bass Rock can also be seen.

The normal route up to the summit of Ben Cleuch is fairly steep, but only in places. This upwards path, via Silver Glen and then Ben Ever, prohibits viewing the summit cone until one is about two-thirds of the way into the journey. When it does come into view, assuming the walk has been undertaken in early March, the viewer will not only see the upmost point of the Ochils, but also the charmingly titled 'Lady Alva's Web' snow patch, as referenced in the opening poem.[13] The origin of this name is lost to history. Was it after Lord Erskine's wife, or is Lady Alva just a female name for the town itself? Whatever its history it lends a nice description to this fragile wreath of snow.

It is formed only in the right conditions. If the predominant weather in winter is from the Atlantic then the wreath will not develop, irrespective of the amount of snow that falls. Unlike most well-known snow wreaths, Lady Alva's web faces south-west. Therefore, it forms most fully when Arctic weather systems predominate. In 2010 it was massive, exposing multiple dead sheep when it finally melted out in May of that year. From that year it has formed only sporadically and – even then – is typically ephemeral because of its height and aspect. But if one can go quickly to it after it forms, the wreath is highly impressive for its bulk. The topography of Ben Cleuch means that big accumulations are formed quickly.

---

13   It has also been called Lady Alva's 'diadem' and 'necklace', but 'web' is best known locally.

Gazing down to the remains of Alva House, it is just possible to imagine the eponymous Lady staring back.

* * *

Creag Meagaidh (snow locations: NN428883 and NN428882)

Harold Raeburn was a man apart. Born in Edinburgh in 1865, he was fascinated by ornithology as a boy. This love led him to climb up and down steep faces and cliffs to seek out eggs and nests. The practice evidently stood him in very good stead, as his list of first climbing ascents in later life would take up the rest of this chapter. His name hovers over Scottish mountaineering even to this day, despite the fact he died almost a hundred years ago. From 1896 to the early 1920s he claimed fifteen of the thirty new routes on Ben Nevis *alone*. His ice axe is one of the two seals of office held by the Scottish Mountaineering Club president. That's how much of a big deal he was. Not content with Scottish hills, he summited the Matterhorn and led expeditions to the Caucasus, as well as being a member of a Himalayan trip. He died and was buried in Edinburgh, leaving a CV that earns him a place in the pantheon of world-class climbers, not just British.

As much as Raeburn was a man apart from his peers, the same case could justifiably be put forward for Creag Meagaidh.[14] In height and scale it is similar to Geal-Chàrn, but it is more complex. A myriad of ridges and corries emanate from the large summit plateau. It presents serious navigational difficulties in poor weather. The hill rises to an impressive 3,700 feet and is marked by a small cairn at the top. Many walkers confuse the summit cairn with the very, very odd, far larger one which sits just a hundred metres or so away. Like the summit cairn, this larger structure is man-made. It goes by the strange name of 'Mad Meg's cairn', and it has an apparently peculiar heritage.

Mad Meg's cairn is reputedly the final resting place of a woman who committed suicide sometime in the last few hundred years. The church would not allow her family to bury the poor soul, on account of her

---

14   Pronounced as 'craig MEHgay', meaning 'the rock (or crag) of the boggy place'.

taking her own life; consecrated ground could not be troubled by such an act. The family, therefore, decided to lay her on the highest ground they could, literally on the other side of the parish boundary (itself marked by Creag Meagaidh's summit cairn) just to spite the landowners. How much truth there is in this fable is unclear, but parts of it make sense. The cairn is a considerable structural undertaking in a very out-of-the-way place, so its construction would not have been an easy or frivolous endeavour. Its very close proximity to the parish boundary is also telling. Maybe one day it will be excavated, but I sincerely hope not. Leave poor Meg to the elements, I say. It is a wonderful, albeit tragic, tale that deserves to maintain its mystique.

I digress.

Of the many corries and ridges, the finest on the hill by far is Coire Ardair. Many consider it to be one of the greatest Highland corries, able to stand on a podium with Beinn Eighe's Coire Mhic Fhearchair and Lord Byron's favourite, Lochnagar. Having seen all three in the flesh I feel Coire Ardair is the finest. Its cliffs of the Post Face are simply magnificent, rising over 1,000 feet from the lochan below. If winter conditions allow, most climbers will tell you there are few places outside Ben Nevis where their sport is better.

But it is not the cliffs that are the object of desire here for the snow hunter. As usual, it is the gullies that concern us. The two main ones are Easy Gully and Raeburn's Gully. The latter is named after Harold, who completed its first winter ascent in the late nineteenth century. At the very top of this gully there lies a medium-sized, but deep, bowl. This place is highly unusual in that – like the Aisre Cham at Geal-Chàrn – it is one of the few places where snow has survived from year to year outside the Cairngorms or Nevis range (it did so in 2014 and 2015).

Life is confined in this hollow to only the most primitive types of mosses. This is in total contrast to the snow-free plateau just a matter of metres from it, where grasses and small plants thrive. In our snowy hollow the growing season is just too short for these. Around the fringes of the snow, out of the way of hungry sheep and deer, alpine plants grow. Of these the most photogenic is the starry saxifrage. This charming flower has five leaves, all brilliant white except for two yellow dots. In the middle, the stigma is a large violet ball from which emanate a number of white filaments, all topped with pink anthers. Google it.

The main gully on Creag Meagaidh is huge. It runs in a crescent-moon shape from just above Coire Ardair's small loch right to the summit plateau: the best part of a thousand feet of a climb, but longer as the fox runs. It is yet another gully with a misnomer. Easy Gully is anything but. It may well be in winter when there is ample deep, even snow to utilise, but ascend it in summer and be prepared for an experience like no other. The ground here is constantly on the move. In 2014, as I ascended the middle section, the whole floor that I was standing on started to slide backwards. It was terrifying, even if easily negotiated. In the past I've been in gullies where the ground was unstable, but never had I experienced a whole *section* of gully sliding away.

Easy Gully, however, is a treasure trove in summer. Apart from all the booty that can be obtained (ice axes, ropes, etc.), the gully is home to massive and weird snowdrifts. These are shaped by the unceasing drip-drip of a thousand springs above on the cliff faces. They are also the reason the gully is so mobile: the stones are lubricated constantly. The snow gathers there after funnelling down from the top of the gully. Avalanches tumbling down the Post Face also add to the great depth, ensuring snow is present right through to August and sometimes beyond. The shapes and scale of the drifts that remain here in late August are worth the walk on their own.

It was in Easy Gully that I managed to replicate a celebrated photograph from September 1899. The picture appeared in the Scottish Mountaineering Club Journal from 1901 and has become something of an iconic image. Taken high in the gully by SMC member Donald Cameron-Swan, it shows a behatted friend of his, Dr Kenneth Campbell of Laggan Bridge, standing beside a huge snow-bridge that spans the full width of the gully. Dr Campbell stands on the left-hand side of the picture and is dwarfed by the left pillar of snow which must be all of twenty feet high. The arch, according to Cameron-Swan, had a span of 'thirty-five feet'. In the foreground of the photo lie the remains of a previous iteration of the arch, now collapsed due to melting. It is altogether a marvellous image, of historical value. It is the earliest photo of a snow patch (or tunnel) at close quarters I am aware of in the UK.

The replication of this photo was in August 2016. Along with another hillwalker, Ben, I set off to see if such a thing would be possible. Ben was looking for interesting stories for a monthly column he wrote for a website. I had long fancied seeking out the precise spot on the hill and had a vague idea of where the photograph was taken. But, given that anyone who might have had knowledge of it had been dead for decades, it was a bit of a crapshoot. Right away, though, when we entered the gully – laminated 1899 photo in hand – it became clear from the skyline in the background we were definitely on the right track. As we climbed higher up the gully we marvelled at how (and why) those walkers from over a hundred years ago carried such heavy photography gear up the backwards-moving defile. We soon hit upon a likely spot for their photo. Luckily for us there were also quite a few drifts of old snow present, which added to the overall authenticity. Then, eureka, we saw a rock feature on the right-hand side of the cliff that matched up exactly with the one in the 1899 photograph. The skyline was perfectly aligned too. We had it. Ben stood where Cameron-Swan did and I where Dr Campbell was. It was as close to a perfect match as one could hope for, and Ben had his story.

* * *

Upper Tweeddale
(snow locations: NT180268, NT170250 and NT151241)

As the years roll on and on, I find myself becoming less and less well disposed towards sheep. Years ago, when I knew far less than I do now, I didn't mind them bumbling around on the hill, eating grass. They seemed like inoffensive, placid and daft creatures. In spring the fields were alive with the calls and sights of little lambs, gambolling across fields which were resplendent in their first flush of growth. What could be more natural? Those days, I'm afraid, are gone. Sheep (and deer) eat absolutely everything in their paths that has the potential to grow into a tree. In Scotland we still have far too few trees and shrubs, and a lot of this is down to the fact that our woolly and antlered friends munch them as they emerge from the ground.

The destruction of large tracts of the upland environment in Scotland lies squarely at the doorstep of mankind. Whether it was the killing of the last predator of the red deer (the wolf), the current over-management of heather moor (the better for people to shoot grouse) or overgrazing due to too many sheep, in terms of flora and fauna, our glens and hills are a shadow of what they once were. Once again, space restrictions – and self-imposed ordinances – prohibit me from delving too deeply into these subjects. There are plenty of well-written books and scientific papers that cover the topics better than I could, especially on the matter of grouse-moor management.

What do I have against sheep? Nothing, *per se*. It's not their fault they do what they do. It's just that there are so *many* of them all over the country that some parts of it feel almost monocultural. The Borders' hills – Upper Tweeddale in particular – are some of the worst affected, as far as I can tell. The wild, sparsely populated and unfrequented hills to the south of Peebles, around the farm of Manorhead, are real jewels. The unproductive soils of the hills would rebound with flora if there were a drastic reduction in the numbers of things roaming around that want to eat them. One only has to look at the trees and plants that grow in the inaccessible streambed shelves to see what *could* be. So it is, then, that whenever I walk in this fine part of the country looking for snow I do so with a tinge of disappointment that it should be so denuded of the life just waiting to erupt were it given the chance.

The three hills I visit most in this area are Dollar Law, Broad Law and Cramalt Craig. The first of the three doesn't hold the longest-lasting patches, but on its eastern slope, just to the east of Fifescar Knowe, a series of deep, north-east-facing gullies fill up readily in a winter of southwesterly snowstorms. This results in some excellent shapes and depths when visited on a warm April afternoon. Until they melt, these snows birth the splendidly titled Ugly Grain burn, which eventually gets gobbled up by the Tweed, miles to the north. Manorhead Farm, which the Ugly Grain passes, is extremely isolated. At well over 1,000 feet above sea level, it has the feel of a Highland bothy.

Broad Law, the Southern Uplands' second highest hill, carries a lot of strange paraphernalia on its summit. A radio beacon station, which looks for all the world like a lunar module, aids transatlantic planes in their navigation. The beacon is one of a global network of VHF Omnidirectional Radio Range (or VOR) stations set up as a radio navigation system for aircraft. Nearby a mobile phone mast adds to the incongruity of the scene. But as interesting as these things are, the real star of the show on Broad Law is its large northern corrie, Polmood Craig. This large scoop is Highland in feel, one of the very few in the region that can lay such a claim. Along its curved edge in spring forms the most impressive snow wreath in the Southern Highlands. Mini-crevasses appear, and it is quite possible to descend into them and be hidden from view to all watching above. The walk to this location is best from Cramalt Craig to the east. From there the finest views of the snow are seen.

Speaking of Cramalt Craig, this is the last in the Borders' trio that is worth a visit. (On a fine spring day, assuming one starts early enough and packs enough sandwiches, it is quite possible to take the snow from all three hills in.) I have yet to visit this hill in spring and not come across avalanche debris below its broad north-east shoulder. The avalanches occur firstly, of course, because of the volume of snow that accumulates on this high hill. Crucially, however, there is an abundance of well-lubricated grass. Any increase in temperature above freezing will see the bottom of the snowpack lose its grip. Eventually a critical mass is achieved and from there the outcome is inevitable: the snow lets go and down it falls. Large chunks of turf and earth are frequently to be seen lying churned up beside a car-sized lump of snow. It is certainly a place to avoid in the winter. A spring visit, though, is very pleasant – assuming the rain and wind stay away. On this exposed, flat ridge it is a very unpleasant place to be when walking into the teeth of a wet gale.

* * *

An Stùc (snow location: NN637432)

The story goes that, in 1878, a large group of about twenty men was led up
to the summit of Ben Lawers, the highest hill in the Southern Highlands,
by a chap called Malcolm Ferguson. They spent the whole day erecting
as best they could a twenty-foot-high cairn. The reason for this eccentric
endeavour was that they wished to bring their beloved local hill above
the magical 4,000-feet barrier. It was previously thought to be this height,
but when measured with specialist equipment by cartographers it was
found to be lacking by sixteen feet. Incensed and highly motivated,
Ferguson – like the Grand Old Duke of York – led the men up the hill
then back down again. Alas for Ferguson and his workers, the Ordnance
Survey ignored the man-made cairn, which has blown away in the intervening years.

Ben Lawers itself is the chief hill in a very fine grouping that encircles
the northern part of Loch Tay. From Meall Greigh ('the hill of the horse
stud') at the range's eastern terminus, the peaks stretch westward until
reaching the summit of Ben Lawers, at an impressive 3,984 feet.

The journey into the foot of Meall Greigh, where one normally starts
when completing a round of these hills, takes the walker up alongside
the Lawers Burn (the derivation of Lawers as a name is disputed: some
say it means 'loud water', while others say it translates as 'hoof' or similar).
Beside the Lawers Burn there are over thirty shielings scattered on well-
drained ground. A shieling is a summerhouse that was often used by
local folk when they grazed their animals at higher level. They are very
small and cramped and would have been, I'm sure, unpleasant in prolonged periods of poor weather. It is not known how old these structures
are, but it's thought they date back at least several centuries.

Despite Ben Lawers' great height, it is not on the hill that the longest-
lasting patch of snow in the range is to be found. That honour lies with its
easterly neighbour, An Stùc.[15] Sounding little like how it's written, An
Stùc is a superb peak and is translated into English as 'the peak'. Nothing
could be more descriptive.

---

15   Pronounced as 'an STOOCHk', with the 'ch' as in *loch*, with a 'k' tacked on at the end.

When viewed from the east, on the summit of Meall Garbh – 'the rough hill' – An Stùc looks majestic. It is shaped like a huge pyramid, with even slopes tapering to an obvious point at the top. Its east face from here looks insurmountable to a walker. Indeed, care is needed when ascending this face, especially in winter. The rock is loose in places (as I know only too well), even if the climb up itself isn't nearly as hard as it looks from afar.

The long-lying snow on An Stùc tends to last until June or July, or exceptionally (à la 2015) August. It is situated in a pronounced hollow on the northern spur of the hill, where it gathers to itself lots of wind-blown snow from around the summit crags. Because of Ben Lawers' relative ease of access and popularity among walkers, photographs of the range are in no short supply in summer. Many hillwalkers become unconscious snow researchers when I gatecrash their social media posts. An Stùc's shapeliness ensures the snow photos keep coming. Consequently, it is yet another of the bellwether or signpost patches that serve as a reliable comparison year on year.

Alert readers will have by now realised that most of the snow patches I visit are out of the way, even if they're on accessible hills. An Stùc is no different. If the basin in which the snow sits is full, which it tends to be even in early summer, then one is obliged to creep very gingerly around its southern rim. The grass surrounding the snow is exceedingly slippery in wet weather. So much so that on my last visit, in July, I used crampons on my boots to get purchase on the grass beside the snow. Never before in my years of doing this had I used crampons on *grass*, only to take them off when reaching the *snow*.

That same day I was fortunate enough to encounter some of the rare and specialist alpine plants that Ben Lawers and its cohorts are renowned for. Among others, I came across a small clump of *Myosotis alpestris*, known charmingly in English as the 'alpine forget-me-not'; a stunning violet-petalled flower. To cap it all off, on the walk back out I was treated to a fine display of soaring from a golden eagle. The perfect way to conclude a memorable day.

nine

South of the wall

There is a curious school of thought in sections of polite Scottish society that all Britain's finest hills are located north of the border with England. This inexplicable theory can only be explained by a lack of familiarity. Many walkers in Scotland would, for reasons that I have not been able to properly ascertain, scarcely think to venture south. Some do, of course, but the vast majority do not. I am by no means a veteran of walking in England and Wales, but I have ambled on, I think, just about every upland part of England and quite a few in Wales. In my experience some of these areas are just as fine as parts of Scotland. Moreover, if we are talking about individual hills then very many indeed in Scotland lack the character and challenges of the best that Wales and England have to offer.

They have little to do with snow, but one of my favourite groups of hills in the whole of Britain is an almost unknown range in South West England called the Quantock Hills. These diminutive peaks rise to about 1,250 feet and are very much overshadowed by their neighbours on Exmoor. Few know that they were England's first Area of Outstanding Natural Beauty, having been designated as such in 1956. The main route to this range for most people is not done in a car or on foot: it is literary. Samuel Taylor Coleridge's most famous works (*The Rime of the Ancient*

Mariner and *Kubla Khan*, amongst others) were completed while the poet lived in the picturesque village of Nether Stowey, at the very foot of the Quantocks. In August the entire top half of the range is covered in the most beautiful carpet of purple heather that I have seen in Britain. One of the few truly wild red-deer populations of old inhabits the serene, wooded dells on the west side of the hills. It is one of the finest places I can think of.

The rest of the chapter is given over to some of the marvellous hills and ranges of England and Wales where snow can lie, or has lain, long. I go to some of these peaks regularly as part of my research for the annual snow-patch paper which appears in the Royal Meteorological Society journal *Weather*.

* * *

The Cheviot

Journeying south from the border between Scotland and England, the first stop is less than a mile from it. The Cheviot is a landmark hill in the Borders, conspicuous for miles from many vantage points. It has always attracted a high level of interest from travellers, and so – apparently – have its snows:

> *Sir* – I have read your paragraph about 'Snow Conditions on Cheviot.' It might be of interest to mention that 'pre-war I have taken parties up the Bizzle and struggled through snow pockets of about six foot depth during the early part of July.' These snow pockets or what used to be called 'Snow Flowers' frequently remain in deep drifts until late summer.
>
> – B.A. Parkes (ex-Lone Wolf),
Berwick-upon-Tweed, June 14th 1951.[1]

This letter, written to the editor of the *Berwickshire News*, was the first in a series penned in response to a newspaper contributor mentioning seeing

---

1   *Berwickshire News*, 26 June 1951.

patches of snow on the Cheviot in early June and remarking upon how late it was. Mr Parkes's letter, followed by several thereafter by different correspondents (including one from Seton Gordon, remarkably), is a charming reminder that people have, for years, been intrigued by these long-lying drifts and moved enough to write to newspapers.

The Cheviot's name is ancient, probably Celtic. It may derive from the Brittonic word *ceμ*, meaning 'ridge', plus the suffix *-ed*.[2] At 2,673 feet it is the highest point on a large tract of upland Northumberland. The range in general gives its name to a breed of sheep, which is perhaps what most people will recognise the name for. (Someone formerly of my acquaintance used to call all sheep, irrespective of breed, 'stupid cheevs' in response to witnessing one of them doing something peculiar. The *cheevs* in question were Cheviots.)

The 'big Cheviot', as it is still known locally, has provoked writers down the years to gush in flowery, whimsical verse when describing even the most mundane of activities. One writer was so taken with the landscape and the simple-living people who dwelt in it that he felt compelled to write the following:

In a word, a shepherd's life may be very pretty and poetical in Arcadia, but a Cheviot Tityrus[3] has other fish to fry, than piping like a bullfinch, or playing 'love among the roses' with some Daphne or Amaryllis, in full costume, as Fentum turns out his shepherdess for Jullien's *bal masqué*, with a straw hat, ribbons that would supply a recruiting party, crook in hand, and everything complete, barring the lamb.[4]

A reader could be excused for thinking, on the strength of the above extract, that the Cheviot is a spectacular hill of breathtaking loveliness. Though beauty rests in the eye of the beholder, I would not describe it as such. It is a large, mostly featureless lump with a very high percentage of

---

2   James, Alan G. (2019), *The Brittonic Language in the Old North*, Scottish Place-Name Society.

3   The shepherd in Virgil's *Eclogues*.

4   Maxwell, W.H. (1852), *Border Tales and Legends of the Cheviots*, London: Richard Bentley.

its slopes given over to heather and peat. Its main selling point is the spectacular views from the summit and the fact it lies right on the Scotland–England border. To the north-west the fertile verdure of the Scottish Border farms stretches out like an Astroturf carpet towards the escarpment of the Southern Uplands. On a clear day it is rumoured one can see Ben Macdui on the horizon, some 124 miles distant. I find this fanciful and have never seen it myself, but respected observers assure me it's gospel.

The Cheviot does possess a few parts which are worthy of note nonetheless, namely College Burn, Bizzle Burn and Bellyside Burn. College Burn faces south-west; the latter duo north. Of the three it is usual for the two north-facing ones to hold snow longer than College Burn, due to the preponderance of snow to blow from the south-west in winter. However, and this is a big 'however', if conditions allow, the College Burn area can have deposits of snow which make it difficult to believe one is in England.

In the aftermath of the so-called 'Beast from the East' snowstorm, which hit the country on 1 March 2018, colossal amounts of snow were dropped all over the country. At the College Burn, upwards of thirty feet was dumped. During a visit there over two months later I was able to walk among the small crevasses that had formed, with the snow still towering over my head. What a spectacle it was, especially given the delicious warmth that the high spring sun was giving off. It recalled to me the adventures of noted ornithologist Abel Chapman, who went to the Cheviot in May 1915:

> On 14th May we had climbed out some 2,400 feet – beyond the head of the Bezzil [Bizzle] ravine – when further progress appeared absolutely barred by upstanding ridges of green ice that resembled nothing so much as a glacier ... The sheer face of green ice exposed stood seven feet in vertical height ... scarcely credible in the temperate zone of our British Isles, especially in mid-May ... it remains to add that, even at midsummer, whole acres of snow still choked the Bezzil ravine.[5]

---

5   Chapman, Abel (1924), *The Borders and Beyond*, Gurney & Jackson.

Top Emerging into the light after ascending a snow tunnel. The roof of this part of the tunnel collapsed some days previously, leaving large chunks to melt on the relatively warm rock. © *Alistair Todd.*

Above Mud, soil and rocks cover snow debris deposited in a huge avalanche.

Top A runner stands beside the relic of the previous winter on Ben Macdui in November 2020.
The old snow was buried the following day, only to emerge once more at some point the following year.

Above This old patch of snow on Ben Macdui lies partially buried by fresh falls in November 2020.
The scalloped shapes are called 'ablation hollows' and are caused by the passage of warm air across
the surface. These edges attract dirt and debris.

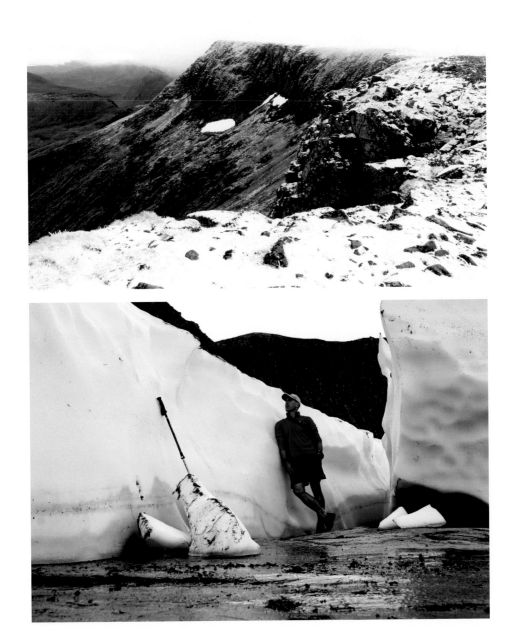

Top A classic view of old snow patches clinging on to the cliffs. Here, at Aonach Mòr in October 2014, the old patches survived until the new snows of the following winter.

Above The author leans against the shear in a huge patch of snow at Aonach Mòr in September 2020. The shear was created by the patch to the author's right sliding down the granite slope, which had been lubricated by meltwater.

Ablation hollows and mesmeric colours at Ben Nevis's Point 5 Gully.
The vista here was described by one commentator as akin to 'the eye of a dragon'.

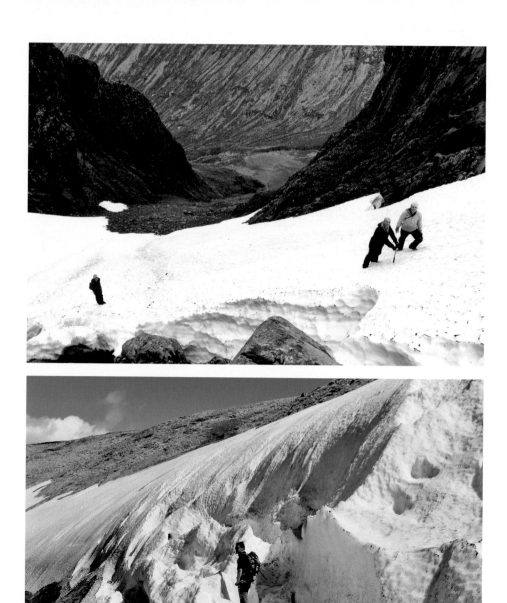

Top Three volunteers assist with the measurement of a massive relic of the previous winter, high on the north-east face of Ben Nevis in August 2015.

Above Huge drifts of snow remain at Hen Hole on The Cheviot in northern England in May 2018. This snow was deposited almost entirely during the so-called 'Beast from the East' in early March of the same year.

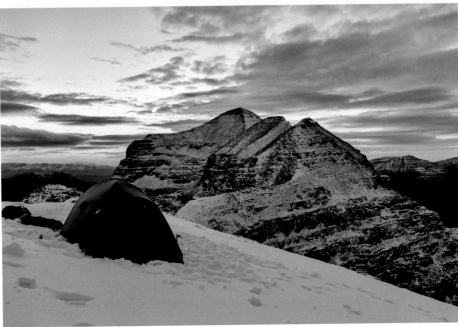

Top Marching towards the snow, high on a Cairngorms hill. © *Murdo MacLeod.*

Above A bitterly cold winter camp on Liathach in Torridon, February 2017. The wind chill was extreme, but the views that evening were some of the most stunning the author has ever encountered.

Top BBC commentator Andrew Cotter with Mabel and Olive, during one of our many walks.

Above Andrew Cotter again, this time on skis with his two dogs.

But the Beast from the East's crevasses melted quickly in the strong sunshine of May 2018, to the extent that they – and the snow they occupied – had vanished by the first week in June. Nearby, on 9 June, an observer went to the Bizzle Burn, just as Abel Chapman had done over a hundred years before, and saw the hill's very last snow. As he ate his sandwiches he noted how the small, dirty relic was the size of a serving platter. It was to succumb to the elements the same day. The observer wrote in a note to me the next day that he found this 'sad'.

* * *

Cross Fell, North Pennines

A mountain that is generally ten months buried in snow and eleven in clouds cannot fail exciting the attention and curiosity of a traveller.[6]

Outside of the Lake District, no hill in England is higher than Cross Fell. At 2,929 feet it commands panoramic views across the fertile Eden valley to the Lake District hills. It is also the highest point in the Pennines, which – along with the Grampians of Scotland – is one of the longest hill ranges in Britain. Anyone who travels to or from Scotland either by car or train has a superb, albeit distant, view of the Cross Fell tops on the east side of the road and railway track. Its subsidiary top, Great Dunn Fell, has a very obvious white golf ball structure on the top which serves as a radar station for the National Air Traffic Services. A necklace of snow wraps around Cross Fell in spring if favourable conditions prevail.

Cross Fell was once called Fiend's Fell. It is unclear exactly as to when, or why, the name changed. A local legend has it that an exorcism of said fiend (credited to St Augustine) was carried out and a cross planted on the summit, giving it a more wholesome aura and its new name. The so-called fiend that St Augustine exorcised was evidently not too keen on travel, as he or she fled only five miles to the north-west, along the Pennine escarpment, to take up residence on another Fiend's Fell. In truth it is likely the original name Fiend's Fell is drawn from the

---

6   Hutchinson, W. (1794), *The History of the County of Cumberland [etc.]*, London: F. Jollie.

shrieking noise often heard on the summit when the famous Helm Wind is blowing.

Like the big Cheviot, Cross Fell's shape is conical rather than craggy. Whereas the former has some interesting geographical features, the latter has rather fewer. It does, though, have very good historical records of snow-related events – good enough to put it near the top of the best-documented hills in England. The snow researcher *par excellence* Gordon Manley gathered many accounts for this area. One of the most interesting was his comment that 'the indigenous use of skis by the lead miners in the Northern Pennines has died out since 1900', which suggests that by this date it was no longer practicable to use them, as it had been in the past.[7] He cautioned, though, that this in part may have been to do with the demise of the industry as much as the reduced snow cover. It is a delightful thought to imagine lead miners in Alston or some other North Pennines village setting off from their houses after breakfast on skis.

Snow in general on Cross Fell is to be found on its northern side where it has melted everywhere else. Crossfell Well, a high spring sitting at grid reference NY685347, just at the foot of the summit bluff, normally carries a good drift into May. Its relatively open aspect, however, means that it is prone to rapid melting.

Remarkably, and perhaps exceptionally in the last fifty years, the ghosts of the old lead miners were roused from their graves on 1 July 1979. On that day, members of the Carlisle Ski Club hiked up to a gully near the summit and skied a large patch that was still plainly visible from the M6 motorway below. A local paper, the *Cumberland and Westmorland Herald*, ran the story a few days later. Over a month and a half later, on 18 August, a farmer by the name of Mr Willan saw two tiny white dots at the same locality while working in his fields, and no doubt they melted a day or two afterwards. This is the latest I have ever heard of snow lying on an English hill in the current era. It is very hard to imagine that date being beaten in my lifetime, alas.

* * *

7   Manley, Gordon (1949), 'The Snowline in Britain', *Geografiska Annaler* Volume 31, 179–193.

The Lake District

It seems rather futile introducing the most famous range of hills in England. So many wonderful writers, poets and musicians have eulo-gised the Lakes over the years that perhaps every native-born English-man or woman has the memory of them running through their veins.

No writers are more associated with the Lakes than Dorothy and William Wordsworth, both of whom were born and grew up in Cocker-mouth. The landscape had a deep impact on them and they wrote extensively on it. It is pleasing to know that these two most famous of Wordsworths were also snow observers.

> It was a Cove, a huge Recess,
> That keeps till June December's snow;
> A lofty Precipice in front,
> A silent Tarn below!
> Far in the bosom of Helvellyn,
> Remote from public Road or Dwelling,
> Pathway, or cultivated land;
> From trace of human foot or hand.[8]

It may not rank as William Wordsworth's best work, but it is highly instructive for the mention of June harbouring snow from the previous year. The 'silent tarn' is, without question, Red Tarn below the north-east-facing cliffs of Helvellyn. The 'cove' is likely to be Brown Cove, just north of the summit. It is almost always the place on the hill where snow lies latest. William's sister, Dorothy, in a publication of 1817, mentions that snow remained on Helvellyn above Red Tarn in late June.[9] Given that both the Wordsworths mention Helvellyn's snow in June, it is possible and, perhaps, likely that William's observation in 'Fidelity' was taken on the same day as his sister's, conceivably during a family excursion. Having sat at Red Tarn myself and looked up at the last chunks

8   Wordsworth, William (1815), 'Fidelity', *Poems in Two Volumes: Part Two*, London: Longman.
9   Manley, Gordon (1952), *Climate and the British Scene*, London: Collins.

of the previous winter's snow that were about to expire, it is heartening to think that I could have been sitting just yards away from where they picnicked and penned their verses.

Helvellyn itself is the finest hill that I have walked on in the Lakes. At the start of the chapter I mentioned that some English hills were more than equal to their Scottish counterparts: Helvellyn is just such an example. Though modest in height, at just 3,117 feet high, it is a complex hill of airy ridges and high cliffs. Striding Edge – a tentacle reaching out from its summit and stretching to the east – is the best ridge-walk in England, offering superb scrambling for those who are so inclined. Swirral Edge provides an excellent descent and horseshoe route back to the 'silent tarn'. The vista from Red Tarn on to Helvellyn's north-east face is one of the classic English mountain views.

On Helvellyn's Brown Cove nowadays, June snow is still possible to find (2010 was a recent example), though it is becoming increasingly rare. This is also true for nearby hills such as Great End, Scafell Pike and Esk Pike, which normally hold snow late into spring. Gordon Manley wrote in 1952 that all the above could expect to have snow till June in a normal year. 'Normal' nowadays means something different. Things have changed remarkably in a relatively short period of time. Like the August snow at Cross Fell in 1979, replicating Manley's observation in 1951, when Brown Cove held a small drift in mid-July, seems unlikely any time soon.

* * *

The Peak District

To Scots, the term 'down south' means everything from Carlisle to Southampton. This echoes how people from northern England playfully mock southerners for calling anything above the Watford Gap 'The North'. It has never been clear to me where the North begins. Leaving aside the Watford Gap stereotype, the question is vexed. The North York Moors, the Lake District, the Cheviot Hills: all these ranges are unequivocally in 'The North'.

So, then, where does the Peak District lie? It surely can't be in the

North as it forms the most southerly spur of the Pennines. It is *undeniably* not in the South. It is also too far north to be in the Midlands. It is, therefore, in no man's land. QED.

Whatever its location, the Peak District National Park is immensely popular, attracting over ten million visitors each year. This is hardly surprising, given its proximity to some the largest cities in England. Manchester, Leeds, Stoke-on-Trent and Sheffield all lie within striking distance of the Peak District's borders. Most of the ten-million-plus sightseers will not ascend the highest hills or even venture beyond the periphery of the roadside. Despite the throng, for those willing to seek it out, there is some wild country to be had.

Kinder Scout is the highest hill in an area that is known locally as the 'High Peak'. At 2,087 feet it doesn't trouble the list-makers of England's highest hills. Nevertheless, there is nothing higher in England farther south than Kinder Scout, not even on Dartmoor. It is, therefore, magnetic to walkers because of that. The same walkers, if they felt so inclined, could walk across its huge and impressive summit table for three miles and not once drop below 2,000 feet. One would be hard-pressed to do that in many places in England, even in the high Pennines. Kinder Scout is a totem in the area, much-loved and visited. It is also much observed and has been for centuries. Once again, snow observers in days gone by do us proud:

> This year, 1615, January 16th, began the greatest snow-storm which ever fell upon earth within man's memory. It covered the earth fyve quarters [45 inches] deep upon the playne. It fell ten several tymes, and the last was the greatest, to the greate admiration and fear of all the land; it came from foure p'ts of the world, so that all e'ntries were full, yea, the south p'te as well as these mountaynes. It continued by daily by increasing until the 12th of March, (without the sight of any earth, eyther upon the hilles or valleyes) upon wch daye, being the Lord's daye, it began to decrease, and so by little and little consumed and wasted away till the eight and twentyth day of May, for then all the heapes or

drifts of snow were consumed, except one upon Kinder Scout, wch lay till Wit-sun-week.[10]

In 1615, Whitsunday fell on 7 June. Old snow visible on any part of the Peak District in June nowadays would be unthinkable. It is just about conceivable that an Arctic blast during the first week of the month could bring a dusting to the range's very highest tops, but it is probable that such snow would be ephemeral. Nowadays one is lucky to find old snow there in April. There are occasional exceptions. One notable example was 3 April 2012. During the early morning, very cold air edged southwards and an intense fall of snow resulted in road closures in the Peak District and Pennines. Three weeks later large drifts were still in evidence across Bleaklow and other high hills in the area. The very last to go that year clung by its fingertips until the first few days in May. Another unusual event happened in 2013. Weather fronts associated with a deep depression to the south-west of Britain collided with cold, easterly air to bring intense falls of snow to the Peak District from 22 to 24 March. That year, the last patch gave up the ghost around 10 May.

All the snow that I have talked about so far in the book has been above ground. This may sound like an obvious thing to state, but in the Peak District there is at least one example of a phenomenon that occurs nowhere else in Britain. This unique feature is a deep fissure called 'Eldon Hole'. This pothole, dubbed the 'fourth wonder of the Peake',[11] is little-known, except to speleologists[12] and locals. Entrance to the pothole cannot be achieved without the aid of specialist equipment, so consequently it is visited only rarely, even in summer.

With its entrance lying in a northwest–southeast alignment, and sitting at 420 metres above sea level, Eldon Hole is prone to collecting snow when winds blow in from a north-through-easterly direction. The nature of the pothole's internal structure means that any snow blowing

---

10 Hall, T. S (1863), 'Derbyshire (England) – description and travel' [the account was transcribed from the Youlgreave parish register by Hall].
11 Cotton, C. (1683), *The Wonders of the Peake*, London: J. Brome.
12 People who specialise in the study of caves.

into its opening is relatively unimpeded until it hits the floor, resulting – given the right conditions – in very deep accumulations. Moreover, because the cave is enclosed, offering complete protection from solar radiation and strong winds, melting is reduced dramatically when snow gathers here. If the accumulations are large enough, the snow will significantly reduce the ambient temperature in the pothole through spring and summer by a process of radiation, thereby further slowing the melting process.

In 2013, I was contacted by a caver who had had to abandon a trip to Eldon Hole during the first week in May because snow blocked the bottom cavern, and it was still present on 4 July when he returned. I noted this event in the *Weather* journal paper I wrote that year. Since that account, other evidence has come to light (personal communications from cavers). Firstly, deep snow was seen in June 2010 by S. Sharp, and the same drift – much diminished but still substantial – persisted in November, five months later. It is a reasonable inference that this snow endured to the first heavy falls of the season, which came only a few weeks thereafter. This is the only example we have in England where snow has almost certainly persisted from one winter to the next. Eldon Hole is a truly extraordinary location.

* * *

The Cotswolds
Nowadays, so little in the way of meaningful snow falls on the Cotswolds that no reports of long-lying patches have been communicated to me or anyone I know. The only reason I am including a section on them is because I live in hope, rather than expectation, that repeats of events past might happen once again.

The Cotswolds are associated more these days with weekend homes that well-off people from London return to after work on a Friday. But this is not a true reflection of the reality. Of its 800 square miles, eighty per cent of this designated Area of Outstanding Natural Beauty is given over to farmland. In a sleepy corner of this farmland lies the pretty village of Snowshill. Our old friend Gordon Manley said that this settlement, sitting at 800 feet above sea level, was 'likely to owe its name to past experience', which is hard to disagree with. Manley went on to say in the same book that after the extremely cold early months of 1947, a shady location of the northern Cotswolds held remnants of a drift until mid-July.[13] Now, were it not for the fact that it was Manley who reported this striking event, I would be inclined to dismiss it out of hand. Mid-July, on the Cotswolds? Surely Manley got that wrong? To my deep regret, Manley didn't say *where* in the Cotswolds this happened. One piece of historical evidence – actually, I'd go further, the *greatest* piece of snow-related historical evidence extant in Britain – casts a certain doubt in my mind about dismissing this, on the face of it, absurd claim.

During research in 2009 for a book I co-authored, I came across a reference to snow being visible in a Cotswolds quarry in August 1634. I thought initially that this was a ludicrous, highly overexaggerated claim. The report came from a book on the Cotswold Hills in Gloucestershire, but the author did not cite the reference. However, Herbert Evans on a cycling tour gave a more precise description:

So past Brockhampton Park, leaving to the left the quarries where in 1634 the snow and ice lay till August, past Sevenhampton.[14]

The large-scale OS map shows disused quarries that fit this description, just east of Brockhampton and at 850 feet altitude. The large expanses of smooth flat exposed ground to the north-east and east would be good gathering grounds for blowing snow. Hmm. Interesting, but still not conclusive.

Unfortunately, Evans did not give a reference either, but it did lead me to a more original and authoritative source in a book by William Dyde of Tewkesbury, published in 1790. On page 99 he wrote:

---

13   Manley, Gordon (1952), *Climate and the British Scene*, London: Collins.
14   Evans, Herbert (1905), *Highways and Byways in Oxford and the Cotswolds*, London: Macmillan and Co.

In January 1634, the greatest snow, that was ever remembered in the memory of man; and, it was attended with such extreme cold, violent, and tempestuous weather, that many people going from this market, were smothered and starved to death. And in the August following, great quantities of the same snow, and ice was to be seen in Brockhampton quarries, notwithstanding it was an extreme hot summer.[15]

Predictably, Dyde did not cite a reference either. I thought that, just maybe, he heard it as a child from an ancient relative who had, in turn, been told of it by their grandmother who had witnessed it first-hand. That turned out not to be the case. I came across some other breadcrumbs, just enough for me to conduct an online search among the dusty archives of the Gloucester Records Office. I found something that I thought likely to be the original source, so made an appointment to go in and view it.

With low expectations, I submitted the reference number to the helpful archivist at the records office that Saturday. It took about ten minutes for her to find it, but when it arrived and I leafed through to the relevant folio, my hair stood on end.

It was clear to me that Dyde had also read the very account I was now looking at, as the wording was almost identical to his own. In wonderful handwriting, inked on vellum, was a description from the early seventeenth century. The account, though difficult to transcribe in entirety, stated that:

on January did fall the greatest snow that was known in the memory of man ... and in August following a great quantity of the same snow and ice did remayne at Brockhampton quares ... yet it was a most extreme hott summer.[16]

---

15   Dyde, William (1790) *History and Antiquities of Tewkesbury from the Earliest Periods to the Present Time*, Tewkesbury: Dyde and Son.

16   *Account book of Giles Geast Charity*, Gloucester Records Office, reference D2688/1: folio 79v.

I was beside myself with excitement upon reading this. Of course, a handwritten entry in a book is proof of nothing except the person who wrote it. The question posed itself, though: *why* enter something like this if it weren't true? Why go to the trouble? Maybe, just *maybe*, given the right circumstances, some of this snow could have held on until August. It is such a tantalising story.

I mention this story at some length mainly because I think it's a tale worth telling, but also because it illustrates the value of accurate recording by the author, as well as perseverance and diligence by the researcher. In this game, these two attributes are not so much desirable as essential.

* * *

The Carneddau, Snowdonia, Wales

As far as this section of the book is concerned, some might contend that leaving North Wales till the end is keeping the best till last. There is certainly some merit in this position. Snowdonia's peaks are rightly lauded for their beauty and the rich variety of walking and climbing available. Tryfan is a pyramidical peak of sublime proportions, easily capable of being included in a photo-finish for the 'handsomest hill in Britain' award. Crib Goch is one of the classic British ridge walks and among the very best in terms of sport. All over this range soaring peaks and deep cwms leave the visitor awestruck at the range and depth of the landscape on offer.[17] Snowdon itself – or Yr Wyddfa to Welsh speakers – is the highest peak in Britain outside the Scottish Highlands and possesses a set of cliffs on its north-east face that many Scottish hills can only envy.

The most northerly section of Snowdonia encompasses a range called the Carneddau – literally 'the cairns' – in Welsh.[18] It has the largest collection in Wales of hills over 3,000 feet, including the one we are concerned with: Carnedd Llewelyn.

In the main, Wales's hills are too craggy and steep to hold on to snow for extended periods. (Snowdon's great height – 3,560 feet – would, if it

---

17 A cwm is the Welsh equivalent of a corrie.
18 There are three distinct sections to Snowdonia's hills: Snowdon, Glyderau and Carneddau.

were more rounded and had gullies rather than cliffs, allow significant accumulations of snow.) The Carneddau are different from the peaks to the south. Carnedd Llewelyn has a whaleback ridge running for over a mile north-east from its summit cone. Some decent-sized cliffs sit above cwms on all sides of the hill. Despite all these glacially carved features, the place on Carnedd Llewelyn that virtually *always* holds snow later than anywhere else in Wales, and often outlasts England's as well, is called Y Ffoes Ddyfn (pronounced, very approximately, as 'uh-FOICE-thuvin', with a 'th' sound as in 'the').[19]

Back in the early part of the twentieth century this location, translated as 'the deep cut', aroused some interest in a local man by the name of J.R. Gethin Jones. A native of those parts since birth, he used to wonder why the 'snow always remained latest on Carnedd Llewelyn' rather than on Snowdon. To resolve this matter he 'undertook a bit of local exploration'. In the resulting paper, which he wrote for the inestimable *British Rainfall* magazine in 1909, he gives a superb overview of the location of Y Ffoes Ddyfn and accompanies it with valuable pictures.[20]

With a theme that has not only been running through this chapter but the book itself when historical records are mentioned, the amount of snow present at 'the deep cut' back in Jones's day was significantly more than is present today. In the paper he described a visit in June 1910. He concluded that the last snow on Snowdon disappeared generally at the beginning of June, but in Y Ffoes Ddyfn at the beginning of July. During the very snowy winter of 2009–10, by comparison, the patch at Y Ffoes Ddyfn persisted until 28 June.

Even this late date by today's standards would appear to be earlier than one might have expected in the mid-nineteenth century, according to writer James Orchard Haliwell. In the late 1850s he walked with a guide on Carnedd Llewelyn and noted:

19 During the period 2010–2019, Wales's snow has been the last to vanish six times, compared to England's five.

20 Gethin Jones, J.R. (1909), 'The Spot in England and Wales Where Snow Lies Latest, with Observations of Snowfall on the Snowdonian Range', *British Rainfall*.

the cold was intense; I should suppose it to be five or six degrees below the freezing point, and this while the sun was shining at noon day ... The snow lay very deep in some situations. We walked over a ravine which my guide informed me was at least forty feet in depth. In this, and familiar places, he said that it would not disappear until the close of August.[21]

As with the account from the Cotswolds, we cannot take one anecdote as proof of anything except itself. Or, as Adam Watson and I state in our conclusion in *Cool Britannia*:

Although much of the above historical evidence is anecdotal and qualitative, it appeared independently with many observers in different years and regions. Because of this, we infer that observations were unlikely to be atypical, such as being concentrated by chance in an unusually snowy year or area. Comparisons involving quantitative observations from recent decades add confidence to this inference. To sum up, it is reasonable to conclude that more snow lay on the hills of Britain during the 1700s to early 1900s than in decades since 1930. Accounts from the lowlands of Scotland and England fit with this.[22]

Today, because of accounts like this being more widely known, many observers keep keen eyes trained on Carnedd Llewelyn's Y Ffoes Ddyfn to see when the last vestige of snow melts. The pioneering onlookers of the olden days fuel the desire to keep these visits and observations going. I find it enormously gratifying that it is not just in Scotland that people are apt to record long-lying snow. It tells me that there is a thread that links like-minded people across Britain, a thread that grows stronger by the year.

21 Watson, Adam and Cameron, Iain (2010), *Cool Britannia*, Bath: Paragon Publishing.
22 Ibid.

ten

Modern perils

Although snow is one of the motivations for people like me going up into the hills, for others the need or desire to do so has remained virtually the same since people started doing it in large numbers some 100-odd years ago. For many it is the re-engagement with nature and the wild; for others it is the challenge. For some others again, it is a combination of all of these. The thing that has really changed in the last ten to fifteen years, however, is the leap forward in technological capability, especially the handheld device. Allied to this change in technology is the now almost ubiquitous use of social media and online applications. Before we delve into that vexed subject, complete with its many pros and cons, it might be useful to look at the modern attitudes to hillwalking in general.

The reasonably observant, societally mindful citizen will have noticed a change in their lifetime regarding risk and the perception of risk. I mean here the general risk, real or imagined, that one comes across via the workings of everyday life. Risk when walking or climbing is just an extension of this. The reader may remember from the opening chapters that my occupation revolves around the management of risk (in the construction industry). In addition to this I've been a Chartered Member of the Institution of Occupational Safety and Health (CMIOSH) for over ten years, so I am reasonably conversant with the subject.

There is little doubt in my mind that society, in general terms, is considerably more risk-averse than it was when I started hillwalking properly in the early 1990s. For many years now, as it seems to me, the public have been looking more and more to the State or some third-party body to provide the things that a generation ago they would have done – in large part at least – for themselves. The reasons for this are many and complex, but a significant one (perhaps *the* most significant in my eyes) was the changing of the law in the 1980s to allow individuals to litigate in an American 'no win, no fee' model. This move by the government, almost certainly to cut down on the legal aid bill, had unforeseen consequences.[1] One of them was the eye-watering expense to the public purse as a result of so-called 'ambulance chaser' solicitors scouring the country for plaintiffs. Thirty years on from these changes, the effects are clear to see (England and Wales figures):

> There were 143,167 practising solicitors in 2018, up 2.7 per cent on the year and 27.3 per cent over the past decade. However, there were 50,684 in 1988, and growth was nearly 50 per cent in each of the following two decades.[2]

The knock-on effects of such a legal-system-heavy society (in relative terms) are everywhere. Daytime TV adverts for 'no win, no fee' law firms pay the salaries of the people making the programmes we watch. Children's play parks that were once concrete-floored are now rubber-floored, surreal environments where the swings are shorter and the climbing frames lower. (I am not saying this change is a good or bad thing, merely observing that it *has* changed.) Disclaimers are put on everything from sandwich wrappers to car park signage. They say effectively the same thing: 'The consumer uses these services at their own risk'. Safety legislation in the workplace is another area that has

---

1   Critics of this move would say that it was not at all 'unforeseen'.
2   Legal Futures. 2021. *Growth in number of solicitors starts to slow – Legal Futures.* Available at: www.legalfutures.co.uk/latest-news/growth-in-number-of-solicitors-starts-to-slow [accessed 13 May 2021].

grown massively since the late 1980s. This expansion is another of the reasons why the legal profession has grown to the bloated level it is now. We look to the law to solve problems that would once have been dealt with by the political process. The citizens of this country now place more faith in judges than they do in politicians. To my mind this is a damaging and unhealthy space to occupy.

Gradually, by osmosis, the general public has adopted this mindset as well. Whether we like to admit it or not, our collective societal toleration of risk has reduced and, consequently, created a void that we look towards others to fill. One example is the attitude of 'If I get lost I'll just phone Mountain Rescue'. I've seen that particular line written on internet forums a few times. In other words, the person appears to be happy to outsource their responsibility to some other poor soul. It is not an effective or societally egalitarian strategy to adopt. The thought of having to call Mountain Rescue fills me with horror. I'd sooner crawl off a hill than put someone else out.

This risk-aversion void is now occupied by a whole host of institutions and companies that barely used to exist: advice bodies, training centres, mountain guides, government-funded information institutions and websites. On and on it goes. None of these organisations or individuals have formal legal competence. But we, the hillgoing public, are expected to follow the advice they give to the letter.

The following list of equipment appears on a popular outdoors website and is deemed 'essential' for summer walking. The list is produced in full:

» Rucksack – about 35 litres
» Boots (with ankle support and soles which will grip on rock, grass and mud)
» Waterproof jacket (with hood)
» Waterproof overtrousers
» General trekking trousers (not jeans or cotton material)
» Thermal top
» Fleece top
» Gaiters

» Warm hat
» Gloves or mitts
» Spare layer, e.g. fleece top
» Compass
» Map (waterproof or in waterproof case)
» Watch
» Torch (preferably a head torch)
» Food and drink
» Emergency survival bag (polythene is OK) and group shelter
» Whistle
» First aid kit (small)
» Mobile phone

The organisation that produced this list is an excellent one. I have walked and worked with them on several occasions. Their members are knowledgeable and – I'm sure – issue advice to the hillwalking public that they feel is sensible and proportionate, no doubt after consultation with various other bodies and individuals. But let's look at that list in some detail. First, there's the cost. Even conservative calculations of the equipment leave little change from £700, and even that doesn't include a mobile phone, which most folk carry as a matter of course nowadays.[3] Multiply twice, assuming a couple want to walk together, and you're looking at something which is borderline prohibitively expensive to average would-be walkers. If I'm completely honest, of the twenty items on the list I routinely carry only half of them, with another four sometimes (normally in winter) and four never. If I have looked at the weather forecast, and seen that it's set fair and warm, I'll carry even fewer: maybe seven or eight. On a fine summer's evening I'll think nothing of walking to the top of my local hill Ben Cleuch carrying nothing on my back except a camera case.

---

3   A decent pair of boots alone will leave no change from £120. A waterproof jacket and pair of overtrousers will set the walker back another £150. Add a cheap pair of trekking trousers in and you're at £300 before fleeces, gaiters, hats, etc. These are all conservative estimates.

Second, the issue of practicality. If one has a settled anticyclonic weather system in summer – with no reasonable prospect of heavy rain or high winds – then taking waterproof jacket and trousers is, as far as I am concerned, an unnecessary burden. Doubtless my profession affects my personal thinking here but in risk management the phrase 'reasonably practicable' is one that is sacrosanct. All this phrase means is the computation of potential risk being weighed against other factors, such as cost.

The third issue I have with lists like this is the charge being levelled at me for being irresponsible should I choose not to take any of the items when I go on to the hill. I would be much happier if websites in general didn't stray into the realms of telling people what to do (i.e. something is *essential*). The term *essential* is subjective.

I remember quite clearly when jeans and trainers were considered acceptable hillwear. Photographs I've seen from the 1980s show kids, some as young as five or six years old, above 3,000 feet wearing wellies, trainers and sometimes sandals. Their 'waterproof' jackets offered only limited protection against wind and rain. Overtrousers, if they were carried at all, were of the same type. Despite this, many responsible parents took their kids up into the hills to try and instil into them the same love for the place that they themselves had. When I say 'responsible', what I mean is parents who were competent hillwalkers and who made sure they'd checked the weather forecast, brought plenty of food and water, and knew their way off the hill in a hurry if things took a turn for the worse. The responsible parent would also limit the scale of their ambition on a given day to the capabilities of the youngest child. It used to be a regular occurrence – as a total percentage of the hillwalking population – to see small children out. Can I be alone in wondering why so few now do it?

No sensible or reasonable person would take issue with a website that gives a handy list of things to look out for when going hillwalking. After all, what is a weather forecast but a guide as to what hills one might want to avoid on a given day? But I think it is unwise and overly paternalistic to advise the general public what is *essential*. If someone must be *told* that they need to carry a map and compass when venturing on to a remote hill the chances are they're not going to know how to use them. Therefore, the authority of self-appointed guardians of our hill safety must always be questioned, and never be allowed to become received wisdom. Attitudes can harden like concrete, so the mix must be stirred to keep it from setting.

Part of the issue, at least for me, is that I feel my personal autonomy is being challenged. I cleave to the dreadfully old-fashioned view that personal responsibility is the single most important attribute a person can possess. From it everything flows. But that is only part of it. For good or for ill, I have always had a contrarian mindset. If the flock flies right, my instinct has always been to fly left. Over the course of my life I think this attitude has helped me, not hindered.

Does anyone, really, like being told what to do?

* * *

In terms of the many specific changes to equipment and practices over the years, some have been welcome, but others less so. I'll speak about a few here.

The first one is the route card. This device is – as the name suggests – something that is filled in prior to going into the hills. At its most basic level it is no more than telling someone close to you where you intend to go and, roughly, what time you expect to get back. Some opt for a more formalised version and write it down. It is then given to a loved one or someone in a position of trust, the better to find you if you fail to return by the allotted time and end up lost or incapacitated. A route card can also be displayed through the windscreen of one's car.

I have two main objections to route cards as 'aids'. The first is that filling in such a thing ties a walker to a set journey, with little scope for deviation to a different glen or hill if the weather turns. I have frequently had to change plans when faced with poor or atrocious weather. By so doing, did I leave myself open to the charge of being irresponsible, or should I have just ploughed on against the weather because that's what I said I was going to do? Did deviating from my predetermined route

162

place an unacceptable risk on those that would seek to rescue me? It also goes against the spirit of being outside. Or, as Mountaineering Scotland state on their webpage on the pros and cons of route cards:

> The imposition from outwith the mountaineering fraternity of an obligation to complete an over-elaborate route card for every expedition however mundane is not acceptable to many mountaineers. Such a requirement implies that hillgoers are in some way accountable to officialdom. This conflicts with the ethos of mountaineering.[4]

The second objection I have to them in the form of physical cards displayed in vehicles is derived from the tongue-in-cheek term *burglar's friend*. Advertising where you've gone and what time you are likely to return is a lip-smacking prospect for a potential miscreant. 'Dear burglar, you have at least two hours available to ransack my car, safe in the knowledge that I'm up on the hill and can do nothing about your criminal act.'

All that said, the *premise* of telling someone where you are going is not really mountain-craft, but good manners and common sense. If travelling somewhere solo I always inform a loved one where *roughly* I'm going, and I'm sure many folk like me do the same. What I am wary of in the future – and I can see a day when it happens – is that people will be obliged to inform a government-aligned or approved website where they intend to go if there's a chance they might need rescuing. Pressure will be put on the hillwalker that if they *don't* do this then their chances of being rescued – and thereby of surviving – will be reduced.

The best and most profound advance in technology over the last ten years when hillwalking and mountaineering, without question, is the ability now to navigate without error. Satellite-based navigation systems, such as the omnipresent Global Positioning System (GPS), have revolutionised navigation and are used by, if my experience is anything to go by, a good deal of the hillgoing population. Gone now is the need to keep an Ordnance Survey map around your neck in a large, transparent pouch.

---

4   www.mountaineering.scot/safety-and-skills/essential-skills/navigation/route-cards [accessed 13 May 2021].

(Having one of these constantly blowing in your face when you're walking into a stiff wind is a pain.) A mobile phone can carry the full 1:25,000 Ordnance Survey Landranger series in its memory and can pinpoint the holder to within feet of their location. If someone is on a plateau in thick mist and wishes to walk in the right direction, then all they need do is remove the phone from their pocket and the application will show them not only where they are, but what direction they are facing.

Some purists eschew using these devices on the grounds that it encourages laziness. Knowing how to use a map and compass, they say, is essential, and preferable to mobile technology. This position is broadly fair, but a tad curious. Technology makes life easier and – ultimately – more enjoyable. We no longer wear hobnail boots and tweed jackets, as we now have better waterproof materials and equipment. It is more lightweight. Does the purist who waves away the GPS also cling to his or her Victorian ice axe and walking breeches? Things move on. They evolve.

'What happens if the battery goes?' is frequently posited by way of a counter. To this question I answer that I always carry a small powerpack in my rucksack in case it is needed (although in ten years it has never emerged from my pack). In any case, my almost obsessive interest in maps over the years has allowed me to have a reasonable internal map in my head wherever I tend to be walking. Prior to going somewhere I've never been before, I have an idea of the hill and its topography based on looking at the Ordnance Survey maps. Personally, I cannot understand anyone who *wouldn't* do this. Maps are wonderful.

So good is the technology available currently that I will confess to not carrying paper maps and compasses on many of the trips I undertake. The brand of mobile device I have used over the last ten years has never once let me down on the few occasions that I have had to use it in anger.

To summarise: unless the current climate of risk aversion reverses or slows down then we will hear increasing clamour for hillwalkers, or people whose pastimes may require them to be rescued, to carry some form of insurance. In effect this will professionalise a cohort of people who currently do it voluntarily. To be fair to the various mountain rescue groups, part of the cohort to which I refer, they have always resisted the compulsory carrying of insurance, and pooh-poohed any notion of people being charged for their services. This is good, and I hope it continues. I fear, though, that the tide of public and political opinion which waxes will be irresistible. We shall see.

In terms of technology there is little doubt that some modern items are a step forward. Lightweight waterproof gear, boots and rucksacks make for a far more pleasant walk or trip compared to the old kit. As discussed, maps able to fit on the palm of your hand *and* which show you the direction you're going in are likely to limit mountain rescue call-outs.

* * *

A couple of years ago, I asked two friends a series of questions during a walk. What would they rather be without, their internet-enabled phones or their cars? The phone lost, perhaps predictably given the value people place on convenience and independence. What about a cooker? There was some debate about this, but in the end the cooker won out three to nil. A washing machine? This is where opinion started to differ. One of my friends stood adamant that the lack of washing machine was sur-mountable, while my other friend and I said not. (Perhaps tellingly, the friend who opted for the washing machine had two children.) What surprised me, when asked the same question about the following items, was the phone won either a majority or clean sweep: TV, chocolate, alcohol, books, laptop or computer.

Now, the three of us are all men in our mid-forties. If we were to ask men and women ten or even twenty years younger the same questions, would we get the same answers? Probably not. What point am I trying to make here? Mobile phones and their associated applications have become absolutely and utterly ubiquitous, to the extent now that many people wonder how on earth they'd manage without them. There's no need to list here their uses. Chances are you have one either in your pocket or lying close to you. From an outdoors perspective, as outlined earlier in this and other chapters, they fulfil a variety of uses, from navigation to camera, from notepad to weather forecasts.

When I am on the hill I very seldom, if ever, look at my phone, and only then to confirm a coordinate or use an application to determine a distant peak that I cannot identify by sight. It gives me peace of mind to know it's there if I should have need of it, but I try never to rely on it. In ten years of carrying it as a navigation aid I have seldom had cause to use it when extricating myself from genuine potential peril.

You can probably bring to mind, without too much trouble, news stories where a mountain rescue team has been called out to escort lost hill-walkers off some Munro, Lake District Fell, or Welsh 3,000er. Almost without exception the excuse for getting lost will be related to not having the correct navigational tool. In multiple cases I've seen and read about, the rescued person or party was at the mercy of Google or Apple Maps, thinking either of them sufficient. And this is the fulcrum of the problem. So dependent have people become on their phones that they are lulled into a false sense of security. 'If I have my phone I'll be all right' seems to be a prevalent attitude. But even the very best and most helpful pieces of equipment are only as good as the person to whom they are attached. The most up-to-date, super-flashy model of mobile phone is useless and worthless if it does not have the correct type and scaling of map on it, or if the user lacks the wit to operate it. Reliance on technology and tech-nology alone is a single point of failure. This is indeed a modern peril.

* * *

As much as over-reliance on the mobile phone as a navigational tool is a modern peril, there are others. The one I have specifically in mind, however, is also a boon. I refer to the pervasive, inescapable spectre of social media.

From the perspective of research, social media – if used correctly – is one of the most useful tools currently available. At a talk in Aberdeen in late 2019 given by Cairngorms National Park Authority, I heard how their staff used an exercise-tracking application to work out where people were going in forests, the better to see how this could affect the mating habits of rare, ground-nesting birds: a really useful way of harnessing a technology in a way that the inventors could not possibly have known at the time. The architects of the social media sites I utilise probably *did* have this sort of thing in mind when they concocted it. The ones that provide the largest number of visuals (somewhat predictably, Facebook, Instagram and Twitter) are the ones that I frequent.

Contained within these three sites is a truly vast repository of photographs and information. To complement these there is usually a refreshingly large store of goodwill. It typically goes something like this: either someone has a day off work or it's the weekend. Intent on making the best of the good weather that is forecast they take a trip to a hill they've been meaning to do for ages. On the way up they'll take lots of photographs, usually on their phones. When they reach the summit, more photographs, and a video of the panorama. Sandwiches are consumed and the happy walker heads back down the hill to the car or whichever mode of transport awaits them at the bottom. Once home and showered, the walker uploads their photographs on to their social media platform of choice, which their friends and other members of the public will 'like' or comment on. This is where I come in.

At the end of a weekend of good weather in, say, late April, it is an absolute racing certainty that hundreds, if not thousands, of people will have ventured to the hills of Snowdonia or the Lakes and uploaded their photos. A very small percentage of these photographs will have displayed in them some of the last patches of snow visible, usually by accident. The trick for me is to find these photographs among the mass. It can be needle-in-a-haystack stuff. But all it takes is one good shot with a decent camera. Bullseye. This stranger's photograph has saved me the trouble and expense of having to go to the same place over the next few days. I will often comment on the photograph to see if the taker has a larger-scale version that he or she can forward to me so that I can better estimate the size of the snow that has almost gone unnoticed by them. Often pleased that someone has commented or taken an interest in their photograph, the owners of the snapshots are (usually) only too happy to send you what you ask for.

Because the work that I am engaged in has gained a larger following over the last few years, I often get unsolicited emails and messages via social media. Appearing on national TV and radio a few times has raised my profile, to the extent that when people do message me it's usually prefaced by words like, 'You're that guy who does the snow stuff' or 'You're that snow-patch guy.' Often the photographs they send are of no use at all. Occasionally, though, I'm sent a cracker; a genuine beauty of a patch that I either thought had vanished or didn't know was there in the first place, though these are quite rare. Whatever the quality of the photograph that is sent, I always thank the senders, who were good enough to go out of their way to send me the fruits of their labours. This is where social media is a terrific and valuable tool. The sharing and dissemination of certain information and pictures among the greater public is highly useful for research.

For every positive, alas, there is a negative.

* * *

23 August 2018

I should have known better. The internet – social media to be exact – doesn't do nuance or self-effacement very well. I (re)discovered this to my cost this evening when, on a popular Facebook outdoor page, I posted a provocative piece on not carrying certain equipment when hillwalking.

Current orthodoxy dictates that venturing above 2,000 feet in the UK necessitates getting kitted out for just about every conceivable eventuality. Things we are exhorted to consider are, amongst others, health, hygiene, first aid, weather, navigation, food, water, having a guide, terrain, footwear, clothing, route cards, map, compass, etc. *ad infinitum*. Not having or doing some of these things is, apparently, tantamount to a death wish, and is placing the walker (and the potential rescuer) in great peril. While there is absolutely no doubt that some of these things are valuable some of the time, the idea that all these things are needed all of the time is a nonsense.

In the post, I suggested a competent walker, familiar with the hill and terrain he or she is on, has absolutely no need for a map and compass.

Moreover, a waterproof jacket and trousers are also unnecessary in summer if you have a settled anticyclonic (i.e. a big high-pressure) weather system. I finished the post by adding that, on top of all this, the hills I had ascended the previous day (Aonach Mòr and Aonach Beag, both above 4,000 feet) were shrouded in thick mist when I climbed them, and I had little visibility. The level of vitriol in the responses was breathtaking.

Despite me pointing out that I was familiar with these hills to the point of saturation and knew exactly where I was going, regardless of the mist, I was classified as reckless, having a death wish, a fanny, arrogant, akin to a camper who leaves his rubbish after breaking camp(!), and various other forms of nasty personal abuse. No matter to what degree I was reasonable or considered with my responses, the bile continued un-abated, to the point that it was impossible to reply to everyone. (At all times I kept in mind Lao Tzu's adage of *responding intelligently to un-intelligent treatment*.) The moderators, not content that I was getting a proper pasting, weighed in and deleted the post, handing me a twenty-four-hour ban into the bargain.

So, what does this tell us?

In some ways it is indicative of the current poverty in general discourse. Intolerance is everywhere and immediate. My post was a reasonable one, albeit provocative and probably controversial. So what? I would have thought that is what online forums are there to cater for. Ninety-five per cent of the contributors, as well as the moderators, thought differently. They clearly imagined it set a worrying precedent. Someone with an alternative and dangerous point of view. One which didn't chime with the current tenet of safety above all else, even if it is to the detriment of one's enjoyment. My views were to be silenced and expunged.

Having the correct equipment for the conditions is essential if you are to go alone on a long day. But being in possession of all the things that you are told you need is completely unnecessary. Take, for example, a map and compass. I was told by virtually everyone in the long list of replies that these were required pieces of equipment. 'Why? What use is there in carrying something that you will never take out of your bag, because there is no need?' I asked. Among the many replies came banal statements like 'You might get disorientated' and 'You could get hit by a rockfall and the landscape changes'. To the latter suggestion I commented that if I were to be hit by a rockfall then not having a map and compass would be the least of my worries.

Once upon a time an opinion like mine would have been commonplace. Before the era of advice bodies, interest groups, mountain rescue teams and overzealous health and safety fanatics, people could be trusted to make their own decisions based upon an apprenticeship of going to the hill. They would build their knowledge slowly and surely, graduating to longer days and higher summits. You knew that there was no realistic prospect of being rescued, therefore you had to be more circumspect with your ambition. Those days are long gone. Nowadays, in an era of instant gratification, those long-held conventions have been bypassed. Too many people are rushing out ill-equipped and inexperienced. Mountain rescue call-outs are ubiquitous. The net result is that every social media poster seems to be a finger-wagging paternalist, chiding their naughty offspring for daring to not follow their advice to the letter.

Too many people, unthinkingly, swallow everything they are told when it comes to what one must do and what one must not do. I, for one, won't have it. If I think the advice I've been given is unnecessary or over the top, then I am absolutely within my rights not to follow it.

What I find most disconcerting of all, though, is the way in which debate, however reasonable it may be, is getting more and more difficult if you have an alternative point of view. Any casual observer of the political scene will know what I'm talking about. If someone disagrees with what you say they will shout and harangue you. They will resort to abuse and to invective. Over the last five to ten years there has been a noticeable breakdown in civility in this arena. There are many reasons for this, which are too long and too numerous to outline here. Suffice it to say that I find it depressing.

I have removed myself from that particular online group, incidentally, and all the online scrappers. It is impossible to wrestle with chimney sweeps and stay clean, so I shan't bother.

* * *

In the eighteen months since I wrote the above piece the situation has not changed. What *has* changed is that I no longer engage in any form of online dialogue. It is just about pointless. As the Argentinian writer Jorge Luis Borges said about Argentina and the UK fighting over control of the Falkland Islands, it is 'a fight between two bald men over a comb'.

On balance, however, social media is – despite its undoubted flaws – an immensely rich and useful seam of information. It often removes the labour- and energy-intensive process of going to a particular hill – energies and attention that can be diverted elsewhere. It is a medium that is here to stay, whether we like it or not.

eleven

Alone again (naturally): meditations on the outdoors

Solvitur ambulando. [It is solved by walking.][1]

'Why isn't it moving?' I say to myself. I crouch down and try to make myself look smaller, less intimidating. Walking slowly and deliberately towards the ball of feathers that I can see no more than fifteen feet away from me, I puzzle at the situation. Some of the feathers are white, but most of them are grey. It's definitely a ptarmigan. If I can just sneak up a bit closer to it I can get a good picture. 'I wonder why it isn't moving,' I say to myself, again. About six feet away now I look at where its head should be. There's no sign of it anywhere. Is it looking the opposite way? Wait a minute. I straighten my back and stand upright, directly over the bird. The reason I can't see its head – and the reason it isn't moving – is because it doesn't have one. Ah. Peregrine. Must've been a peregrine. An eagle or fox would've taken the whole thing. A peregrine ripped the head off and flew away. A tasty treat to finish breakfast, maybe.

It's a cold wind up at 3,700 feet. The T-shirt on the walk in seems a long time ago. At least the sun is out, though. It takes the worst of the chill out

---

1   Popular theory has this quote credited to St Augustine, though its origins may be much older.

the air. Below, at the bottom of the glen, the wind is whipping up white horses on the loch. I shall be glad to have that on my back walking home.

I pick my way carefully down the loose, steep slope. The rocks are as well bonded to the ground as my superglue repairs tend to be (i.e. not very), so care is needed. At this altitude it is still much too high for heather, whose root systems help knit the soil together and bond it. The vegetation here is a light brown grass that is about as hardy as it gets. I'm making good progress down it to about 3,000 feet, whereupon I hit the heather, which starts patchy and sparse but gets deeper with every few downward steps I take. I place my foot near to where I guess the ground will be best able to support it without turning an ankle. I repeat the process hundreds – thousands – of times descending from the plateau. You get used to it. In truth I enjoy it. The idea of it is worse than the reality. At least to me it is. The key is to keep the joints loose but poised. An unseen stone can roll your ankle any which way. A few of these can be anticipated on any given walk. It's unavoidable.

Every now and again, usually when I'm within about six feet or so of it, a brilliantly camouflaged grouse springs up from the heather like a ghost in a horror film, and it almost causes me heart failure. This, too, can be expected as a matter of certainty. It doesn't make it any less of a fright.

The mat grass that provides such a verdant hue to the stream courses in summer has changed colour to the yellowy-brown shade of autumn. There are many of these silvery water veins cutting down the hillside, creating a natural block between great swathes of heather. Nature's firebreaks. Heather itself does not like waterlogged ground. It is true to say that, as a general rule, if one walks on heather, one's feet will remain largely dry. The pay-off is the risk of turning an ankle. But heather also dislikes snow. If snow lies in one location on a hillside for any longer than a few months unbroken, year after year, the heather will struggle to take hold.

Anyway.

As I descend the hill, through the heather and past the streams, and as my heart returns to normal after the grouse episodes, I spot a large object moving in the sky. It is often said that if you see a big raptor overhead and you can't work out whether it is an eagle or a buzzard, it'll be the latter.

But there is no mistaking this bird. It is huge and soars on winds that are being whipped up by the ridge to the west. It's spotted something, I think, but seems hesitant to do anything about it. After a few minutes it turns towards the wind, without as much as beating or flapping a wing, and disappears over the crest of the hill. No one who has seen a golden eagle can ever forget it. More than any other living thing in Scotland, the red deer and wildcat included, it is the embodiment of the wild. In many churches, lecterns are fashioned from wood in the shape of an eagle. From ancient times it was believed that the bird could stare straight at the sun. Christians, with their apparent ability to gaze fearlessly at the revelation of the divine word, adopted the eagle as a metaphor for their faith. Roman legions used them on their standards. Eagles have long been admired by us.

I reach the bottom of the slope and pick up the path, at last. After several hours of traversing uneven ground, it is good to once again be on a semblance of something that is kinder to the limbs. The wind is now at my back as well. Only a few miles left to get back to the car. I walk for a few minutes. Out of the corner of my eye, off to the side of the path and semi-buried in deep peat, the bleached remains of ancient pine tree stumps and roots lie. Some may be thousands of years old. With their white root systems spreading out in all directions they look like massive octopus carcasses that have been petrified by staring at some Greek monster or other. Back in a time when this land was warmer and less wet, the trees thrived here. But Scots pines dislike waterlogged ground, so eventually these ones succumbed to a changing climate and no shoots could germinate from their spread-out cones. Once the trees died and had fallen, the peat overtook the stumps and buried them. Only now, as we live through a comparatively drier spell, the peat diminishes, and the stumps re-emerge.

I look up, high on the side of the hill to my east. The descendants of these bleached skeletons stand on better drained ground. They sprout seemingly from between large boulders. The terrain around them is rocky, enough to deter the sheep and deer from feeding on them. Small pines grow at their parents' feet. A good sign. On the second-from-last tree, jutting out at an angle, a huge pile of twigs and sticks is visible through the binoculars. It can only be an eagle's nest. An eyrie. Nothing else could be so large. The tree must be over 200 years old. Its thick trunk has withstood scores of years of weather but shows no sign of being defeated. Perhaps the eagle's nest is ancient as well? Some nests can be used for generations by the bird.

The canopy of green contrasts with the sun-bathed pink granite of late afternoon, and as I round a corner I hear the whoosh of the river as it tumbles off a small ledge. A dipper, keen to exploit the turbulent, fast-flowing water, sits on a rock by the riverbank, bobbing up and down. It springs from its perch and disappears into the noisy cauldron of water. After emerging a few seconds later, I marvel at how the water drops off its down as though it were coated in wax. The white bib it wears stands out against the dark water. Luckily, I see two red squirrels corkscrewing around the trunk of a nearby large pine tree that sits close to the river's edge. They seem to be playing. Another treat on a trip that has already provided plenty.

As I cover the final few hundred yards on brittle gravel, the sun has just set behind the large hill to the west. The taller of the hills to my east are still bathed in its glow, but in the river valley it is already getting darker. At this time of the year the temperature drops quickly once the sun has departed. I can see my breath by now and a frost is sure to form. The gravelly path crunches hard as I arrive at the car. Then, silence. A few birds chirp somewhere close. The forest in the distance catches the last of the light, its canopy emerald-coloured by the horizontal rays of the sun. The snow I had been to many hours ago is now a memory. So much of the day was spent walking to and from it. The snow is the reason for the trip, but it would be so much less of a day without the memorable, if brief, sideshows that go on all around.

* * *

The devotion
Although this book is almost exclusively about snow and the hills where it reposes, there is more to seeking it out than just the act itself. In going

to these shrines of winter, walkers spend an inordinate amount of time in a semi-wilderness environment that they would not normally visit. That being so, they are disproportionately exposed to – and influenced by – the unusual and thought-provoking things they see, hear and feel around them. For many this may be a profound experience; for others not. Lots of folk are content just to walk to the top of a hill and back down again. A day's exercise such as this is a fine thing on its own and need be examined no further. There are some of us, however, who struggle to leave the memory of it behind so easily.

Hills and mountains speak to people like us in a way that few other things are capable of. They invoke a primeval memory within. In bygone days Palaeolithic and Neolithic man *generally* avoided these lofty places. Little in the way of food could be gathered there, and the weather was often far less reliable than at ground level. Therefore, mountains were, in the main, to be admired for their beauty but, for their unpredictability, kept at arm's length. This mindset remains with us today, I submit, through a process of atavism. A good deal of the public avoids going to the high places, in part because of an association with danger. (It may be postulated that humans' innate fear of being in forests at night is similarly atavistic. It is no coincidence that writers and filmmakers through the ages have used wolves in forests as a plot device to invoke in the reader or viewer a primordial fear.) It cannot, surely, be any kind of accident that some of the language certain people use to describe climbing hills or mountains is so adversarial in its tone. Peaks are not climbed but *conquered*, as if they were foes to be triumphed over. Hills are *bagged* as though shot for sport, like grouse in August. Walkers talk of a *plan of attack* when contemplating the best route up the hill. Part of the reward for many walkers is the overcoming of this physical and psychological barrier. Reaching the top of a high hill can become addictive – an obsession, almost.

As much as some walkers seek to *conquer* a hill, I believe that in some measure it is a conquering or overcoming of *themselves* that is the foundation for their desire. But a hill can never be *conquered* by a walker. For a while he or she may stand on its summit and gaze out to the view that will have changed little since the last Ice Age, but the moment lasts only fleetingly. When that walker goes home and to bed, as they will do that night and every night for the rest of their lives, the hill will endure, even after the memory of the *conqueror* is long forgotten. The notion that one can *conquer* or *bag* a hill is an anthropocentric view, short on humility.

For some others, the act of going on to the hill is akin to devotion. These people are, perhaps, less concerned by the act of ticking hills off a list. They are content just to climb a hill – any hill – for no other reason than its own sake.

Church attendance figures in the UK may have declined dramatically in the last fifty years, but our desire for some form of spiritual fulfilment has not diminished at all. Mankind, uniquely, has evolved with a divine curiosity. Since forever, all over the world, *Homo sapiens* has looked around the land it occupies to imbibe meaning from it. But not just the land. Gods of the sky that represent the sun, moon and stars were common in many countries for thousands of years. Tree deities, water nymphs, mountain gods: these historical attempts to understand and give context to that which we are surrounded by were invented by our ancestors to be suitably reverential deities. Moreover, our devotional pull was so strong that we built places where it could be better practised. Whether Neolithic man's erection of stone circles, Ancient Egypt's construction of the pyramids, or the creation of the great Gothic cathedrals of medieval France, our yearning for answers gave rise to us assembling not only a set of belief systems, but huge physical structures to validate and propagate them. Do those without religious beliefs worship the mountains and hills in the same way? Are the people who hitherto would have gone to church now adopting a different form of service when they walk up a hill as opposed to kneeling in a pew?

Humanists and others who evangelise for the removal of religious studies in schools imagine that children freed from its shackles will grow into adulthood without passing on any inherited spiritual beliefs to their own offspring and, so, eventually, it will be weeded from our society entirely, possibly within a generation or two. From this religious extinction, they suppose, a new, fact-based humanity will emerge, with science at its heart and reason as its God. I believe this to be fundamentally faulty thinking. What they are displaying is faith, not scientific reality. A moment's reflection on the history and nature of our species shows we are incapable of relinquishing the need for a sense of mystery or spiritual fulfilment.

Even if science does work out the genesis of the tiny particles that life or the stars are constructed from, people will continue to look for answers and a sense of the divine. Not necessarily in organised religion – which is just a formalised version of devotion – but also in the tangible, everyday terrain that surrounds them. No scientific theory or proof can eradicate this.

Few if any of us now believe in Olympian gods atop mountains, or water nymphs that leap up and down streams. Plenty of folk, however, still look to the outdoors to quench their spiritual thirst. Though many of these people may not necessarily be religious, they are simply acting on a profoundly devotional impulse that is as old and strong as any of the major religious tenets that steer others in the direction of the mosque, church or synagogue. There can be few human needs whose roots run so deep. They will be part of our DNA for a long while yet.

* * *

The island

If one looks at a topographical map of Britain, several features are obvious. The clearest of these is that the landmass itself is homogenous. From Dartmoor to Torridon there is an irrefutable interconnectedness. Several other features are distinct too. The south and east parts of England are largely flat, having only minor hill ranges to ruffle the otherwise smooth and contourless landscape. Wales and Scotland are the opposite, being predominantly mountainous but pocketed with fertile land along river courses and in the lee of high hills. (The Wye Valley in Wales and the Findhorn in Scotland spring to mind.) The hills of northern England rise near Manchester and extend in an almost unbroken chain, via the Pennines and Southern Uplands, to within striking distance of Edinburgh.

One of the consequences of my long love affair with maps was that, from a very early age, I didn't really understand the political concept of borders. I could tell easily enough where county and country borders lay, but they didn't really mean a great deal to me. The national borders on an Ordnance Survey map were so small and insignificant they barely registered. I was interested in natural features: where rivers originated, cliffs, standing stones, chambered cairns and old places with Gothic writing. Arbitrary lines on a map that were man-made and, in any case, modern, didn't interest me. Streams and rivers crossed borders uninhibited, such as one of northern England's rivers, the Tyne, whose ultimate root lies in Scotland.[2] Hills lay in two countries at once. Place names reflected groups of people, not which country or county they resided in.

Not that I had the necessary critical faculties to understand any of the significance of these observations at the time, of course. Only through subsequent reflection did I come to understand that part of my philosophical and political outlook has been shaped by an interest and love of physical geography.

The very northernmost hills in England are indistinguishable from those in southern Scotland to the extent that few people could tell the difference. The Lammermuirs in south-east Scotland would sit quite happily on the Welsh Marches and no one would think them out of place. The Lake District and the Trossachs are cut from so similar a cloth as to be hard for all but a resident of either to tell them apart.

There is a *Britishness* to all these landscapes that is immediately recognisable to me. I say this without a trace of jingoism or chest-beating patriotism. I believe in neither. My sense of what constitutes 'Britishness' is not rooted in a confected, man-made idea of nationhood. (I recoil at flag-waving in general and don't buy into loyalties being dependent on what side of a line on the map one lives on.) The Britishness I am thinking of is purely a convenient term with which to describe a landscape that is wrapped up in its, amongst other things, geographical, zoological and

---

2   The Kielder Water, which is the main tributary of the River North Tyne, has its origin on Hartshorn Pike in Scotland, whereupon it is called the 'Green Needle'.

geological history. (The name Britain, with reference to the island containing Wales, England and Scotland, has been with us in some form since Ancient Greek times.) Britain will seldom be mistaken for anywhere else. There are, certainly, times when it can look more continental (as when the Highlands are blanketed in thick snow, or a Cornish beach in summer looks like a Greek island). In the main, though, our hills lack the height of the great world ranges such as the Alps, Pyrenees, Rockies, Andes, Himalaya or even the Atlas Mountains. Though they may not possess the magnificence of these splendid ranges, they have a character way out of all proportion to their modest height.

Walking in the Lake District about ten years ago, I passed a couple of what looked like experienced hillwalkers, probably in their mid-forties. Being in late May, I was heading for Great End, a hill that falls just short of 3,000 feet, but which carries long-lasting snow in most years. The couple were from Bolton. Neither they nor I seemed to be in a particular hurry to go anywhere, as the day was hot and time was no barrier. As we talked, as hillwalkers who are not in a rush are apt to do, I could tell they were seasoned outdoors folk. We discussed some of what we considered to be the finest places in Britain. They spoke fondly of Torridon in Scotland, as well as Dartmoor in the South West of England. We spoke for about ten to fifteen minutes before going on our respective ways. As they hopped down the flagstone path back towards their car, something very odd happened. Just for a minute, I literally had to do a double take as to where in the country I stood and where I was going to. In the short space of time I had been speaking with the couple I had lost the sense of my location and which hills I stood among. This moment passed within a few seconds, of course, but it was a genuinely peculiar feeling. That night, as I sipped on a drink and pondered the day's success (I found some snow on Great End as hoped), my thoughts looped back to the instant after I chatted with the couple from Bolton.

In that moment, standing there in the pleasant afternoon warmth, it really didn't matter *where* I was. In that few-second gap I could have been anywhere in upland Britain. There was something remarkably familiar about it all. So accustomed was I to the different smells, vegetation, rocks, humidity, temperature, streams, trees, walls, noises and wildlife of so much of upland Britain that it all became interchangeable. The uplands of Britain are connected by being on the same land mass, but there is more to it than that. Something intangible, perhaps, but no less real to me for that.

My feelings about this have, over time, been reinforced because of where I have lived. For a number of years, I dwelt on the small Scottish island of Iona, just off the west coast of Mull – itself an island near Oban. What became patently obvious to me whilst living there was the sense that the islands of the Inner and Outer Hebrides were their own country. Whenever I left Iona to visit family or friends on the mainland, it felt like I was travelling to a different nation; the same being true when coming home. More than the journey itself, there was – and is – something *different* about the islands. Something un-Scottish, un-British. Just *island-ish*. I have an enduring and deep love for Iona and Mull, and might still be living there yet if circumstances had been different, but they and other islands are, to me, apart. Maybe it is to do with the act of crossing water acting as a gateway to somewhere else. An invisible threshold, maybe. I'm sure that is part of it, but by no means all. Some Orkney and Shetland folks say much the same thing.

Britain, the island, has a range of upland areas through which a thread runs. This thread is made up of and spun from the strands each of them have in common, such as the flora and fauna. As far as I am concerned there is no difference between Exmoor and the Pentlands. They are all equal parts in the wonderfully and colourfully varied island we live on.

* * *

The realisation

Most of the walks I undertake, either as part of my research or in general, are done alone. There are two very good reasons for this. The first reason is that often the places I go to for snow research are not easy to reach, as will be amply clear by now. I generally do not wish to inflict the experience on people who are unused to difficult conditions. Even seasoned hillwalkers and mountaineers that I've been with voice surprise at just how tricky the terrain can be. I confess that, within reason, I enjoy walking on such ground. I don't know if this is partly masochistic or what. All I know is that the more challenging the terrain, the more fulfilling I find it. That's part of it. The second reason I tend to walk alone is, on reflection, less complex.

As a child, maps fascinated me. I couldn't get enough of them. Whether it was old AA road atlases or second-hand Ordnance Survey maps it really didn't matter. I pored over every one of them, intrigued at place names, rivers, hills, ancient monuments and everything else in between. It mystified me why no one I knew really shared this passion. Maps were my friends and therefore precious to me. In geography classes I excelled at anything to do with them. One of my several mediocre teachers over the years used to eye me nervously if anyone in class asked a question that had any cartographical connotations, knowing that I'd pounce on any form of slip-up.

One of the consequences of this passion for maps was a realisation that I was *alone* in it. I understood quite quickly that it was something else I seemed to like that none of my peers did. It's not as if I was a loner, either. I was a socially gregarious boy when young. I enjoyed football and other sports. I ran fast and enjoyed being boisterous. But I soon realised that the things I found truly meaningful and which had any form of inner worth were to be found by doing them alone. Nobody else *got it* the way I did, as it seemed to me. As a teenager I used to lie in a dark room listening over and over to the same piece of music for hours, not wanting to share it because I felt embarrassed and I thought nobody else would understand. I would read my atlases and maps alone and not tell anyone, for the same reason.

When I got older and started venturing out into the hills, or just walking some place in general, I found that I enjoyed it more when I was alone. I didn't have any distractions, or people to hold me up. I could come and go where I pleased, whenever I liked. The story remains pretty much the same right up until today. But that is not to say that great trips are done exclusively when one is alone. Some of the very best and most memorable days have been with others. Good companionship on the hill is a precious thing, especially sharing a special moment with someone who understands what is unfolding around them.

I suppose, then, that when I walk to a fast-disappearing patch of snow by myself part of it is me reminding myself of being young. The boy who found interest in the things that other people didn't; the nine-year-old who looked across to Ben Lomond that day in May 1983 and saw a small drift sitting alone. Maybe, even, by seeking out the last drift of solitary snow, what I am really trying to do is revisit a part of my childhood that I've never really been able to get over or come to terms with.

But, equally, it might just be that I'm a hopeless geek.

epilogue
The A9 and the canine

The first fall of snow is not only an event, it is a magical event. You go to bed in one kind of a world and wake up in another quite different, and if this is not enchantment then where is it to be found?

– J.B. Priestley[1]

Quite a few years back, when I had posted something or other on Twitter about the depth of snow on a Highlands hill, I started to correspond with another Twitter user. It transpired that, like me, he was an avid hillwalker and very keen specifically on winter treks. His desire to get on to the hill during that season was, to use his own subsequent words, 'Because snow makes the world feel very different. It makes everything wilder, far more dramatic, perhaps even with an edge of danger and risk. It covers up all man-made things. It makes a noisy world quieter. You feel totally lost in nature when out in the snow.'

After some to-ing and fro-ing with tentative suggestions to meet up, we aligned our calendars and arranged to go for a snowy, midweek sprauchle on Ben Lawers, a huge hill reaching almost to 4,000 feet, right in the heart of the Southern Highlands. The weather on the day we met

---

1   Priestley, J.B. (1928), *Apes and Angels: A book of essays*, London: Methuen & Co Ltd.

in the car park was, at best, patchy. Low cloud swirled all around us, with mist threatening to reduce the walk to a semi-blind stumble. But, still, there was plenty of snow to go at and the forecast suggested that an inversion was possible.[2] For their beauty and transformative effects on the landscape, inversions are highly prized by walkers in winter and spring. Emerging from a cold 'pea-souper' of a day into brilliant, relatively warm sunshine and clear blue skies is one of the great joys of hillwalking. It lifts the spirits and provides views that can, literally, take the breath away. Anyway, when we left the car park that morning there were no guarantees on the weather, only the promise of snow. That was enough.

On the walk that day with us was a dog: a chocolate Labrador, called Olive. With her unquenchable enthusiasm and seemingly boundless energy, Olive made a memorable companion. I could have had no idea at the time that this unassuming, placid and hungry hound would go on to become – it is no exaggeration to say – internationally famous as one half of the Olive and Mabel Labrador pairing that quite justifiably deserves the handle of 'internet sensation'. Olive's (and Mabel's) owner, and my walking buddy that day, was, of course, BBC commentator Andrew Cotter. During the 2020 lockdown, Andrew put his energies and, frankly, abundant Covid-related spare time into making a series of videos that went viral, and which have since been viewed literally tens of millions of times. But on that quiet, midweek April day at Ben Lawers, it was just Olive, Andrew and me. Mabel was only the proverbial glint in the eye of her doggy parents. But what a day we had. Ben Lawers, resplendent in miles of unbroken spring snow, was an absolute treat. The inversion that we hoped for duly arrived halfway up the hill, exposing only the top 1,000 feet of the surrounding hills in their naked glory. As we climbed higher still, we saw the thick cloud hide the cols that conjoined the neighbouring hills, giving them the appearance of lumps of ancient rock emerging from a grey sea. Nothing could have been finer.

---

2   An inversion is when low cloud sits in the valley below, due to the cold air sinking to the bottom of it. But as height is gained, the temperature rises slightly, so that one bursts through the cloud and looks down on top of it, not unlike when an aeroplane emerges from the gloom of a wet and miserable day into clear, unbroken blue skies.

Watching the interaction between Olive and Andrew that day was fascinating. As someone who didn't grow up in a dog household and had never owned one, I was always puzzled as to why so many people invested so much emotional (not to mention financial) capital in them. To me they were nice playthings, but they seemed to be awfully hard work – as well as being quite a tie. I was never a fan of the *wet dog* smell either. Also, when I was young I saw a couple of dog-attacks on children, so I had a bit of reticence about their behaviour. However, this was the first time I'd ever been in the company of a dog for such a long snow walk. And, I must confess, by the end of it I was smitten. Watching Olive that day, bounding around with absolute, unadulterated joy, was a real tonic. Glissading, or *bum-sliding* to give it its less technical name, down a steep bank of snow was of particular delight to Olive. She didn't seem to get it. 'Human Is Moving Without Moving His Legs' seemed to flash through her head – so far as anything can flash through a Labrador's head. This activity elicited a lot of barking and jumping up and down. All of this, of course, was met with sheer adoration in Andrew's expressions and actions. Many chicken-y treats were dispensed that day, each of them eaten more quickly than I thought physically possible. For a first walk, it was definitely memorable.

It can be a risk meeting someone in the flesh on the strength of their social media profile and limited interactions you might have had online. Fortunately, though, Andrew and I seemed to have a good deal of outdoors interests in common, and the day passed all too quickly amid stunning scenery, steep banks of pristine spring snow, and dog-related fun.

Over the next couple of years, we met up regularly in winter and would go for midweek snow walks in out-of-the-way places, with Andrew always keen to avoid the honey-pot spots that tended to draw the crowds. In terms of fitness and passion for the hills we made good companions. There was, and remains, a tacit, unspoken respect for each other's abilities and capacity to get up at 4.30 a.m. to drive 200 miles just to grab the opportunity of a good day out. The latter is absolutely essential if one is to seize the very short weather window opportunities that Scotland offers in winter.

* * *

I think I can remember most of the walks that I have done over the years. Of course, I don't recall all the details of everything seen or done on these days out, but there is usually something that happens which leaves an imprint on the memory. An eagle sighting, maybe, or the remains of an avalanche. It might only be something fleeting, but that can be enough. One of the very best recent trips was in January 2018, and it was the first time I had the pleasure of meeting Andrew's new Labrador, Mabel. On this particular day, however, the amount of memories experienced means that this is likely to last longer than, and be remembered just as fondly as, any walk I've done for quite a long while.

We had arranged to meet up early in the morning, as usual. The weather promised to be as good as you could ever hope for during a Scottish winter. As I drove north that morning the pitch-black sky was pierced by the pinlight of a million stars. The thermometer on the car showed sub-zero temperatures, and the snow that lay on the ground was covered in a hoary, deep frost. As I crossed the northern Perthshire boundary into Highland district the sun started to rise, turning the inky-black sky slowly to the colour it would remain all day: a glorious, deep metallic blue. The sun's low winter trajectory meant that it took a while for it to climb above the low hills to the east. But rise it did, just as I pulled into the car park to meet Andrew, who had arrived not more than a few minutes before. As I stopped beside his car, I noticed the silhouette of two dogs in the boot, both with tails wagging so hard that they might shatter the windows. As he popped the tailgate, the familiar figure of Olive jumped out first, closely followed by a blonde Labrador. This was Mabel. Perhaps worried that one of them might be hit by a lorry on this busy Highland road, Andrew wanted to get going as quickly as possible. Apart from anything, days were short at this time of the year, and we had a big, ten-mile round ahead of us in deep snow.

The four hills ahead of us were coated in the most beautiful white powdery snow. Thankfully, though, a strong crust formed on top, meaning that, once a bit of height was gained, it took our weight, thereby avoiding the need to do the incredibly tiresome and tiring job of breaking trail. The higher we got, the less vegetation we saw. Even the long grass and tall heather were invisible beneath this huge volume of snow. All the while Olive and Mabel were going absolutely bananas playing in the snow. If Andrew or I stopped for a breather or to take a picture, creating a small gap between us, the dogs would expend untold energy darting back and forth almost as if to make sure we knew that they were looking out for us.

I recall the futility of Mabel trying to chase a mountain hare up the side of a hill. This latter animal, supremely adapted over thousands of years to evade the stealthiest and quickest pursuers, made Mabel look like a malfunctioning robot. Never have I seen such a one-sided contest. By the time Mabel had reached the bottom of the slope where the hare had been, it had long disappeared over the top and was probably half a mile distant. The older, and infinitely wiser, Olive watched on impassively and incredulously.

Our walk in, along the Perthshire–Highland boundary, was done in the shadow of the magnificent hill called the Boar of Badenoch. It gazed lustily across the floor of the glen to the Sow of Atholl. These two fine hills of old marked, as their names suggest, the outer edge of Atholl and the start of Badenoch: two ancient regions of Highland Scotland.

In the end, Mabel's pursuit of hare, grouse, ptarmigan, and anything else that moved that day, meant that by the last summit she was whimpering and tired. At one point she tried to fashion herself a little nest out of the snow. Once again, Olive, who had seen it all before, stood stoic and impassive. As with Ben Lawers, I could really see how the presence of dogs could add to the experience of being on snow-capped peaks. The sheer unbridled joy and pleasure of the dogs in this environment was heart-warming. I was beginning to come round to the idea that perhaps I had, hitherto, been too hasty when dismissing them as more hassle than they were worth, irrespective of their loveliness. Maybe the time is nearing when I get one of my own.

Apart from the dogs, what made this trip so memorable was a combination of weather, hills, and the deep snow. Not once in the entirety of the walk did the sun threaten to disappear behind a cloud.

The deep blue winter sky could not have contrasted more starkly with the gleaming white of the snow. The views from the summits, across much of the Central Highlands, were unsurpassable. As pretty as these hills are in summer, they cannot compete when covered with snow. As I say to most people when out walking, snow makes everything better.

General terminology and Gaelic pronunciation

Place names in the Highlands of Scotland are overwhelmingly of Gaelic origin. This is *Scottish Gaelic* as opposed to *Irish Gaelic*. The former is pronounced as 'GAAH-lick', the latter 'GAY-lick'. It's an important distinction that native speakers can become irked at.

It is a long-held and generally established fact that Scottish Gaelic was brought to these shores via Irish settlers around AD 500.[1] The language spread eastward for the next few hundred years until it eventually became the spoken tongue for much of the country. In the twelfth century, virtually the entire population spoke it, save for parts of Lowland Scotland (including Lothian and Caithness), who spoke variants of Scots and Norse.

As is normally the case throughout history with languages in a time of social and political instability, Gaelic's dominance began to slip over the centuries. Lowland Scots, now spoken by the landed class and increasingly at the royal court, started to penetrate westward through the glens and river valleys. The drop in numbers of people speaking Gaelic accelerated after the 1745 Jacobite uprising, when it was proscribed by the British government. Children were censured in schools from

---

1   Jones, Charles (1997), *The Edinburgh History of the Scots Language*, Edinburgh University Press.

speaking it up until relatively recent times, which accounted for further significant falls in numbers. (It was seen as a backward language by some.)

Records from the last 250 years paint a sobering picture. In 1755, twenty-three per cent of the Scottish population spoke Gaelic (290,000 speakers). In 2011, it was just 57,602 (1.1 per cent of the total population).[2] Though substantial efforts are being made to revive the language, both in cities and in the country, as a spoken, community language it is to a very significant extent restricted to the Outer Hebrides. Some parts of northern Skye and other Inner Hebrides have small, isolated communities, but with over half the native speakers living in Glasgow, the future for the language as an everyday one is very unclear.

It is true that some of the very earliest strata of names, such as those for rivers like the Dee, Clyde or the Tay, pre-date even Gaelic and are probably thousands of years old. Most names though, especially hills, are Gaelic in origin.

Pronunciation

Like the dialects of England (Brummie, Scouse, Geordie, Yorkshire, East Anglia, etc.), Gaelic is not a homogeneous language. Though words may be written the same for places in Mull as they are for Perthshire, they often have different pronunciations. Trying to negotiate this is a minefield, and to their credit some websites make attempts.[3] However, often fine institutions and well-meaning websites do make mistakes. This is no truer than in the Cairngorms, where local pronunciation is quite different to other parts of the country.

Throughout the book I have attempted to give rough translations and meanings wherever possible. The list below is more authoritative, giving the names of the main hills contained therein. The name of the hill comes first, in bold, followed by an approximate sound in Scots/English, indicated by italics. Emphasis of syllables is given in upper case. Then follows the meaning, with an explanatory note if necessary.

---

2   MacAulay, Donald (1992), *The Celtic Languages*, Cambridge University Press.
3   https://getoutside.ordnancesurvey.co.uk/guides/the-gaelic-origins-of-place-names-in-britain/ [accessed 13 May 2021].

Aonach Beag *eunuch BAKE* – 'little ridged hill'. The *ao* sound in Gaelic has no English equivalent. The best way of mimicking it is to try and say *oo* without rounding your lips.

Aonach Mòr *eunuch MORE* – 'big ridged hill'.

Beinn a' Bhuird *been-a-board* – 'table hill'. The 'a' is very short, and hardly pronounced. It indicates a euphonic (for ease of pronunciation), not the article. The designation 'Bhuird' is incorrect. Older folk pronounce the hill, approximately, as *board*. Recently *voord* or even *voorsht* have become common.

Beinn Bhrotain *been VROHtan* – 'hill of the mastiff'.

Ben Avon *ben AAHn* – meaning unsure. Possibly from *athfhinn* ('very bright one'), or from *abhainn* ('river').

Ben Macdui *ben macDOOi* – 'hill of Macduff'. The highest hill in the Cairngorms and the second highest in the UK.

Ben Nevis *ben NEVis* – meaning unknown. An ancient name, possibly pre-Gaelic. One interpretation is 'hill with its head in the clouds', though this is disputed.

Braeriach *bray REEach* – 'the brindled upland'. The first syllable is generally pronounced as 'bray', but – properly – is 'bry'.

Cairn Gorm *karn GORom* – from *An Carn Gorm*, meaning 'blue hill'. This is the hill that gives its name to the whole range.

Cairn Toul *karn DOWel* – 'hill of the barn' (from *Carn an t-Sabhail*).

Geal-Chàrn *g-YAHL charn* – 'white hill'.

Liathach *LEE-a-huch* – 'the grey one'.

Sgòr an Lochain Uaine *scorn lochan OOan* – 'peak of the green tarn' (small loch). Also known as the Angel's Peak, a name given by Alexander Copland as a counterbalance to the nearby 'Devil's Point'.

Sgòr Gaoith *scor GOOee* – 'peak of wind'.

Sgùrr na Lapaich *scoor na LAHpeech* – 'hill of the bog'.

Get involved

Though there are now many people who are actively involved in the study of snow patches and snow in general, there is always room for more. The uplands of the UK are huge, with many nooks and crannies where relics of winter can cling on unseen, just waiting to be discovered by a walker, climber, cyclist or caver. To that end, I encourage people who stumble across these old patches to get involved in their reporting. The best way to do this is via two specific pages for Scotland, England and Wales on Facebook:

www.facebook.com/groups/snowpatchesscotland
www.facebook.com/groups/snowpatchesengland
(for England and Wales)

There are now several thousand people on these pages, many of whom are very knowledgeable and keen to engage with new members.

The other medium is Twitter, specifically via my own page. I can be found at @theiaincameron all year posting updates and interesting snow-related pictures. Please feel free to say hello on there if you've enjoyed this book.

Acknowledgements

Not in any particular order, but I would like to thank the following people for helping to bring this book to print:

Laura – for her unceasing and helpful advice on what worked and what didn't when I was trying to knit the story together.

Jenny Brown – my agent (www.jennybrownassociates.com) for her excellent and constructive suggestions and having faith in what I was trying to say.

Vertebrate Publishing – for the belief they showed in taking on something that was outside of their natural comfort zone. To Commissioning Editor Kirsty in particular.

Murdo MacLeod – the master photographer. His photos from various trips he and I have completed in the last few years have helped to bring this subject to a wider audience.

Andrew Cotter – for his permission to allow me a shameless addition of a doggy epilogue.

Finally, to the late Adam Watson. Despite it being over two years since his passing, I cannot help but say to myself, 'I must send that photo to Adam,' when I see one of the Cairngorms covered in snow. His knowledge and passion remain a daily inspiration.

ECONOMIC PLANNING AND TOWN EXPANSION

A case study of Basingstoke

by

JOHN H. DUNNING

Senior Lecturer in Economics, University of Southampton

with the assistance of Eric Stokes and
a W.E.A. study group at Basingstoke

SOUTHAMPTON:
WORKERS' EDUCATIONAL ASSOCIATION
(Southern District)

Printed in Great Britain for Workers' Educational Association
by W. & J. Mackay & Co. Ltd., Chatham, Kent

216711

By permission of Basingstoke Town Development Joint Committee

EXPANDING TOWNS

RAILWAYS

ROADS

THETFORD

BURY ST. EDMUNDS

HUNTINGDON

HAVERHILL

LETCHWORTH

LUTON

AYLESBURY

LONDON

STAINES

GUILDFORD

EDENBRIDGE

TUNBRIDGE WELLS

ASHFORD

OXFORD

READING

HOOK

BASINGSTOKE

ALTON

SWINDON

NEWBURY

ANDOVER

WINCHESTER

SOUTHAMPTON

PORTSMOUTH

BIRMINGHAM

100 miles

50 miles

25 miles

10 miles

thames valley

berkshire downs

hampshire downs

north downs

south downs

isle of wight

WORCESTER

GLOUCESTER

BATH

SALISBURY

salisbury plain

new forest

BRISTOL

EXETER

YEOVIL

RAILWAY

ECONOMIC PLANNING AND TOWN EXPANSION

PREFACE

EARLY in 1960 Mr. F. E. Foden, a member of the Committee of the Basingstoke Branch of the Workers' Educational Association, proposed that the Branch should set up a group to study the expansion of the town. During the previous two years the Branch had already arranged public meetings at which the implications of town expansion were discussed. The formation of the group seemed an appropriate step to take, since it would enable the Branch to further its policy of providing opportunities for the objective consideration of the many factors involved. The proposal was therefore adopted by the Branch at its Annual General Meeting on 20 May 1960.

At the suggestion of Mr. P. J. Pitman, District Secretary of the Southern District of the W.E.A., the proposal was discussed in June 1960 with two members of the Faculty of Economics (now Faculty of Social Sciences), University of Southampton–Dr. J. H. Dunning, an economist, and Mr. M. Broady, a sociologist. The Branch was much encouraged by their promises of help, as also by the interest shown in the early days of the project by the Town Clerk of Basingstoke, Mr. L. Womersley, who became a member of the sociological section of the study-group.

The original intention was that the group should consist of three sections–historical, economic and sociological–and that each should, from time to time, produce reports on the work they were undertaking. Eventually the three sections were reduced to two–the economic and the sociological; these were later organised as classes under the aegis of the Department of Adult Education of the University of Southampton.

The economic section began its work in the autumn of 1960 with a series of occasional meetings, and met as a class from September 1961 until June 1962, with Dr. Dunning as tutor.[1] At various times during its work it received visits from the Mayor of Basingstoke (Councillor J. B. Peat), Mr. R. W. Steel (Director of Town Development), Mr.

[1] The names of the class members are given at the end of this Preface.

1

PREFACE

A. G. McCulloch (Chief Architect/Planning Officer) and from Mr. H. S. Gray (Youth Employment Officer). Members of the section also visited Aylesbury, where they were kindly entertained by the Town Clerk (Mr. R. D. W. Maxwell), who gave a useful survey of the problems and opportunities presented by a town expansion scheme.

It is appropriate here to say that the work of the section would have been impossible without the co-operation of a large number of people, who helped directly by the provision of material, or indirectly by the interest shown in what we were trying to do. This opportunity is therefore taken to thank all who contributed in various ways to the work.

Individuals whom we especially wish to thank for their help are Mr. John Parker and Mr. Brian Harvey of the Faculty of Social Sciences for statistical computations; Mr. J. Corfe, for his work in preparing part of the population projections; Mr. H. Higgs, for details about the work of the Basingstoke Development Committee; Miss M. C. MacFarlane, Curator of the Willis Museum, for permission to consult old directories; Mr. J. E. G. Pearce, who provided material about the early history of the bus services in the neighbourhood, and Mr. T. Pettle, who helped with the loan of a history of the Southern Railway.

We also wish to thank the following, who provided historical and statistical material: the various departments of the Basingstoke Borough Council; the Board of Trade; the British Transport Commission; the County Library; the Development Corporations of the London New Towns; firms located in the Basingstoke Employment Exchange area; the Ministry of Labour, both locally and nationally; the Registrar-General; the Area Electricity and Gas Boards; and the Town and Country Planning Association.

We are particularly indebted to the Editor of the *Hants and Berks Gazette* for allowing us to make frequent consultation of the files of his paper. Our very sincere appreciation is due to Mrs. P. Powell for typing a difficult manuscript so well; to Mr. E. Poole for valuable advice on printing specifications; and to Miss D. Marshallsay for compiling the index, designing the dust jacket and for her very considerable assistance in preparing this work for publication. Finally, we thank Mr. C. J. Thomas, Senior Lecturer in Economic Statistics at the University of Southampton, for his helpful comments on Chapter IV and its Appendix.

In the event, the results of the work of the economic section appear

2

not as a series of reports but in the form of a book. That they do so is due to the enthusiasm that Dr. Dunning has shown for the project and to the encouragement and financial support afforded by the Southern District of the W.E.A. and the Department of Adult Education of the University of Southampton.

E. G. STOKES,
Chairman,
Basingstoke Branch
Workers' Educational Assn.

The names of the class members were:

Mr. A. Aslett	Miss A. Hunt
Mr. A. Attwood	Miss M. Sadler
Mr. L. Bone	Mrs. B. Swift
Mr. A. Cumberbirch	Mr. D. Toohig

Mr. E. G. Stokes (*Secretary*)

CONTENTS

5

LIST OF TABLES

7

8

INTRODUCTION

THIS book is a case study in economic planning. More particularly, it attempts to look into some of the problems of interest to the economist which are likely to face a town which plans to expand its population from 25,000 to 80,000 in the next two decades. The subject of our case study is Basingstoke, which is one of the towns now undergoing a programme of expansion under the Town Development Act of 1952. In the fifteen years between 1961 and 1976 some 37,000 Londoners are expected to migrate from the capital to Basingstoke and by 1981 the population of the town[1] is scheduled to rise to something approaching 85,000, two and one-half times its current (1962) level.

An expansion of this magnitude within twenty years brings with it a host of problems and challenges of interest to the economist, sociologist and town planner alike. Some of these concern the implementation and programming of the scheme, e.g. how is the growth of population to be synchronised with the expansion of housing and employment; what industry and shops will be needed by 1970; how is the scheme to be financed, and so on. Others are more long term in character and relate to the effects of the scheme when completed; its likely repercussions on the life of the inhabitants of Basingstoke and the town's role in the economy of the South. What sort of age-structure of population is likely to result in 1976 and 1981 and what are its implications for education, housing and conditions of employment? What types of industry and commerce are most likely to suit the future pattern of Basingstoke's resources? What kinds of firms will want to site their plants in the town and is the town likely to become a more or less attractive location in the future? What are likely to be the service requirements of an urban centre serving a population of 110,000 or more? What of the future of Basingstoke as a regional economic centre and its relationship to Southampton and Reading?

[1] More particularly the area contained by the (1962) Town Map.

11

Many of these issues involve much wider considerations than the purely economic, but the economist can often make a useful contribution by bringing to bear his own particular skills and training. This we have sought to do in this study, conscious that our task is not so much to propose specific courses of action as to probe and advise on the possible economic consequences of such courses of action.

Our analysis is thus deliberately confined in scope and argument. We have tried not to trespass too far beyond the boundaries of our subject, but this should not be taken to imply that we are unaware of the non-economic implications so often involved in decision-taking on economic matters; indeed, we fully appreciate that economic betterment is but one aspect of social welfare which has to be taken into account in planning for town expansion: at the same time no one would deny that it is often a very important aspect.

Basically, our task is to review the pattern of resources likely to be available at different stages of Basingstoke's expansion programme and to suggest ways in which they can be co-ordinated and effectively utilised. Since, apart from agriculture, the region's resources consist almost entirely of its building land, labour supply and the capital equipment of its firms – it produces virtually no raw materials of its own – our attention is first directed to the likely provision of the basic services necessary for a town of 80,000 inhabitants, and secondly, to the size and composition of the population as it affects the supply of labour and the demand for goods and services which such basic services make possible. In Chapter IV we pay some attention to population growth and the factors which determine the rate of growth and the pattern of age-structure. What are the implications of the town development scheme in this connection? What rate of migration must be achieved per annum to achieve the proposed target and what are the chances of achieving this rate?

But perhaps the most important economic problem of town expansion surrounds the provision of adequate employment opportunities. This is more than seeing that there are sufficient jobs for the available labour force: a town can maintain full employment (at least for a while) by various patterns of resource allocation. However, each of these possible patterns involves the satisfaction or sacrifice of certain objectives crucial to economic welfare, e.g. stability of employment, growth of production, balance of male/female labour, economy of space, and so on. We thus seek in Chapter V to examine the particular needs of Basingstoke and the possible ways in which these aims may be achieved.

Having analysed the provision of work in the new community, we then seek to examine the repercussions of the increased wealth and income of Basingstoke on the various consumer expenditure trades. What are the implications of the Town Development Scheme on retail distribution – the main medium through which income is spent? How far ahead can one plan in estimating demand for particular types of shopping facilities? What, too, of the implications for education and housing? Such questions as these we deal with in the latter part of Chapter IV and Chapter VI.

In addition to examining the significance of town expansion, we have also sought to analyse the likelihood of such expansion being achieved in the time envisaged, i.e. we have paid attention to the phasing of the development programme. In doing so we have drawn heavily on the experience of other communities faced with similar problems to that of Basingstoke. While recognising that such comparative analysis has its weaknesses, particularly when looking to the future, it is nevertheless the best guide we have to assessing the problem in hand.

We have not touched on the social implications of the development scheme, as these are currently being examined by another study group of the Basingstoke W.E.A., and we have virtually ignored the whole problem of local government politics and the integration of the 'old' into the 'new' Basingstoke. Neither must the merit of our work be judged on its success in making predictions: rather we have sought to examine the economic basis of alternative predictions and their likely implications for economic welfare.

CHAPTER I

HISTORICAL BACKGROUND

IN their report to Parliament in 1837 the Commissioners for Municipal Corporations Boundaries said of Basingstoke:

> The town is represented to be gradually increasing in size and importance. It contains no manufactories, but is a place of very great thoroughfare and is situated in the centre of a rich agricultural district. The markets held here are well attended, being resorted to by persons from Andover, Winchester, Newbury and Reading.[1]

This report was made two years before the coming of the London and Southampton railway to Basingstoke. But for some years previously the town had been an important stopping-place on the great coaching road between London and the South-West, and in 1796 a canal, now disused, was completed which not only facilitated the transit of agricultural produce from the outlying countryside but also linked Basingstoke with the Thames via Weybridge. As Table I/1 indicates, the rate of growth of population of the Borough fluctuated in the first half of the nineteenth century, but in general was below that of the national average, due chiefly to the sweeping industrialisation of the North. With these developments Basingstoke had little to do; the town had few industrial raw materials of its own and was some distance away from the centre of technological developments. But a rising population and national income consequent upon such industrialisation did mean an increased demand for food, and Basingstoke expanded as a regional market centre, providing regular grain and cattle markets for the farmers of the district, together with the traditional crafts and trades associated with such an economy.

Moreover, even if large manufactories, such as those characteristic of the industrial north, were lacking, there were in the town several establishments engaged in various forms of manufacturing activities. The importance of the malting industry was mentioned by a local

---

[1] *Municipal Corporations Boundaries. England and Wales.* Royal Commission, Reports, Pt.I; 1837 (238) xxvi.

15

TABLE I/1

POPULATION GROWTH IN BASINGSTOKE* AND
ENGLAND AND WALES 1801–1921

	Basingstoke	decennial % increase	England & Wales (000s)	decennial % increase
1801	2,589		9,000	
1811	2,656	2·6	10,000	11·1
1821	3,165	19·2	12,000	20·0
1831	3,581	13·1	13,897	15·8
1841	4,066	13·5	15,914	14·5
1851	4,263	4·8	17,928	12·7
1861	4,654	9·2	20,066	11·9
1871	5,574	19·8	22,712	13·2
1881	6,681	19·9	25,974	14·4
1891	7,960	19·1	29,003	11·7
1901	9,793	23·0	32,528	12·2
1911	11,540	17·8	36,070	10·9
1921	12,723	10·3	37,887	5·0
1931	13,865	9·0	39,952	5·5

* Municipal Borough.
Source: *Census of Population.*

directory in 1804 and again by the Boundary Commissioners in 1835.[1] Other firms included those engaged in coach-building and repairing and foundry operations. One foundry, established by Aaron Caston in 1820 to produce agricultural equipment, later became the firm of Wallis & Steevens, which is still manufacturing today in Basingstoke. This company and the railway appear to have been the largest employers of labour in the town in 1851, by which time the population had reached 4,263. Thereafter the pace of population expansion quickened, and by the turn of the century it was increasing at nearly twice the national average. Perhaps the chief impetus to this resurgence of growth was the widening of markets and opening up of new sources of supply made possible by the establishment of Basingstoke as an important railway junction. In addition to its links with Southampton and London, a connection with the Midlands and North via Reading and Oxford was established in 1856, and with Salisbury and the West via Andover in the following year. The town might have developed into an even more important railway centre had there been sufficient suitable land to allow the accommodation of the carriage

[1] *Municipal Corporations in England and Wales.* Royal Commission. 1st Rep. App. Pt. II. pp. 1106–7; 1835 (116), xxiv.

16

and locomotive works of the London and South Western Railway which were eventually transferred from Nine Elms to Eastleigh in 1891 and 1909.[1]

With the improved transport facilities, and the growth of a wide variety of consumer goods industries less tied to the site of their raw materials, several manufacturing firms set up in Basingstoke in the late nineteenth century. In 1878 the firm of Gerrish, Ames & Simpkins was established to carry on the manufacture of clothing begun in the town in 1868 by a Mr. Burberry. Mr. Burberry retained the made-to-measure side of his business, but later disposed of it to Messrs. Mares, so that he could concentrate his attention on the manufacture of gabardine cloth.

But the most significant industrial development in the town occurred in 1898, with the transfer from Chiswick of the Thornycroft motor vehicle works. Within four years the firm was employing 200 workers, mostly recruited from Basingstoke and its surrounding area. Further additions to manufacturing industry were made in 1913, when a leather manufacturing firm transferred from Godalming, and in 1926, when Auto Tyre Services set up its factory at Worting. The 1921 Census of Population included particulars of the distribution of employment. The figures for Basingstoke Borough show that of a total population of 12,723 there was a working population of about 5,800. Primary industries (mainly farms and market gardens) employed 300, manufacturing 2,400 and the service trades 3,070. Agriculture, however, continued to remain the mainstay of the local economy, accounting for 75 per cent of the labour force in the rural districts around Basingstoke. The main employers of male labour in the Borough were the metal-working industries (877) and transport (592, including 254 railway workers). Of the female labour force, some 545 were engaged in the manufacture of textiles and clothing, and 534 in personal services (of which 335 were employed as domestic servants).

There was little new industrial development in Basingstoke during the 1920s, but the trade and commerce of the town received a significant stimulus in 1926 with the establishment of the Venture Bus

[1] The difference this made to Eastleigh's economy may be illustrated from the following population statistics of the civil parish taken from the Census of Population—1891: 3,613; 1901: 7,779; 1911: 13,056; 1921: 13,389. See also H. P. White: *A regional history of the railways of Great Britain*, 1961, Vol. 2, p. 140. With reference to the prospect of railway extension at Basingstoke some years earlier, see *Hants and Berks Gazette*, 2.7.1881, p. 4.

Company.[1] By 1933 this firm had become the largest of several local operators connecting Basingstoke with its immediate hinterland and neighbouring towns of Andover, Alton and Newbury. The rapid increase in this and other forms of vehicular traffic, both local and long-distance, soon brought congestion to the roads which was not relieved until the building of the by-pass to the south of the town in 1930.

Meanwhile, several ideas were being formulated as to the pattern of development likely best to suit the resources of Basingstoke. There were, in general, two schools of thought. The first, epitomised by Mr. C. G. Clark in a talk given to the Rotary Club in the town in 1930, entitled *The Architectural Development of Basingstoke*, was that the future of the town lay in an extension of its manufacturing facilities and an improvement in its communications.[2] The second, put forward by Mr. P. M. Crosse in 1937, also to Rotary, argued that Basingstoke was particularly well placed to become a residential or outer-dormitory town for London, and emphasised the advantages of a 'clean, fast and frequent electric train service to London'.[3]

In general, development in the '30s proceeded on the former lines. Towards the end of 1931 the Finance Committee of the Borough Council reported that it had received a letter from the Basingstoke Chamber of Commerce, which asked that the Council should form a Development Committee. The Finance Committee recommended that a conference should be convened to discuss ways and means of attracting new industries to the town. This recommendation was adopted and, as a result, the Basingstoke Development Committee was set up, consisting of representatives of the Borough Council, the Chamber of Commerce, the Trades Council and the Rotary Club of Basingstoke. Eventually a booklet, 'Basingstoke for Industry', was produced. This listed the sites available for factories, the names of the owners of the sites and the approximate acreage of each site. In the booklet, Basingstoke was stated to possess 'many special features as a provincial industrial town on account of its geographical position'. Reference was made to its suitability in respect of transport facilities and also to the availability of labour, both within the Borough and from the neighbouring villages. Initially, little resulted from the efforts of the Basingstoke Development Committee to attract new

[1] Taken over in 1945 by Red and White Motor Services Ltd. (a South Wales company) and by the Wilts and Dorset Bus Company in 1951.
[2] *Hants and Berks Gazette*, 25.1.30.
[3] *Hants and Berks Gazette*, 4.6.37.

industry to the town, which for most of the 1930s continued to depend largely on the prosperity of the two important engineering firms, Thornycroft[1] and Wallis & Steevens, a number of small clothing establishments and market-gardening carried on at the Cranbourne Lane nurseries. The farming of the surrounding region also continued to exercise an important influence on the economy of the town, in its function as a market and shopping centre. Basingstoke was not un-affected by the depression of the '30s, but unemployment was not as serious as in the heavy industrial areas of the North and South Wales.

Indeed, locational forces at this time were moving in favour of towns in the Midlands and South of England. This was a period of rapid growth for the new industries based on new sources of power (electricity cf. coal) and methods of transport (road cf. rail). Such industries were much less influenced by the site of their raw materials in their choice of location than by the proximity of their chief markets. For many trades a location in Southern England was also an advantage from the standpoint of administration and of access to financial and technical services.[2] Basingstoke shared in this development to a limited extent, but it was not until just before the Second World War that any major expansion in new industry occurred. First, in 1937, a 38,000 sq. ft. factory was opened on the Winchester Road estate by the Glasgow firm of Kelvin, Bottomley & Baird, to manufacture air-craft instruments; this company has since been taken over by Smith's Instruments. A year later Airpak Ltd., producers of quilted fabrics, established a plant at Overton, in the Basingstoke employment dis-trict, and in 1939 the third important development occurred when the American firm, Eli Lilly, set up a factory on the Houndmills Estate to supply pharmaceutical products. Other companies already established also expanded in the late '30s, and by 1939, out of a total insured population of 8,160 in the Basingstoke Employment Area, some 2,839 or 34·9 per cent were employed in manufacturing industry.[3] The population of the Borough at this time was estimated by the Registrar-General at 14,920.

Meanwhile, during the period we have just been discussing, ques-tions involving the national distribution of industry and population were brought very much to the fore. Two related problems emerged,

---

[1] The largest employer, having 1,800 employees in April 1930.
[2] G. C. Allen: *The Structure of Industry in Britain*, 1962, p. 158.
[3] For further particulars, see Chapter III, p. 77 ff.

viz. (i) the continued presence of areas of severe localised unemployment and (ii) the accelerated drift of industry and population into a number of urban conurbations, such as London and Birmingham. The unemployment problem arose for two reasons; first, the decline in Britain's export trade and of her competitive position overseas, and second, the changing structure of U.K. industry at home coupled with the substitution of electricity for steam power as the main source of industrial energy. The industries most affected by these developments were highly concentrated in certain geographical areas, e.g. the coal and steel industries in South Wales, the cotton industry in Lancashire, the shipbuilding industry on the north-east coast and Clydeside. The falling prosperity of these, the so-called depressed areas, led to the migration of people to other areas in search of work. At the same time, the London and Home Counties and the Midlands area conurbations underwent a marked expansion. Whereas between 1923 and 1937 population fell in the depressed areas by up to 11 per cent it rose in the Midlands by 10·4 per cent and in London and the Home Counties areas by 16·4 per cent. Most of this latter increase was centred around the cities of London and Birmingham. An important factor in this expansion of population was that within these areas there was a high proportion of the new expanding industries.[1]

To look into these two problems, the Government appointed in 1937 the Royal Commission on the Distribution of the Industrial Population (the Barlow Commission).[2] The Commission was asked first to examine the causes and effects of the existing distribution of industry in Great Britain and any changes thought likely to occur in that distribution, and second, to consider the various social and economic disadvantages which arise from large concentrations of population and industry in urban areas. In its report, published in 1940, the Commission recommended that congested urban areas should be redeveloped where necessary, and that industry should be dispersed from these congested areas; it also considered that there was need for a more balanced development of industry throughout the country, while within each region industry should be appropriately diversified.

With the recommendations of the Barlow Commission in mind,

---

[1] For further particulars, see S. Pollard, *The Development of the British Economy 1914/50*, E. Arnold, 1962, pp. 125–34, and J. H. Dunning and C. J. Thomas, *British Industry*, Hutchinson University Library, 1963, pp. 68 ff.

[2] *Distribution of the Industrial Population*. Royal Commission Report; 1939–40 Cmd. 6153, iv.

the Greater London Plan (the Abercrombie Plan) was drawn up in 1944.[1] In an attempt to check the growth of industry and employment in the capital and suburbs, several measures were suggested, one of which was the dispersal of over one million people from the continuously built-up metropolitan area. The decentralised population was to be redistributed as follows:

1. Quasi-satellites—immediate post-war housing programme (L.C.C. and County Borough of Croydon) — 125,000
2. Additions to towns outside the continuously built-up area within the Greater London Plan boundaries, largely in the Outer County Ring — 261,000
3. Additions to existing towns outside the Greater London Plan boundaries, mostly between forty and fifty miles from the centre of London — 163,750
4. Eight new satellite towns outside the Green Belt Ring — 383,250
5. Dispersed wholly outside the metropolitan influence — 100,000

1,033,000

This programme, adopted in principle by successive post-war Governments, and encouraged by such legislation as the New Towns Act, 1946, and the Town Development Act, 1952, has been only partially successful. While it is true that in the last sixteen years more than 200,000 people have been dispersed from the Metropolitan Area to the new towns, 60,000 to London County Council estates beyond the Greater London Plan boundaries, and 20,000 to the expanded towns, the growth of London, as an employment centre, has hardly been halted. New industry, particularly service industry and office employment, has continued to expand in and around the capital; between 1952 and 1959 the London conurbation gained 45 per cent of the increase in employment in England and Wales, even though it had only 27 per cent of the increase in population: during this period the number of new jobs provided in the central area grew by 20,000 a year.[2]

Basingstoke was included in Section 3 of the above table, together

[1] *Greater London Plan 1944*; 1945 Non-Parl. Ministry of Town & Country Planning.
[2] A. Goss, *British Industry and Town Planning*, Fountain Press, 1962. W. Thomas, 'The Growth of the London Region', *Town & Country Planning*, May 1961. A. G. Powell, 'The Recent Development of Greater London'. *The Advancement of Science*, No. 65, Vol. XVII, May 1960.

with towns such as Aylesbury and Bletchley: others have since been scheduled under the Town Development Act. The Plan proposed that Basingstoke should receive 20,000 people, compared with a proposed 35,000 for Aylesbury and 22,500 for Bletchley. Of Basingstoke's suitability for expansion it was said:

> This is a railway junction and is also on the Exeter Road. It is a thriving market and shopping centre serving a wide district. Of recent years, the town has tended to expand industrially and is well located for the reception of certain types of industry, including aero-engineering, food factories and other concerns needing a clean atmosphere.[1]

Two further points were made about the town: first, it was suggested that it should become a regional goods distribution centre; second, that the railway line from London to Basingstoke should be electrified.

In May 1947 the Borough Council received a report from its Finance Committee about the Greater London Plan to include Basingstoke among towns to be expanded. Reference was made to the need to ensure a careful selection of the type of industry, so that there might be suitable diversity of employment opportunity in the town. The report also dealt with the necessity for a building programme which would ensure that houses and factories would be erected concurrently and for an appropriate expansion of the various service industries.

The Hampshire County Council referred to the expansion proposals for the town at its meeting in November of the same year, and asked that it should be given an opportunity of joining with the Borough Council in any discussions which might take place. Further discussion of the proposals to expand Basingstoke took place at a meeting of the North-East Hampshire Planning Committee held early in January 1950. Reference was made by the Area Planning Officer to plans to enlarge the town's population to about 35,000, and although this was welcomed by the Borough Council, some anxiety was expressed lest it might grow to 60,000. The Town Clerk emphasised that Basingstoke would not become a new town, but would merely undergo accelerated expansion. He added that the Borough Council had acquired 90 acres of land off the Kingsclere Road (the Houndmills area) to provide accommodation for factories and housing for the decentralised population.

[1] *Greater London Plan 1944*, p. 167.

HISTORICAL BACKGROUND

Early post-war legislation relating to the overall distribution of employment and industry centred around three pieces of legislation:

(i) The Distribution of Industry Act, 1945, which, by the designation of 'Development Areas' (the successors of the pre-war depressed areas) into which new industry was to be encouraged, aimed to deal with the problem of localised unemployment.

(ii) The New Towns Act, 1946, which created the necessary machinery for the establishment, administration and financing of new towns in Britain, and the dispersal of population and industry from congested urban areas and

(iii) The Town and Country Planning Act, 1947, a comprehensive enactment which dealt with the control of land use in general.

By the provisions of this latter act, each County Council in the country was required to make a survey of the resources in its area and to publish a report and a plan for the future use of land within its boundaries. The Hampshire Development Plan 1950, published by the Hampshire County Council, revealed that the area covered by the Basingstoke Employment Exchange[1] had a total population of about 26,000 in 1948. Of this number, the working population accounted for roughly one-half, viz. 8,775 men and 4,002 women. The pattern of employment in that year was as follows:

TABLE I/2

EMPLOYMENT IN BASINGSTOKE EMPLOYMENT
AREA, 1948

	Male	Female	Total	%
Agriculture, fishing and forestry	1,438	199	1,637	12·8
Mining and quarrying	8	1	9	0·1
Manufacturing	2,743	1,253	3,996	31·3
Building and service industry	4,370	1,875	6,245	48·9
Miscellaneous	216	674	890	6·9
Total	8,775	4,002	12,777	100·0

Source: Ministry of Labour.

The manufacturing industries of the area were stated in the survey to be: agricultural implements, steam-rollers, lorries and motor-coaches, scientific and aeronautical instruments, clothing, fancy leather goods, pharmaceutical and biological products.

[1] Basingstoke Municipal Borough, together with most of Basingstoke Rural District. For further particulars see Chapter V.

23

Part II of the 1950 Hampshire Survey contained an outline Development Plan for Basingstoke. This made detailed proposals for the accommodation within the town of the overspill population envisaged by the Greater London Plan. Assuming that this would be the 20,000 proposed and also that there would be a natural increase of 3,230 over a period of twenty years, it put the 1971 population of Basingstoke at 40,000. Of this number, one-half would constitute the total working population, of which, it was estimated, 10,000 would be employed in secondary or manufacturing industry. To accommodate this industry, it was recognised that more land would have to be made available, and the Plan proposed to increase the industrial acreage within Basingstoke from 80 to 225 acres by 1971.[1] The main areas set aside for secondary industry, in addition to the existing Thornycroft and Brook Street areas, were to be:

1. The northern or Kingsclere Road area (Houndmills).
2. The southern or Winchester Road area.

For both of these, light industrial projects were proposed. Long-term proposals included the transfer of industry from Station Hill and Worting to the new industrial areas of the town.

The original proposed development plan was later modified to take account of the need to preserve land of high agricultural value in the immediate neighbourhood of the town. The main modification was the reduction of the size of the immigrant population, and a revision of the target population for 1971 from 40,000 to 32,000. A new Development Plan drawn up in 1955[2] repeated this new estimate of population increase; it also allocated a further 5 acres of land for industrial purposes. In the meantime the Town Development Act, 1952, had been passed which made it possible for local authorities of small towns some distance from large and congested urban areas to receive financial assistance towards expansion and development where required. Tentative approaches were made almost immediately by the London County Council to several towns, including Basingstoke, and in a progress report published in October 1955 by the Housing and Town Planning Committee of the London County Council, it was intimated that the London County Council had concluded overspill agreements with Bletchley, Daventry, and Swindon, and that negotiations were in progress with thirteen other towns, of

---

[1] Calculated on the basis of 35 persons per industrial acre.
[2] Hampshire County Council. *Development Plan, 1955, 1957.*

24

which Basingstoke was one.[1] As a result of these negotiations an agreement was signed in October 1959 under the Town Development Act, 1952, between the London County Council and Basingstoke Borough Council by which some 3,500 houses were to be built in Basingstoke over the following ten years to accommodate an over-spill population from London of 12,000.

During the 1950s several new firms set up manufacturing units in Basingstoke, despite the efforts of the Board of Trade to attract industry to the Development Areas. One of the most important of these was that of Lansing Bagnall Ltd., which in 1949 began manufac-turing industrial trucks and materials handling equipment. It is interesting to record that this firm first approached the town in 1946 through the Basingstoke Development Committee; its expansion over the years has been particularly impressive and at the end of 1961 it was the largest industrial employer in the Borough. Firms engaged in bookbinding and printing and the production of electric fork-lift trucks and industrial diamonds also set up production units between 1946 and 1960 and further added to the region's industrial diversity and stability. Further details of these post-war developments are given in Chapter V; accompanying them there have been substantial in-creases in the numbers employed in the service trades. A further impetus to the growth of the town was provided by the building of some 380 houses on an estate to the north-east of the town to accom-modate the families of employees working at the Atomic Weapons Research Establishment set up at Aldermaston in 1952. In 1960 it was estimated that around 1,000 people employed at the Establishment lived in Basingstoke.

For the most part, the growth of employment and population in Basingstoke have kept pace with one another in recent years. In 1959, for example, in a report to the Basingstoke and District Employment Committee, reference was made to the fact that local industry had been able to absorb most of the workers who had moved into the district. A warning was given, however, that the time must come when existing firms would be unable to meet the already increasing demand for employment opportunities.

By the eve of the new (and present) projected Town Development Scheme the town of Basingstoke had developed into a light industrial

[1] The others were Ashford, Aylesbury, Bury St. Edmunds, Grantham, Haver-hill, Huntingdon, Kettering, Letchworth, Luton, Dunstable, Peterborough, Thet-ford, and Wellingborough.

25

centre of some importance. Throughout the post-war period, un-employment had been maintained at below the national average[1] and the local rate of national savings above the national average. The population of the Borough in mid-1961 was 26,010 and increasing at the rate of 6 per cent per annum, of which the larger part was accounted for by immigration.[2] Of the working labour force in the Employment Area of 19,581, some 8,431 or 43·1 per cent were engaged in manufacturing and 9,571 or 48·8 per cent in the tertiary industries.

[1] See Chapter V, p. 84.
[2] Between 1951 and 1961 the total number of people immigrating into Basing-stoke was 7,162.

THE TOWN EXPANSION SCHEME

THE BASIC AGREEMENTS

ON the 31st October 1961 an agreement was made under the Town Development Act, 1952, between the London County Council, Hampshire County Council and Basingstoke Borough Council for the erection of 11,500 dwellings in Basingstoke by 1976 to be occupied by tenants nominated by the London County Council, as part of the scheme to relieve congestion and overpopulation in the County of London.[1] In addition to this housing development, it is intended to provide all appropriate public services, facilities for public worship, recreation and amenity buildings, as well as ensuring that there will be sufficient industrial, commercial and other employment opportunities for the new population. Assuming the size of the average household migrating to Basingstoke is that of the country as a whole, viz. 3·1 persons, the scheme would allow an intake of overspill from London of 36,800 people; if it is the size of the average New Town household, viz. 3·5 persons, the population accommodated would be nearer 40,000. Assuming for the moment that the lower figure is valid at the time of entry of such households into Basingstoke, and taking into consideration (i) the present civilian population in the (amended) Town Map area (approximately 32,500), and (ii) the expected natural increase and private migration into the town, it is estimated that the total population of Basingstoke by 1981 will reach 86,000,[2] and by the close of the century nearly 100,000.

[1] Any of the following may be nominated as tenants:
   (a) persons who are employed by an industrial, commercial or business concern which moves from London to Basingstoke and who transfer with the firm;
   (b) tenants or licensees, or persons on the housing waiting lists, of London County Council or any metropolitan borough who find employment in Basingstoke under the Industrial Selection Scheme;
   (c) if suitable persons in the preceding category are not available for employment, any other person resident in London who accepts employment in Basingstoke;
   (d) persons resident in London, or on the London housing lists, who are self-employed or retired and who wish to move to Basingstoke.

[2] Basingstoke Town Map Amendment Map, 1961, Hampshire County Council, 1962. For comments on this expected target see Chapter IV, p. 47 ff.

The present, i.e. 1961, scheme supersedes an earlier and less ambitious agreement arising out of the 1950 and 1955 Development Plans, signed in October 1959 between the Basingstoke Borough Council and the London County Council to build 3,500 dwellings for transferring population from London. It also forms part of the proposal submitted by the Hampshire County Council for major town expansion schemes at Andover, Basingstoke and Tadley under the 1952 Act, as an alternative to the London County Council project for the building of a completely new town of 100,000 people at Hook.[1]

The Hook project had many advantages on paper. The site, one of seventy investigated south-east of a line from the Wash to the Solent, was a reasonable distance from London, with good road and rail communications; water supplies were adequate and the land was of only moderate agricultural value. But it also had serious drawbacks, which in the end, after considerable local opposition, led to the abandonment of the scheme. Three of these may be mentioned: first, the western edge of the Hook site was only six miles east of Basingstoke, where, as has just been mentioned, a town expansion scheme for absorbing 11,000–12,000 Londoners was already in hand; second, the scheduled area was at the centre of a whole polygon of other growing towns, viz. Aldershot, Farnborough, Camberley, Fleet, Wokingham and Reading; and third, it was felt that the dispersal of Londoners could be achieved more quickly by the expansion of existing communities than by a new town starting from scratch on a virgin site. Instead, the London County Council agreed on the plan of the Hampshire County Council for settling about 67,000 Londoners in Basingstoke, Andover and Tadley. Basingstoke, it was decided, would take about 37,000 of these and the other two towns the rest.[2]

ADMINISTRATIVE ARRANGEMENTS

This new expansion of Basingstoke is to take place within the provisions of the Town Development Act 1952, which, like the New Towns Act 1946, is specifically designed to relieve congestion and overpopulation of the major conurbations. However, while the objective of the two Acts is the same, there are important differences in their administration. Whereas, for example, the Town Development

---

[1] See *The Planning of a New Town*, London County Council, 1961.
[2] Andover (present population 16,974) is scheduled to increase its size by 31,000 (20,000 overspill and the balance private migration and natural increase), and Tadley (present population 4,600) by 12,500 (10,000 overspill and the balance private migration and natural increase.)

Act places the responsibility for the implementation of town expansion on the shoulders of the local authorities concerned, in the case of new towns, most of the initial planning and programming is in the hands of specially constituted Development Corporations, financed by Treasury loan. Each form of administration has its own particular advantages. The Development Corporations, for example, exercise complete control of land use in their designated areas and in policy and decision taking are largely independent of local authority jurisdiction. On the other hand, the expanded towns legislation tends to encourage a fuller and more co-ordinated use of existing resources such as the building capacity, and the technical skills of planning staffs of engineers, architects and surveyors of both the receiving and exporting local authorities. Moreover, the 1952 Act appears to offer more latitude and advantages to the receiving authority in respect of the financial provisions than does the New Towns legislation.

The particular administrative vehicle chosen to plan and guide the future of Basingstoke is the Town Development Joint Committee. This Committee, which issued its first report in the summer of 1962,[1] consists of representatives of the three interested authorities, viz. the Basingstoke Borough Council, Hampshire County Council and the London County Council, and its specific function is:

> 'to co-ordinate, manage and execute on behalf of the three authorities a scheme of town development at Basingstoke.'

It is interesting to observe that this is the first time a county council has become an active participant in town expansion in this particular way.

The Joint Committee is required to prepare and submit to the authorities for approval:

(a) a master plan for the scheme;

(b) from time to time, more detailed plans and estimates of individual projects forming part of the scheme;

(c) an annual programme of operations and an estimate of expenditure and income for each financial year; and

(d) supplementary programmes and estimates, if it appears at any stage that the approved programme and estimate for any year is likely to be exceeded.

[1] Basingstoke Town Development Joint Committee. First Annual Report 1961/62.

The Committee is also required to issue an annual report giving details of the progress of the scheme together with financial and other statements. The day-to-day executive work of the Committee is undertaken by the Basingstoke Development Group headed by a Director of Town Development and a Chief Architect/Planning Officer. Both these officers have been engaged by the Hampshire County Council.

To assist the execution of the above scheme, financial assistance is to be provided for the local authority from two main sources: (i) Central Government funds under the Town Development Act, 1952, and (ii) the Hampshire County Council and London County Council in accordance with the specific agreement signed.

As regards the former, while Exchequer assistance is given for housing in the form of a subsidy payable for sixty years on each approved dwelling provided as part of the town development scheme, the Ministry of Housing and Local Government is also empowered to give financial aid in respect of:

(*a*) expenses of acquiring land for development;

(*b*) expenses of site preparation and other works;

(*c*) expenses of providing, extending or improving main water supplies, main sewerage and sewage disposal services;

(*d*) payments by the local authorities to suppliers of water in connection with making supplies available for the purposes of development;

(*e*) payments to a river board or drainage authority in respect of works rendered necessary by the development.

Attention is now paid to the financial arrangements entered into for Basingstoke.

FINANCIAL ARRANGEMENTS

The financial arrangements specific to the Basingstoke scheme are described in detail in the First Annual Report of the Joint Committee, but may be summarised as follows:

(1) *Initial Capital.* The initial capital for land purchase, except that of the comprehensive area, will be provided by the London County Council, as will the capital for the development of housing and industrial areas. It is intended that the Basingstoke Borough Council will repay such capital as the scheme proceeds and development becomes productive.

(2) *Development Capital.* The actual cost of development will be shared by Hampshire County Council and the local authority. For the first two years of the scheme the County will bear one-half of the net additional expenditure incurred by the Borough Council on town development and from the third to the fifteenth year one-half of the excess of the 'true borough rate over the average true district rate of all boroughs and urban districts in Hampshire (plus five per cent)'.[1] In addition, the London County Council will pay half the amount contributed by the Hampshire County Council throughout the fifteen years mentioned above.

(3) *Housing Subsidies.* A third form of financial assistance is the contribution by the London County Council towards the housing revenue. For a period of ten years the London County Council will pay half of the estimated deficiency in the housing revenue account for each dwelling provided for a nominated tenant, and for the first fifteen years from 1st April 1961 one-quarter of the net annual deficit of the housing revenue account as a whole.

Finally, provision has been made in the agreement for any unforeseen contingencies. Thus in the event of any 'unreasonable' financial burden falling on Basingstoke Borough Council, both the London County Council and the Hampshire County Council undertake to review their contributions. In any event, there is to be a review at the end of the eighth year of the share of housing subsidies and developmental expenditure provided by the various authorities.

THE TOWN MAP

Consequent upon and arising out of the 1961 agreement, a new Town Map and provisional Master or Development Plan for Basingstoke has been prepared covering the twenty years to 1982. Both Map and Plan were published in September 1962. The area of the Town Map approved in 1955 under the Town and Country Planning Act, 1954, has been extended to include the whole of the Borough of Basingstoke and a number of parishes (or parts of parishes) in the Basingstoke Rural District.[2] It has been prepared by the Hampshire County Planning Committee and is subject to the approval of the

---

[1] See para. 24 (3) (ii) of First Annual Report of Joint Committee.

[2] Viz. parts of the parishes of Basing, Cliddesden, Deane, Dummer with Kempshott, Farleigh Wallop, Mapledurewell and Up Nately, Monk Sherborne, North Waltham, Oakley, Sherborne St. John, Sherfield on Loddon, Steventon, Tunworth, Winslade, and Wootton St. Lawrence.

Ministry of Housing and Local Government before it has statutory significance.

The object of the Development Plan, which is the work of the Joint Development Group, is to make provision for:

(i) the marked increase in population resulting from the reception of overspill from London, private migration and natural increase;

(ii) the 'extension, redevelopment and improvement of the Central area in consequence of the expansion of the town', and

(iii) 'a road system adequate for the needs of the enlarged town area as regards efficient communications, road safety and visual amenity'.[2] In allocating land within the town to different uses, e.g. housing, industry, schools, shops, etc., the Committee states that 'every effort will be made to keep the best existing landscape features of the town and also to create new treatments both within and on the edge of the town'.[1]

THE SCOPE OF THIS STUDY

This study is concerned with some of the more important economic and planning issues raised by the 1961 agreement and the new Town Map. First we deal briefly with the prospects of supplying the basic services necessary to meet the demands of a town of 80,000–90,000 people. Second, in the light of population statistics at our disposal, we examine in some detail the likely pace and pattern of population growth in Basingstoke over the next twenty years, particularly as it affects the size of the school population and working labour force. Chapter V outlines the main features likely to influence the structure of economic activity in the town and pays detailed attention to manufacturing industry. Chapter VI deals similarly with the pattern of service industry and includes an estimate of Basingstoke's shopping requirements by 1980.

Throughout the analysis which follows we would emphasise that our aim has been not so much to project trends or forecast future events but to apply such data as is available on the subject of town development to the particular challenges and problems facing Basingstoke. Economic predictions are difficult enough to make six months or one year ahead: twenty years ahead they are little more than crystal-gazing. So much of the art of forecasting depends upon the

[1] Basingstoke Town Map, op. cit., p. 3.

initial assumptions which one makes concerning the value and be-haviour of certain variables – assumptions which though valid now may so easily be proved wrong in the future. Yet, in so far as some measure of planning is necessary and accepted as such, the authorities concerned cannot avoid making decisions today which will affect the lives of thousands of people tomorrow; land has to be allocated in 1962 for use in twenty years' time and housing and industrial pro-grammes scheduled well in advance of implementation. The more comprehensive the information available to guide planning decisions, the more reliable such decisions are likely to be. The fact that such problems forbid any precise answers and that conclusions are often highly conjectural, in no way detracts from the usefulness of under-taking research of this kind.

Our task, then, is to examine and draw together such data as we think useful in helping to answer a number of questions relating to the economic future of Basingstoke; to bring ordered reasonableness and consistency into the making of planning decisions; to suggest one or two models which may be used as a basis of such decisions; and in the light of such models and the data derived from them to offer ten-tative predictions of the ordering of economic events.

Finally, an important point of definition. In this study we shall have cause to use the name of Basingstoke in three separate senses: (a) as the present *Borough* of Basingstoke; (b) as the *1962 Town Map Area* of Basingstoke, and (c) as the area of the Basingstoke *Employment Exchange*. This triple definition is unfortunate but unavoidable; employment figures, for example, are not available for the same area as those of population, and the 1962 Town Map creates a new set of boundaries altogether. Unless otherwise stated or qualified in one way or another, however, 'Basingstoke' means, for the purposes of our study, the present Borough of Basingstoke. In Chapter IV most of our population projections are expressed in terms of the Borough–or more particularly, in terms of the existing population within the Borough. In Chapter V, however, most of our employment data relates to the area covered by Basingstoke Employment Exchange–a much wider area than the Borough. Though we would like to have standardised our analysis with respect to the 1962 Town Map area, this has proved impossible, as we know so little about its population and employment composition. We thus ask the indulgence of the reader in our terminological difficulties and for his care in interpreta-tion of the pages which follow.

BASIC SERVICES

INTRODUCTION

THE planned expansion of Basingstoke described in the previous chapter will make substantial new demands upon the basic services. It is not our purpose to discuss roads in this study–an economic evaluation can hardly be given until the detailed framework of road design, as conceived by the planners in the light of overall town development, is made known. Something (though regrettably not much) can be said about the likely development of the main public utility undertakings; the supply and demand for these services is more easily predictable as they depend less upon conceptions of town planning in general, than upon specific economic considerations. Naturally, certain assumptions need to be made concerning such things as the structure of prices, as between gas and electricity, for example, and the rate of technical development–but these are inherent in any programming of this sort.

In general, there would appear to be no insuperable technical difficulties in the provision of the necessary services for sewerage disposal and the supply of electricity, gas and water. Only two of the fifteen new towns designated before 1957 ran into serious difficulties over water supplies and only one over sewerage; gas and electricity installations were constructed without serious delays,[1] although sometimes programmes had to be re-phased. The most serious problem which may confront Basingstoke is the shortage or non-availability of the necessary labour or capital at the time of the actual construction of works.

1. SEWERAGE

The Basingstoke Borough Council already has in hand the work of expanding the existing facilities, which are inadequate even for present requirements. It is estimated, however, that when the pumping station to be built in Swing-Swang Lane is completed it will be

[1] J. H. Nicholson, *New Communities in Britain*, National Council of Social Service, 1961.

sufficient to cover the needs of a population of 36,000. At the same time, planning permission has been granted for the construction of a new treatment and disposal works at Wildmoor in the Loddon valley, which will meet the needs of a population of 80,000. Work has already started on this project which, it is expected, will be completed within some three years. A long trunk 42 in. sewer is also to be installed which can be immediately incorporated in the present system. Within the Borough itself a network of sewers will be constructed and there will be a new main sewer for Kempshott. No technical problems are envisaged in the undertaking of this programme, but some difficulty is contemplated in securing the necessary labour for the Wildmoor scheme, unless the services of a national contractor can be obtained, in which case labour would be imported from outside the area. The cost of the above sewerage schemes has been provisionally put at £1,784,000:[1] of this amount the Ministry of Housing and Local Government will provide £871,250.

2. GAS

Introduction

The supplying authority – the Southern Gas Board – has already studied the plans for the development of Basingstoke, as given in the Draft Development Plan and shown on the Draft Review Town Map, and made tentative forecasts of the likely future demands. In 1961 some 1·1 m. therms were supplied to 5,200 consumers in the Basingstoke area: by 1976 with a population of 75,000 it is estimated that the number of consumers will rise to 21,000 and total gas purchases to 4·4 m. therms: by 1981 assuming a population of 85,000, it is thought that the consumers will number 25,000 and gas purchases 5·9 m. therms.

Plans to meet these increases over present rates of gas consumption have been made as flexible as possible so that changes in the rate of expansion of the town or in the areas to be developed can be accommodated. Flexibility is also necessary to cater for changing living standards and improved industrial techniques, which might well produce a much higher rate of increase in gas consumption than the rate calculated from the physical expansion of the town alone.

Present gas distribution system

Basingstoke is at present supplied with gas made at Southampton through a 16 in. main which links Southampton with Reading and

[1] *Hants and Berks Gazette*, 28.9.62, p. 11.

Oxford. From this main there are two branches into Basingstoke, one 12 in. from Hannington through which a bulk supply of gas is transmitted to the Basingstoke, Farnham and Aldershot area, and an 8 in. branch from Kingsclere which supplies Aldermaston and Basingstoke. A gasholder of 1,000,000 cu. ft. capacity is fed from either or both of the branch mains at a controlled rate and Basingstoke town is supplied from this gasholder through a station governor controlling gas to the local low pressure mains.

There are two principal mains from the gasholder station, one 18 in. diameter feeding the north, north-west and central areas and one 12 in. diameter feeding the south and south-west areas and the village of Oakley. A network of smaller mains links these two principal feeder mains.

This system can deal with gas outputs up to 1965.

Alterations to distribution system

Most of the early development will be in the southern area served by the 12 in. main, and to cater for the increased gas load the main will be isolated from the low-pressure system and operated as a medium-pressure feeder main. Seven district governors will feed gas from the main into the ends of the low-pressure mains.

To supply this medium pressure a four-stage centrifugal booster driven by an electric motor will be installed at the gasholder station.

Demands for gas from the factories to be built in the Kingsclere Road area can be met from the existing mains for a time but, as they increase, two governors will be fixed to take gas from one of the incoming mains to Basingstoke direct to the district mains. This may be necessary by 1965.

Supply to gasholder station

The present method of supplying the gasholder station is adequate for about ten years. Later it will be necessary to increase the capacity of the control equipment at the gasholder station, but it will not be necessary to lay additional mains in the foreseeable future.

Land has been reserved for the erection of a second gasholder of 1,000,000 cu. ft. capacity. It will be required in about ten years, but the need for additional storage will be reviewed as the pattern of hourly variations in the combined new domestic and industrial loads is revealed. It is important that the rate of intake of gas to the gasholder station is maintained constant throughout the day.

36

Long-term proposals

By 1981 the pressure in the 12 in. medium-pressure main will need to be 7 lb. per sq. in. and provision is being made for a compressor to be installed in an extension of the existing governor house.

It is also planned to lay a new 12 in. main to link the medium pressure main with the 8 in. main entering Basingstoke from Kingsclere. This would complete a pressurised ring main around the town with alternative feeds from the Southampton main and the gasholder station.

Eventually a new low-pressure main to the eastern side of the town will be required and its route can be planned to conform with the development plan for the central area.

No difficulties are anticipated in increasing gas supplies, since Southampton uses by-products of the Fawley oil refinery.

Comparative analysis

Data obtained from Regional Gas Boards supplying a number of the new towns suggest that there were no serious technical difficulties encountered in providing the necessary services in the time programmed. Economic problems did arise, however, as building conditions in the period of most pronounced development (1948–54) were difficult. Many materials, e.g. steel and cement, were in short supply and labour was hard to come by.

The expansion in the consumption of gas in a selection of the London new towns from the time of designation to 31 March 1962 is given in Table III/1 below. No estimate is available of the current maximum or peak load demanded by consumers, although a general formula of 1/230 of the annual consumption per day is used by the various Gas Boards. As the last column of the table reveals, the *per capita* consumption of gas by the 'new' population averages 51·2 therms per annum, ranging from 34·3 in Hemel to 65·8 in Harlow.

When estimating Basingstoke's future requirements account must also be taken of trends in average consumption. Over the past ten years the annual consumption of gas *per capita* in Great Britain has risen very slightly from 50·1 therms to 50·5 therms in 1961 with a peak consumption in 1955 of 52·1 therms. If a linear trend of the rate of growth over this period is extrapolated up to 1976 and 1981, this would result in a small increase of consumption of 0·5 per cent in both cases.[1] An alternative projection based on the average rate of

---

[1] A curvilinear trend would suggest a slight *decrease* in consumption of between 1 and 2 per cent over this period.

increase over the past ten years would give a rise of consumption of 2·1 per cent and 2·8 per cent. If we take the mean of these projections (1·3 per cent and 1·6 per cent) and express this as an increase over the 1961 *per capita* consumption of Basingstoke (42·3 therms), we obtain a *per capita* consumption of 42·8 therms in 1976 and 43·0 therms in 1981. Multiply these by the probable population in these years (75,000 and 85,000) and we arrive at total consumption figures of 3·2 million and 3·7 million therms. If one calculates one's projections on the average of the new town's consumption in 1961 (instead of that of Basingstoke) the total likely consumption in Basingstoke works out a 3·8 million therms in 1976 and 4·3 million therms in 1981.

TABLE III/1

GAS CONSUMPTION IN SELECTED NEW TOWNS

	At Time of Designation		Population*	Total	Col.4−2
	Population	Con- sumption (therms)		Con- sumption (therms)	Col.3−1
Basildon	43,000	2,013,000	89,000	4,422,000	52·4
Bracknell	17,000	657,000	32,000	1,309,000	43·5
Crawley	9,500	35,000	55,400	2,809,300	60·4
Harlow	4,500	150,000	52,400	3,422,966	65·8
Hatfield	8,500	426,000	21,500	1,001,278	44·2
Hemel Hempstead	21,200	1,004,000	56,500	2,214,911	34·3
Stevenage	7,000	380,000	44,000	1,745,192	36·9
Welwyn	18,500	1,600,000	36,000	2,525,700	52·9
All towns	129,200	6,265,000	386,800	19,450,347	51·2

*As at 31.12.62, except for Basildon and Bracknell, given at 31.3.62.

Source: *Town and Country Planning*, January 1963, and information kindly provided by Area Gas Boards. Basildon covers respectively the local authority areas of Billericay and Easthampstead.

The above figures, which clearly suppose a less pronounced increase in gas consumption than that envisaged by the supplying authority, must not be taken as other than a very tentative forecast; but assuming the relative contribution of the two main secondary fuel industries in Basingstoke to be broadly comparable with the average of the London new towns and the price ratios of gas and electricity follow the trend of the past decade it is suggested they are quite plausible.

Costs and labour

No data on the expected costs of the above proposals are available;

neither will they be until detailed plans of housing and industrial estates have been worked out. Capital will be allocated to the Board by the Gas Council under arrangements approved by the Ministry of Power: no difficulties are envisaged in this respect, since new housing developments always have a high priority. It is expected that the Southern Gas Board will put all work out to contract; successful programming of building work is thus substantially dependent on an adequate supply of labour being available at the right time.

3. ELECTRICITY

Introduction

Electricity supplies also appear to be adequately catered for and no major problems are anticipated in expanding them. At present the approximate maximum demand for electricity in Basingstoke is 20 mW.: the supplying authority considers a consumption in the order of 70 mW. by 1976 and 90 mW. by 1981 is not improbable.

Present and projected electricity distribution system

At present Basingstoke's electricity is provided by lines of 33,000 volts fed from 132,000-volt sub-stations at Alton and Reading. Within a year a further point near Fleet will be drawn on, and in about three years' time a grid supply point will be put in Basingstoke itself. No particular technical difficulties are foreseen, assuming that all the necessary sub-stations sites, way-leaves, etc., required to house the necessary equipment are available.

Costs and labour

The capital cost of expansion, the exact amount of which will not be known until further details of the Town Map are published, is likely to be between £2,000,000 and £2,500,000.[1] This cost will initially be met by the Southern Electricity Board, the amount being spread over the whole area covered. The subsequent payments by consumers are expected to cover this capital expenditure except in certain cases where the consumer has to pay towards installation costs. No labour problems are envisaged as there is a mobile force of both skilled and unskilled labour which is recruited from Portsmouth, Aldershot, etc., to serve the whole Southern Region.

[1] Estimated in the light of the capital expenditure in the new towns, viz. (up to 1960) Stevenage £900,000; Harlow £1·6 m.; Basildon £862,114; Crawley £1·2 m.; and allowing for rising labour and material costs.

Comparative studies

The South Eastern Electricity Board supplied the following information relating to the provision of electricity to Crawley New Town between 1946 and 1960:

During the whole of the development we met with no particular economic problems in providing supplies either to the domestic or industrial areas. This was in a large measure on account of the very close cooperation and frequent meetings we had with the Crawley Development Corporation whereby we were supplied with details of all projects in their very early stages. It was therefore possible to plan ahead and order cables and material in advance and at the same time prepare expenditure budgets. In this way we met with no difficulty in meeting capital requirements.

By this collaboration it was also possible to arrange for the provision of suitable sites for the Board's sub-stations, the re-routing of certain existing mains and cables and the provision by the Corporation of housing accommodation for the Board's staff. The schedules of the original programme were all well maintained though labour shortages at times did represent a serious problem because the district itself did not have large enough engineering resources and it was necessary to import staff and labour from other parts of the area.

The unskilled labour position was further complicated by the demands made on it for water, gas and sewage mains laying in addition to the major requirements for building work.

Apart from the rather rapid rise in the demand for electricity made by one or two of the new industries on the industrial estate, no unexpected problems arose. The maximum demand in Crawley in 1946–47 was approximately 2·47 mW. In 1959–60 it was 34·2 mW. rising to 41 mW. in 1961–62.

Data in respect of the growth of facilities in four London new towns show that the *per capita* consumption in March 1960 varied from 1,363 units in Basildon to 2,005 units in Hemel.[1] The maximum or peak demand ranged from ·37 kW. *per capita* in Basildon to ·75kW. *per capita* in Crawley. The two main factors making for these differences were:

(1) the proportion of industrial to domestic consumers, and
(2) the relative importance of gas and electricity as sources of motive power in industry (e.g. Harlow industry uses considerably more gas but less electricity than Hemel industry).

For estimating future demand the data in respect of the new towns are too varied to be of much use. A better indication may be gained by

[1] One unit = the consumption of 1 kW. of electricity for one hour.

using Basingstoke's present electricity consumption as basis. Extrapolating (i) the linear trend and (ii) the average increase over the period 1951/61, electricity consumption *per capita* in Great Britain is likely to rise between 60·4 per cent and 109·2 per cent by 1976 and between 81·3 per cent and 146·4 per cent by 1981. Taking the mean of these projections and applying this to Basingstoke, this would mean that the *per capita* peak consumption of electricity may well rise from (approximately) 0·8 kW in 1961 to 1·5 kW by 1976 and 1·7 kW by 1981. Multiplying these figures by a 75,000 and 85,000 population respectively, we arrive at total peak demands of 112·5 mW and 144·5 mW. Both these figures considerably exceed the estimates cited earlier.

4. WATER

Basingstoke's water is at present supplied by the Mid-Wessex Water Company and is likely to continue to be so as the town expands. No serious difficulties are anticipated in increasing supplies, although a new source or sources of water will have to be found and additional storage facilities and brick mains provided. The existing supply in the Borough is being developed, while a new source at Woodgarston ($3\frac{1}{2}$ miles north-west of Basingstoke) is expected to yield sufficient water for a population of at least 45,000. Since no definite plans yet exist of the layout of the expanded distribution system, no estimates of probable cost are possible. The water company has no information as to who will bear the cost of the necessary works. The company itself raises capital by offering shares by public tender and they anticipate no difficulty in raising any additional finance. Neither is any difficulty expected in obtaining labour. Expansion is intended to precede increase of population, with the object of keeping ahead of demand all the time.

Consumption of water in Basingstoke in 1961 averaged $1\frac{1}{4}$ million gallons daily, the peak summer load being $1\frac{1}{2}$ million gallons. It is likely that these needs will be increased four to five times in the next fifteen years.

CONCLUSION

To sum up, all public utility undertakings appear confident of a smoothly running expansion programme once the Plan is finally agreed, although electricity requirements may well prove considerably higher than has been envisaged. Sewage disposal alone appears to present a less optimistic picture, though once present difficulties have been overcome this, too, should proceed satisfactorily. Provided the economic situation is favourable, electricity and gas as nationalised concerns should have no difficulty in providing the large capital outlay; the other undertakings are covered by the tripartite agreement.

CHAPTER IV

POPULATION EXPANSION

THE population of Basingstoke in April 1961 was 25,940. This figure, taken from the provisional results of the Census of Population represents a 52·8 per cent increase over that of the 1951 Census, when the population numbered 16,978. The growth in the population of the town since 1921 is illustrated in Table IV/1; comparative trends for Hampshire and England and Wales are presented in Tables IV/2 and IV/3. From these tables three distinct phases of growth are discernible. First, up to around 1946 the population of Basingstoke expanded steadily at the rate of about 0·9 per cent per annum. There was a small net migration into the town, particularly during the latter part of the inter-war period, but most of the increase, the rate of which was con-

TABLE IV/1

POPULATION GROWTH IN BASINGSTOKE 1921/1961

	Total Population	Annual Net Increase (+) or Decrease (−)	Natural Increase (+) or Decrease (−)	Apparent Immigration (+) or Emigration (−)
1921	12,723			
1931	13,865	+ 114		
1939	14,920	+ 132		
1946	15,850	+ 133		
1947	16,070	+ 220	+126	+ 94
1948	16,260	+ 190	+123	+ 67
1949	16,780	+ 520	+ 85	+ 435
1950	16,720	− 60	+107	− 160
1951	16,978	+ 258	+ 62	+ 218
1952	17,220	+ 242	+116	+ 104
1953	18,170	+ 950	+129	+ 821
1954	19,320	+1,150	+184	+ 966
1955	19,610	+ 290	+197	+ 93
1956	19,930	+ 320	+146	+ 174
1957	20,830	+ 900	+200	+ 700
1958	21,810	+ 980	+212	+ 768
1959	23,130	+1,320	+236	+1,084
1960	24,500	+1,370	+302	+1,068
1961(a)	25,940	+1,440		
1961(b)	26,010	+1,510	+368	+1,142

Source: 1921, 1931, 1951, 1961(a) Census of Population; other dates Registrar-General mid-year estimates.

TABLE IV/2

PERCENTAGE RATE OF GROWTH IN POPULATION:
BASINGSTOKE, HAMPSHIRE AND ENGLAND AND
WALES, 1921/1961

	Basingstoke	Hampshire	England and Wales
1921–31	9·0	11·0	5·5
1931–46	14·3	2·4	1·6
1946–52	8·6	17·6	8·3
1952–56	15·7	3·9	1·6
1956–61	30·5	6·9	3·4

Source: As for Table IV/1

TABLE IV/3

RELATIVE GROWTH IN POPULATION: BASINGSTOKE,
HAMPSHIRE AND ENGLAND AND WALES, 1921/61,
1946 = 100

	Basingstoke	Hampshire	England and Wales
1921	80	88	93
1931	87	98	98
1939	94	106	102
1946	100	100	100
1947	101	103	103
1948	103	107	105
1949	106	114	106
1950	105	115	108
1951	107	117	108
1952	109	118	108
1953	115	119	109
1954	122	120	109
1955	124	121	109
1956	126	122	110
1957	131	123	111
1958	138	125	111
1959	146	127	112
1960	155	128	113
1961	164	131	114

Source: As for Table IV/1.

siderably above both the national average and that of Hampshire,
consisted of the excess of births over deaths. Second, between 1946
and 1952 the annual rate of increase in population quickened to 1·5
per cent. Of the total growth in population of 1,370, some 701 (52 per
cent) represented natural increase, and 669 (48 per cent) net migration
into the town. Most of this migration reflected a general trend southwards of the national population, though there was some expansion
of employment opportunities in Basingstoke, particularly in 1949,
which added to the town's attractions.

44

The impetus behind the third phase of population growth which began in 1952 and has continued to the present day was twofold—first the setting up of the Atomic Weapons Research Establishment at Aldermaston, and the development (under Ministry of Works contract) of the Oakridge estate in Basingstoke to accommodate some 380 workers and their families,[1] and second, a marked expansion in private-house building[2]. Net migration into the town accelerated to 820 in 1952–3 and 966 in 1953–4. On the completion of the Oakridge estate, net migration fell to nearer its earlier level, an increase in private-house building being more than outweighed by a slowing down in the rate of construction of local-authority dwellings.

Between mid-1956 and mid-1961 the population of Basingstoke expanded at an average rate of 5·2 per cent per annum—nine times the national average and more than four times that of Hampshire. The beginning of this period coincided with the third wave of post-war migration into the town—a wave which gathered momentum following the Borough Council's agreement with the London County Council, in October 1959, to accommodate some 12,000 Londoners in Basingstoke in the following ten years. In fact, however, by far the greater part of the net migration into Basingstoke since 1956 has taken the form of private or non-organised migration. About three-fifths of the 4,500 migrants in question came from Greater London or South-East England and the balance from other parts of the country.[3] While employment opportunities in Basingstoke and the surrounding areas have, in general, kept pace with the needs of a growing working population,[4] there has also been a substantial rise in the commuting labour force. Although this latter trend is mainly accounted for by those working at Aldermaston, the number of season tickets (weekly and other) issued to those travelling from Basingstoke to London, Reading and other districts rose from 3,083 in 1951 to 5,761 in 1960.

[1] The building of this estate took place between December 1952 and February 1954. At the same time there was a considerable local recruitment by A.W.R.E. of semi-skilled labour. This build-up of workers from Basingstoke has continued over the years and by 1960 roughly 1,000 people from the Borough were working at Aldermaston.

[2] Between 1946 and 1950 only 77 new houses were built by private enterprise in Basingstoke. With the easing of Government restrictions on such building from 1951 onwards the number of new privately constructed dwellings completed rose from an average of 54 in the years 1951–4 to 258 in the years 1955–8.

[3] *The Social Implications of a Town Development Scheme.* An interim report of the Basingstoke W.E.A. Study-Group on Town Development, Sociology Section, ed. M. Broady, 1962. (Private circulation only.)

[4] For further particulars see Chapter V.

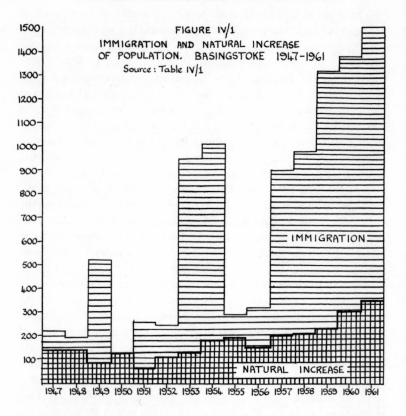

FIGURE IV/1
IMMIGRATION AND NATURAL INCREASE
OF POPULATION. BASINGSTOKE 1947-1961
Source: Table IV/1

It is, however, unlikely that Basingstoke will develop much further as a dormitory suburb, as the current scheme for expansion envisages the growth of population and employment opportunities in the town side by side. This scheme, which was discussed in some detail in Chapter II, has necessitated some revision to the Basingstoke Town Map drawn up under the Town and Country Planning (Development Plans) Direction, 1954, and anticipates a number of changes in boundary composition.

The new, i.e. 1962, Town Map covers a wider area than that of Basingstoke Borough, but not as wide as the Borough and the Rural District. The estimated size of the population in this new area in mid-1959 was 30,130, compared with 23,130 in the Borough and 40,730 in the Borough and Rural District combined. The proposed population of the Town Map Area and its composition in 1982, (i.e. at the end of

46

the twenty-year planning period) as set out by the authors of this map[1] is given in Table IV/4. Unfortunately, for the purposes of analysing the possible implications of the Town Development Scheme the area covered by the Town Map, (or more accurately, the data we have on the population within that area) is inappropriate. This is because there is no way of estimating the age-structure of the population living *outside* the Borough but *within* the Town Map Area in 1959, which is essential if even approximate estimates of population trends and movements in age structure are to be made.

TABLE IV/4

POPULATION IN BASINGSTOKE TOWN MAP AREA

1959/1982

(As estimated in Draft Town Map)

		Total	Population net of new intake area
1. Population mid-June 1959		30,130	23,130
2. (a) Intake of overspill from London	36,800		36,800
(b) Private migration	6,300		6,300
Total immigration		43,100	43,100
3. (a) Natural increase of 1 above	4,200		2,950
Natural increase of 2 above	8,800		8,800
Total natural increase		13,000	11,750
4. Population 1982		86,230	77,980

In the discussion which follows, then, we exclude from our calculations the 7,000 people living in this new intake area in 1959 *plus* the natural increase expected from such a population over the next twenty years (estimated at 1,250).[2]

Adjusting the figures in Table IV/4 to allow for such an omission, the projected population for the town of Basingstoke in 1982 works out at just under 78,000. This is somewhat less than that earlier envisaged – or at least than that popularly supposed to have been envisaged.[3] In our opinion, there are three reasons why the estimate errs on the low side by 8–10 per cent. First, the figure given in the table for natural increase (see 3.(a)) is almost certainly a considerable underestimate; on the basis of projected national birth- and death-rates adjusted for Basingstoke's present age composition, an estimate

[1] See Part I, Basingstoke Town Map, Amendment Map, 1961, Hampshire County Council, 1962, p. 3.
[2] Calculated on the same basis as that of the expansion of the population within the Borough, but adjusted downwards to allow for an 'older' age structure.
[3] See, for example, comments in *Hants and Berks Gazette*, 4.11.60 and 9.6.61.

47

of 7,000 (5,750 after deduction of population in the new intake area) is more realistic than 4,200 (2,950 after deduction). Second, the rate of private migration into Basingstoke thought likely, viz. 274 persons per annum (or approximately ninety families) would seem too low a figure when viewed both in the light of past experience (for the decade 1952–61 it averaged 500 per annum and in 1960 and 1961 net immigration totalled 2,228)[1] and that of a much larger total population. Admittedly, private immigration is largely a residual in the sense that it is primarily a function of the excess of the rate of house building and growth in employment opportunities in the town over the requirements of the existing population and sponsored migrants, but we consider a figure approaching 8,000 (or 348 persons per annum) a more plausible assumption to make.

Third, we also feel the figure anticipated for the natural increase of the migrating population is unduly low. We shall explain our reasons for this opinion in more detail later but, if the provisions of the overspill agreement with the London County Council are adhered to, it follows that the intake of population from London will be completed by 1976. Assuming an age structure of the incoming migrants basically similar to that recently experienced by the New Towns, and a flow of immigration evenly spread up to 1976, a more likely figure for the natural increase of migrant population (including an estimate for that of private migration) by 1982 is within the range 11,000–11,500.

Making due allowance for the above adjustments, it would seem that the total population of the Town Map area of Basingstoke may be expected to rise to between 93,000 and 94,000 by 1982, and that of the Town Map Area minus the adjustments mentioned above to at least 85,000. Our detailed estimates are presented in Table IV/4a. It is recognised that if these are correct, a further enlargement of the Town Map Area may be necessary.

In the analysis which follows we proceed on a more general assumption, based upon the 1961 agreement, that the town will reach a population of 80,000 sometime between 1976 and 1981. In assuming this, only two sets of data – a pre-Agreement (i.e. 1961) population of 25,940 and an intake of overspill from London of 36,800 are taken as given. This latter figure is obtained directly from the Town Map 1962

[1] Most of this latter figure represented private migration. It was reported in the *Hants and Berks Gazette* on 9 November 1962 that the 100th family to be accommodated under the 1959 'overspill' agreement was welcomed to Basingstoke on 3 November 1962. This would mean that about 85 per cent of the migration in 1960 and 1961 was privately arranged.

TABLE IV/4a

POPULATION IN BASINGSTOKE TOWN MAP AREA
1959/1982
(As estimated by author)

		Total	Population net of new intake area
1. Population mid-June 1959		30,130	23,130
2. (a) Intake of overspill from London	36,800		36,800
(b) Private migration	8,000		8,000
Total immigration		44,800	44,800
3. (a) Natural increase of 1 above	7,000		5,750
(b) Natural increase of 2 above	11,500		11,500
Total natural increase		18,500	17,250
4. Population 1982		93,430	85,180

and is itself derived from the 11,500 dwellings which are scheduled to be built under the overspill agreement. The ratio of 3·1 persons per dwelling may, of course, turn out to be slightly high or low, according to trends in housing standards, the character of the houses built and the family structure of the immigrants. The difference between the total of these two figures and the increase in the projected population assumed is based on our own estimates of three variables:

(a) the natural increase (i.e. excess of births over deaths) of the existing population;

(b) private migration;

(c) the natural increase of the migrant population.

The purpose of the exercise which follows is twofold: first to examine the possible pattern (or patterns) of population build-up in Basingstoke over the next fifteen to twenty years, and second, to predict the likely age composition of the population at five-yearly intervals between 1961 and 1981. These latter projections are particularly important in so far as they give some guide to the likely growth of the school and working population over the planning period.

Because of the speculative nature of any population projections – there are so many variables, the value of any one of which may be affected by unforeseen circumstances – we have made *five* separate estimates, each of which is based on a slightly different hypothesis (shortly to be discussed) concerning the size and speed of organised migration into Basingstoke. However, each assumes that a population

49

of 80,000 will be reached at some point between 1976 and 1981, and that the age structure of the incoming migrants will be basically comparable to that experienced by other towns recently faced with similar challenges and problems to those now facing Basingstoke.

Our five projection series (A–E) for individual years up to 1981 are set out in Table IV/5. As can be seen, the population projected for 1976 (the last year of the Town Development Scheme) varies considerably from 67,503 (A) to 80,192 (E): by 1981, the difference between the extreme estimates has narrowed to less than 4,000, viz. 85,129 (C) and 88,957 (E). It will be observed that each of the projected figures for the end of 1981 is considerably higher than that envisaged in the Town Map.

TABLE IV/5

BASINGSTOKE: POPULATION PROJECTIONS* FOR INDIVIDUAL YEARS 1961/1981

Projections

	A	B	C	D	E
1961	25,940	25,940	25,940	25,940	25,940
2	27,496	27,750	27,750	28,000	28,251
3	29,073	29,582	29,582	30,096	30,610
4	30,667	31,435	31,435	32,226	32,977
5	32,279	33,308	33,308	34,360	35,371
6	33,910	35,201	35,201	36,518	37,792
7	36,624	38,184	38,436	40,022	41,544
8	39,372	41,204	41,713	43,571	45,365
9	42,159	44,268	45,036	47,170	49,240
70	44,989	47,378	48,409	50,823	53,173
1	47,867	50,541	51,838	54,537	57,171
2	51,672	54,390	55,707	58,955	61,632
3	55,543	58,305	59,643	63,449	66,170
4	59,475	62,228	63,641	68,015	70,779
5	63,456	66,308	67,689	72,638	75,447
6	67,503	70,401	71,804	77,338	80,192
7	71,037	73,725	74,394	79,000	81,898
8	74,623	77,099	77,021	80,685	83,626
9	78,261	80,521	79,685	82,394	85,378
80	81,951	83,992	82,387	84,126	87,154
1	85,696	87,514	85,129	85,883	88,957

* For the Town Map Area *minus* the existing population of the new intake area (and its natural increase).

THE PROJECTIONS ANALYSED

There are two main components of the increase in population projected:

(i) net immigration;

(ii) natural increase, viz. the excess of births over deaths.

We will discuss each of these in turn.

(i) NET IMMIGRATION

The rate of organised migration into Basingstoke will principally depend, on the one hand, upon the rate at which new houses can be constructed, and on the other, on the growth of employment opportunities in the neighbourhood. In turn, the phasing of growth of these two components will be influenced by the speed at which the basic services can be expanded and the ease with which the town can absorb additional population. Thus, if the experience of other expanding towns under the Town Development Act and New Towns Act is anything to go by, the initial period of the development scheme is likely to be devoted to site preparation, road construction and the provision or expansion of essential services, e.g. the various sewers and public utilities. In the case of each of the London new towns, between four and five years elapsed between their designation date and the time when they received their 1,000th migrant from London.

Likewise, the ease with which additional people can be accommodated depends on the size of the initial population. Thus Swindon, currently expanding under the Town Development Act, already had a population of 69,760 in 1954, when the Government approved a development scheme to increase its population by 23,000. It is clearly better able to absorb this additional population than is Aylesbury, faced with a similar expansion scheme, but with an initial population of only 25,000. Of the new towns, Bracknell, East Kilbride, Harlow and Stevenage were little more than villages at the time of designation and, on the whole, were slower to complete their initial phases of development than Hemel, Welwyn and Crawley, each of which had a substantial nucleus of population and services.

Though it would be inappropriate to view the prospects of Basingstoke entirely in the light of the results achieved by the new towns or those currently expanding under the Town Development Act, the experience of such towns in recent years provides the most useful pointer we have to the likely realisation of the planned migration into the Borough over the next fifteen to twenty years. We thus propose to examine the average rates of net migration into the five new towns most comparable in size and character to that of Basingstoke at the time of their designation,[1] during the first fifteen years of their experience. We shall also give separately the rates of migration for a town which, at the time of designation, was the most comparable to

[1] viz. Hemel, Basildon, Stevenage, Crawley, and Harlow.

51

Basingstoke in population, size and structure, viz. Hemel Hempstead, which in 1946 had a population of 21,530 and over fifteen to twenty years aimed to expand its population to 80,000. Table IV/6 gives these figures.

Two general features stand out from this table:

1. The gradual build-up of migration to a peak in the second or third quarters followed by a slight decline.
2. The peak intake of new people, through the 'best' quarter averaged 4,054 for all new towns, ranging from 2,515 in Hemel to 4,750 in Harlow. The average rate of annual immigration once the main build-up phase has been completed ranged from 1,301 in Crawley to 4,343 in Basildon.[1] For the fifteen-year period as a whole the average net migration totalled 2,395 a year.

TABLE IV/6
ESTIMATED NET ANNUAL MIGRATION INTO SELECTED NEW TOWNS* 1947/1961

	Average	Low	High	Hemel
1947–1950	233	63	648	160
1951–1954	2,188	1,338	3,125	2,499
1955–1958	4,054	2,515	4,750	2,515
1959–1961	2,960	1,301	4,343	2,290
Annual (average) total				
1947–1961	2,395	1,838	2,817	1,838

* Basildon, Harlow, Hemel Hempstead, Stevenage, Crawley (since 1957).

TABLE IV/7
HOUSES COMPLETED ANNUALLY IN SELECTED NEW TOWNS 1950/1960*

	Average	Low	High	Hemel
1950†	87	0	152	152
1951–1954	901	448	1,307	965
1955–1958	1,320	938	1,629	937
1959–1960	953	595	1,171	917
Annual (average) total				
1950–1960	1,000	673	1,313	883

* As in Table IV/6.
† No houses were built in the new towns (under the new towns legislation) until 1950.

Table IV/7 gives similar particulars in respect of the number of houses completed since 1950 and reveals a pattern of build-up and

[1] Although Basildon was a little slower than other new towns to reach its period of maximum expansion.

stabilisation comparable to that of migration. The average number of houses constructed in any one year in the peak period was 1,320; the rate of building has now slowed down and in the period 1959–60 ranged from only 595 in Crawley to 1,171 in Basildon. Let us now look briefly at the main features of each quarter in turn.

1st quarter 1947–1950

Basically this was a period of preparation in which the necessary administrative machinery was created and attention was focused on the provision of basic services. The first new corporation house was not built until early in 1950. Though quite a high rate of private immigration was achieved early in the period, the numbers migrating under the Industrial Selection Scheme[1] were still very small. In Hemel the net migration recorded in 1950 was only 108 people.

2nd quarter 1951–1954

This was the quarter of a gradually accelerating build-up of new population. The provision of basic services was now complete or in the final stages of completion. With the gradual expansion of industry, population moved in and house construction was considerably speeded up. By 1953, the rate of migration in all the new towns was exceeding 2,000 a year or more: by 1954 Crawley had completed 4,303 houses and Harlow 5,343.

3rd quarter 1955–1958

By now, the fifteen-year Development Plan was in full swing. The rate of immigration reached its peak in the middle of this quarter for the new towns scheduled in 1946 and 1947 and in 1957–8 for Basildon, which was scheduled in 1949. The rate of house construction also achieved its maximum; industry expanded rapidly and towards the end of the period attention was focused on the provision of a central shopping area and ancillary service facilities.

4th quarter 1959–1961

This was the beginning of the running-down period for most new towns, although following a change in the target population of Hemel (60,000 to 80,000) and Basildon (from 80,000 to 100,000) there was an upsurge in migration in 1960 and 1961. More noticeably, the proportion of the increase in population accounted for by the excess of births

[1] For further details see p. 126.

over deaths is becoming increasingly important. Realising the tremendous growth in population likely to follow as a result of the age-pattern of migrants, some new towns, e.g. Crawley, have reduced their rate of immigration much sooner than had been anticipated to allow natural increase to make up the balance.

Total

The estimated average total net migration of the five new towns which planned to expand their population to between 50,000 and 80,000 over the fifteen-year period[1] was 34,900; the maximum migration achieved was 42,500 by Harlow, and the minimum 27,500 by Hemel.

In the light of these figures, the Basingstoke scheme to rehouse between 36,000 and 40,000 Londoners in fifteen years (let alone migrants from other parts of the country) might be thought somewhat ambitious. Put alongside the earlier experiences of Dagenham, however, which in the eleven years between April 1921 and March 1932 recorded a net increase in its estate population at Becontree of 103,328,[2] the project seems perfectly feasible.

Moreover, there are several differences between the situation facing Basingstoke today and that facing the new towns fifteen years ago. First, the general economic position is much improved. Between 1946 and 1951 or thereabouts there was an acute shortage of most raw materials, a stringent control on industrial location and the allocation of capital expenditure, and direction of labour. The country was not only in the midst of a transition from a war to a peace economy; it was entering a new phase in world economic development and evolving new policies of domestic planning and control. In their first ten years the new towns had to weather a series of economic crises which curtailed their housing programmes and rate of factory building; they were unfortunate, too, to be caught in the mesh of a scheme of priorities just at the time when the lack of normal amenities was being acutely felt there.[3] Second, the uniqueness of the new towns experiment militated against a quick fruition of results. It took time to set up the machinery of administration, resolve the inevitable conflicts of local and national interests involved, and prepare the groundwork

[1] Thirteen in the case of Basildon.
[2] See *Becontree and Dagenham*, by Terence Young, Becontree Social Survey Committee, 1934.
[3] J. H. Nicholson, *New Communities in Britain*, National Council of Social Service, 1961, p. 39.

for subsequent planning and growth; industrialists were reluctant to set up plants in areas devoid of the usual urban amenities and facilities. With those towns expanding under the Town Development Act such problems are lessened by the fact that a nucleus of administration, in the form of the local authority, already exists and can be built upon and adapted to the requirements of town expansion. There is no division in responsibility as in the new towns, where the undertaking of house and factory building and certain public utility services is in the hands of specially constituted Development Corporations, and that of most communal services, e.g. refuse disposal, street lighting, libraries, etc., is in the hands of the local authority.

On the other hand, there are certain problems which the expanded towns have to face which are not experienced by the new towns. In the case of Basingstoke, for example, the administrative body responsible for town expansion, the Town Development Committee, consists of representatives from three separate councils, and the interests of each have to be considered and co-ordinated before major policy decisions can be taken. There are also important differences in the financial sphere. Rather more responsibility rests on the local authority for raising capital. In the first fifteen years of development in the new towns all money is provided by the Treasury. In Basingstoke's case the local authority has to pay a larger share of the initial cost or, at least, assume the responsibility for raising the necessary capital. Assistance is given by the Hampshire County Council and London County Council in respect of housing and provision of basic services. It is up to the local authority to arrange or decide the means by which finance for factory construction is raised. Thus the success of Basingstoke's expansion programme will be more dependent upon the local authority's ability to borrow capital or (temporarily) increase the rates, and to this extent at least they will be rather more susceptible to the vagaries of the market than were the new towns. In addition, the cost of borrowing is likely to be much higher than that originally faced by the new towns, as these were in part constructed in times of cheap money and/or benefited from below market rates of interest.

Because it is difficult, if not impossible, to attach a numerical evaluation to these variables, we have postulated five possible rates of migration for Basingstoke–each of which corresponds with slightly different assumptions concerning the economic growth of the town and forms part of the population projections A–E (see p. 50) earlier described. These *include* private or unorganised migration, i.e.

migration other than that sponsored under the tripartite agreement, the limit of which may well be set by the rate of house construction and employment opportunities. Inasmuch as first priority will be given to building under the development scheme and the rehousing of Basingstoke's own population, it is expected that the rate of private migration will fall to less than one-half its current figure of 1,000 or more each year.

TABLE IV/8

PROJECTED ANNUAL RATES OF NET MIGRATION
INTO BASINGSTOKE 1962/1980

	A	B	C	D	E
1962/65	1,250	1,500	1,500	1,750	2,000
1966/70	2,250	2,500	2,750	3,000	3,250
1971/75	3,000	3,000	3,500	3,500	3,500
1976/80	2,500	2,250	1,500	500	500
Total to 1976	33,750	35,750	38,750	40,000	42,250
Total to 1980	43,750	44,750	44,750	42,000	44,250

Projection A envisages a build-up of migrant population in Basingstoke more or less comparable with that achieved by Hemel Hempstead, but adjusted in the early years to allow a more speedy growth. It should be noted that these are *average* figures spread over the four- or five-yearly periods in question. To the extent that immigration into Basingstoke over the last three years has averaged 1,050, the rate of migration assumed for the first quarter is probably an under-estimate. The whole scheme envisages an initial construction of about 400 houses a year rising to 950 by the early 1970s. Under this projection, the target of re-housing 36,800 Londoners would not be reached until well after the target date of 1976, even allowing for the very minimum of private immigration.

Projection B allows for a slightly increased initial intake of people into Basingstoke, rising to something in excess of 3,000 per year in the third quarter. The speed of build-up is still fairly conservative, reflecting more or less the pattern achieved in most of the new towns. Again, on the assumption of this projection the town development scheme would not be completed until 1977 or later.

Projection C suggests an accelerated rate of net migration into the town after 1965, which reaches a higher peak than the previous projections, but which slackens off more markedly in the fourth quarter. This projection considerably exceeds the intake achieved by most new towns, but even so the net immigration assumed is only just sufficient to meet the programme of town development and allow for the barest minimum of private migration.

Projections D and E envisage a much more intensive and rapid build-up of immigrants but more accurately depict the pattern of expansion necessary for the acceptance of 36,800 Londoners and a reasonable intake of private migrants by 1976 or thereabouts. For either of these projections to be realised in the case of Basingstoke, it is essential that first, the scheme should be put into operation immediately and a substantial rate of migrant intake be attained by the end of 1963 and second, that there are a minimum of delays and obstacles (such as difficulties in obtaining the necessary resources or fluctuations in the economic climate) throughout the planning period. Both are extremely ambitious programmes in the light of what has been achieved both by the new towns and other expanded towns in recent years. For a period of six/eight years the net annual intake of migrants must average around 4,000–and possibly 5,000 in the peak years. This entails the building of at least 1,300 new houses each year, which, when added to requirements necessary to satisfy the increasing number of potential householders already living in Basingstoke[1] and to re-house part of the Borough's population (e.g. due to slum clearance or the replacement of dwellings requisitioned under the development scheme),[2] means a total building programme three to four times that at present.[3]

Comment

When one also considers the very substantial claims on the building industry likely to arise from industry, commerce and education, it may well be that the limiting factor on the achievement of the population target will be the availability of building labour. In turn this will depend on the extent to which national contractors can be employed. Local builders can supply only a small amount of the services required (probably about one-fifth). To prevent vicissitudes of the national economic climate affecting the availability of materials and labour at the right time, the local authority might do well to enter into as many long-term contracts with national builders as possible.

(ii) NATURAL INCREASE

This is of two kinds–first, excess of births over deaths of the existing population and second, the natural increase of the new immigrant population.

Our method of projection here is first to assume that the natural increase in population in Basingstoke is equal to the (expected) national average and second, to calculate the deviation from this national rate resulting from the present age-structure of the population and the expected age pattern of immigrants. Thus to the present deviation of the natural rate of increase in population in Basingstoke from that of the national average, we must add a projected deviation likely to arise from the divergence between the town's rate of population growth and changing age structure and that of the national average.

[1] Thus, the post-war population 'bulge' will reach marrying age within the next five/ten years. Even if there was no migration into Basingstoke, the number of new houses required by such new householders (less those vacated by others due to death, etc.) is likely to average at least 100 a year from about 1966 onwards.

[2] Draft Comprehensive Development Area Map, 1962, Part 1, p. 3 ff.

[3] The number of residential properties completed in Basingstoke Borough since 1950 is given below:

1950	176	1954	145	1958	525
1951	165	1955	170	1959	527
1952	133	1956	240	1960	353
1953	272	1957	448	1961	386

Data to assist this latter calculation may once again be obtained from the experiences of the new towns over the past fifteen years. Table IV/9 gives the average deviation of the excess of births over deaths (per 1,000 of population) in six new towns from that of the national average for the period 1947–61. Since, however, within these towns there are considerable variations due to the size and age composition of the initial population and rate of increase in immigration, we also show figures for the new towns with the highest and lowest (average) annual deviation for each quarter, and also particulars for Welwyn and Hemel separately.

TABLE IV/9

DEVIATIONS OF NATURAL RATES OF INCREASE*
FROM THE NATIONAL AVERAGE IN SELECTED
NEW TOWNS† 1947/1960

	Average	Low	High	Hemel	Welwyn
Prior to designation	− 0·6	−3·8	+ 6·8	+ 0·1	+ 6·8
1947–50	+ 0·9	−2·2	+ 5·2	+ 1·1	+ 5·2
1951–54	+ 5·9	+2·1	+11·0	+ 3·8	+ 6·3
1955–58	+15·9	+8·7	+25·4	+10·7	+13·0
1959–60	+14·4	+9·9	+19·6	+ 9·8	+12·1

* i.e. births minus deaths per thousand of population.

† i.e. Basildon, Crawley, Harlow, Hemel, Stevenage and Welwyn Garden City. Harlow and Crawley figures are, however, only available after 1955.

Looking at the figures at the time of designation, we see that Welwyn was adding to its population much faster than the national average – a reflection on its earlier development as a privately sponsored new town. Hemel, Basildon and Crawley were expanding much the same as the national average and Stevenage slightly less; Harlow was little more than a village at the time. Because of these initial differences in size and age pattern and the consequent differences in the proportion of migrants to total population, the natural rate of increase in the new towns differed considerably in the early years of growth and is only just beginning to even out. The nearest case to Basingstoke appears to be that of Welwyn,[1] but this new town was exceedingly slow to expand under the new legislation. What is apparently common to all new towns, however, is that the rate of

[1] In 1960 the deviation in Basingstoke's natural increase in population from the national average was 6·7 per thousand. This deviation was due principally to the younger age composition of the town (see Table IV/11), itself a reflection of the substantial volume of immigration since 1957.

deviation steadily increases up to between the tenth and twelfth years of development and then begins to decline. As might be expected, the towns with the smallest initial population and the fastest intake of migrants show the greatest deviations. If one examines the deviation of the three new towns with an initial population between 18,500 and 25,000, i.e. nearest to Basingstoke at present, one sees that the maximum deviation reached for the four years around the peak are much the same, viz. 10·9 for Hemel, 11·1 for Basildon and 13·4 for Welwyn. We have thus assumed that, for peak quarter 1971–5 the deviation in Basingstoke will average 11·0 per thousand and that the build-up to this figure is fairly steady from the 1960 figure of 6·7 per thousand. We further assume that in the final quarter the deviation will drop slightly to 10 per thousand.

It should, however, be noted that this projection assumes that the age pattern of those moving from London to Basingstoke is roughly the same as of those who earlier moved from London to the new towns. It is, of course, possible that the average age of immigrants may be slightly different from this; thus it is estimated that only one-third of new migrants into the new towns originated from London housing lists, the balance being composed of those who were previously employed with firms which have transferred from London. Since it appears that the heads of households of this latter group of migrants are younger than those of the former group, any change in policy towards tenant selection may well affect the age balance of the town. At the moment, experience suggests that the age composition of migrants settling in the other expanded towns, e.g. Thetford, Swindon, etc., is younger than the average of new town migrants.[1] If this proves to be true in Basingstoke, our estimates of the deviations may prove to be on the low side. On the other hand, the London County Council might well be able to influence the age pattern of incoming migrants rather more than the Development Corporations, due to their more direct interest in the administration of the scheme through the Joint Development Committee. On the whole, our estimates probably lean towards the conservative view.

The formula we have adopted for estimating the rate of natural

[1] This experience would seem confirmed by the findings of the Sociology section of the study group on Basingstoke. Of a sample of 174 heads of household who have migrated to the town since 1952, some 39 (22·4 per cent) were under 30 and 66 (37·4 per cent) under 40, at the time of interview (1961). Some 53·4 per cent of the sample moved to Basingstoke for 'occupational' reasons. See *The Social Implications of a Town Development Scheme*, ed. by M. Broady.

increase in population in Basingstoke over the next twenty years is, then, as follows: first we obtain the expected rate of the natural increase for England and Wales and then add to this rate by the amounts listed below:

1962–65	7	per 1,000 of population
1966–70	8·5	per 1,000 of population
1971–75	11	per 1,000 of population
1976–81	10	per 1,000 of population.

SUMMARY

Our detailed population projections as contained in Table IV/5 have been computed as follows:

1. For each year from 1961 to 1981 we have first calculated what Basingstoke's population would be if it increased at the rate forecast for England and Wales. The data for such a projection were obtained from the Registrar-General's estimates,[1] though these have been rounded to give a continuous function. Thus the population in England and Wales is expected to rise from 45·9 million in 1961 to 50·4 million in 1980. On a similar basis, Basingstoke's population would rise from 25·9 thousand to $25·9 \times \dfrac{50·4}{45·9} = 28·4$ thousand.

2. Next, we add to the figure so derived the cumulative expected migration into the town for each year from 1961 onwards. This will vary according to which of the migration projections A to E is used and will range from 5,000 (projection A) to 8,000 (projection E) for 1965 (i.e. representing the total migration in the period 1962–5) to between 33,750 (projection A) to 42,250 (projection E), for 1976. (See p. 56.)

3. Third, we adjust the resulting figures according to our estimate of the deviation of the rate of natural increase in population from the national trend, i.e. by 7 per 1,000 in 1962–5 up to 11 per thousand in 1971–5.

Illustrating for the year 1961–2, and for projection C we have:

Pop. 1961 → 25,940 × 1·00445 + 1,500 × 1·007 = 27,750 ← Pop. 1962

Estimated growth rate in national population   No. of immigrants   Basingstoke deviation

---

[1] As estimated at 30 June 1960.

Further details of the method of calculating this data are contained in the Appendix to this chapter; a picture of the projected population at five-yearly intervals is given in Table IV/10. The main conclusions which we derive from our calculations is that, if these are right, it would seem there is some inconsistency between the population requirements of the Town Development scheme and those outlined in the Town Map. Thus we have observed (p. 56) that for 36,800

TABLE IV/10

BASINGSTOKE: POPULATION PROJECTIONS FOR
FIVE-YEARLY INTERVALS 1961/1981

	A	B	C	D	E
1961	25,940	25,940	25,940	25,940	25,940
1966	33,910	35,201	35,201	36,518	37,792
1971	47,867	50,541	51,838	54,537	57,171
1976	67,503	70,401	71,804	77,338	80,192
1981	85,696	87,514	85,129	85,883	88,957

Londoners to be rehoused by 1976, fairly liberal assumptions need to be made concerning the speed and build-up of immigration, such as our projections D and E in Table IV/5. But when we add to these projections our forecast of the natural increase in population in the interim period, we obtain total population projections of between 77,338 and 80,192 for 1976–figures respectively just below and well above the forecast contained in the Town Map for the (adjusted) population in 1982. Indeed, making the lowest assumptions concerning private migration (5,200 in projection D) and the speed of migration build-up, the population at the end of 1981 is still more than 8,000 higher than that projected in the Town Map. This difference in estimate is clearly accounted for by the different assumption made concerning the rate of natural increase, which in our opinion has been seriously underestimated by the authors of the Town Map.

AGE COMPOSITION

What of the likely age structure of Basingstoke's future population? It is clearly desirable to make some forecast of this at various stages of the development scheme, and particularly of the size of the working and school population. Once again the task of projection is not an easy one; so much depends upon the particular assumptions made and especially, in this case, on the age-pattern of incoming migrants.

E.P.–E   61

One initial difficulty which has to be overcome is that, as yet, there is no information available relating to the age distribution of Basingstoke's population in 1961.[1] The pattern of the distribution for the Censuses 1921, 1931, and 1951 is given in Table IV/11. For comparative purposes we also give the age structure of the population of England and Wales for this latter year; as can be seen, the two distributions are almost identical.

TABLE IV/11

BASINGSTOKE: AGE DISTRIBUTION OF POPULATION
1921/1951
(1951 figures also given for England and Wales)

Age Group	1921 No.	1921 %	1931 No.	1931 %	1951 No.	1951 %	England & Wales 1951 No. 000s	England & Wales 1951 %
0– 4	1,127	8·3	1,007	7·1	1,474	8·7	3,722	8·5
5–14	2,221	17·5	2,192	15·8	2,216	13·1	6,016	13·7
15–24	2,250	17·7	2,318	16·7	2,066	12·2	5,594	12·8
25–34	1,932	15·2	2,315	16·7	2,483	14·6	6,341	14·5
35–44	1,942	15·3	1,934	14·0	2,537	14·9	6,708	15·3
45–54	1,488	11·7	1,812	13·1	2,364	13·9	6,014	13·7
55–64	944	7·4	1,300	9·4	1,880	11·1	4,592	10·5
65+	819	6·4	987	7·1	1,958	11·5	4,813	11·0
	12,723	100·0	13,865	100·0	16,978	100·0	43,800	100·0

Source: Basingstoke–Registrar-General.
England and Wales–Registrar-General's Quarterly Return, March 1952.

To estimate Basingstoke's population structure for 1961, we have adopted the following procedure. First, from the population of 1951 we have projected forward all age-groups of 5–14 onwards for a period of ten years, making the necessary deductions for deaths based on national life tables.[2] Second, from the Registrar-General's returns of live births in Basingstoke for the period 1951–61, and the national life tables, we have estimated the approximate number of children born in these years and still living in 1961 in the 0–4 and 5–14 age groups, *apart* from those already born to incoming migrants at the time of their arrival in Basingstoke.[3] The total number of people in

[1] This is not expected until November 1963, when the Hampshire volume of the Census 1961 is due to be published.
[2] See *Annual Abstract of Statistics 1961*, Tables 34/35.
[3] The result is probably a slight overestimate, as we have not been able to make any allowance for migration from Basingstoke during this period.

these groups when deducted from the *actual* population of Basingstoke in 1961 (viz. 25,940) will give us the approximate number of people who have migrated into the town between 1951 and 1961 and are still living. Basing our calculation of the age distribution of this latter group of people on comparative figures for the new and expanded towns,[1] and adding the result to that already derived, we obtain an age distribution of the total population of Basingstoke in 1961. This is given in the table below and compared with the national average. It can be seen that Basingstoke is not only a somewhat younger town than ten years earlier – 68 per cent of the population are now under 45 compared with 63·5 per cent in 1951; its age composition is also younger than the national average.

TABLE IV/12

AGE DISTRIBUTION OF POPULATION IN
BASINGSTOKE AND ENGLAND AND WALES 1961

	Basingstoke		England and Wales	
		%	000s	%
0– 4	2,811	10·8	3,665	7·9
5–14	3,974	15·3	6,921	15·0
15–24	2,988	11·5	6,088	13·2
25–34	3,976	15·4	5,806	12·6
35–44	3,870	14·9	6,291	13·6
45–54	3,148	12·2	6,445	13·9
55–64	2,678	10·3	5,429	11·8
65+	2,495	9·6	5,521	12·0
	25,940	100·0	46,166	100·0

Source: Basingstoke–Estimated by author.
England and Wales–Registrar-General's Mid-Year Estimates, 1961.

In our estimates for the period 1961–81 we have adopted a rather different procedure. This consists, first, of assuming that the existing population will increase at the rate of the national average, adjusted to allow for the particular age distribution of Basingstoke, and second, of calculating the age distribution of the incoming migrants and any children born after arrival in Basingstoke. Stage I of our calculations is thus similar to that used in projecting the 1951 population forwards to 1961, save that while in the former case allowance was made for immigrant births, this factor is now taken into consideration in Stage II, which is an estimation of the age distribution of the new

[1] For further details see Appendix I, p. 71–6.

population *and* children subsequently born to migrants in Basingstoke, i.e. all additions to population other than that following the natural increase of the existing population.

Any age distribution of population computed will naturally vary according to (*a*) the particular population projection chosen and (*b*) the year being considered. We have calculated figures for each population projection (A–E) at five-yearly intervals, viz. 1966, 1971, 1976 and 1981, and subdivided these into four main age groups, 0–4, 5–14, 15–64 and 65+.

We now describe the basic assumptions of our analysis, further details of which are contained in the Appendix. First, Stage I of the projection. In projecting the existing population forwards, we have based most of our calculations on national birth- and death-rates adjusted to Basingstokes' particular age distribution in 1961. First, we assume births in the town will increase by 21 per 1,000 for each year up to 1966:[1] from 1967 to 1971 a figure of 20 per thousand has been taken; for 1972 to 1976 19 per thousand and for 1977 to 1981 18 per thousand. The fall in birth-rates naturally follows the slight 'ageing' of Basingstoke's existing population.[2] In turn these birth-rates have been calculated on the assumption that the net excess of births over deaths of the existing population will fall slightly from 12 per thousand between 1962 and 1966 to 11 per thousand between 1967 and 1971 to 10 per thousand between 1972 and 1976 to 9 per thousand between 1976 and 1981: again, these rates are falling for the reason given above. Next, from the national life tables, we have forecast the numbers of those already born and those still to be born who will be alive in 1966, 1971, 1976 and 1981. When the total 'old' population so derived is deducted from the particular projection (A–E) chosen, we obtain (approximately) the size of the new, i.e. migrant, population.

The big unknown in Stage II of our calculation is the age distribution of the migrant population. We shall assume that the difference between the population projected for any given year and that resulting from the natural increase of the existing population is net migration plus subsequent births minus deaths of incoming migrants. Our calculation of the age distribution of this latter is derived primarily from

[1] In 1960 the birth-rate in Basingstoke was 22·0 per thousand and in 1961 was 22·6 per thousand.
[2] While the total number of women of child-bearing age is unlikely to rise more than 10 per cent in the next 20 years, the total population will increase by about 20 per cent. We have had to assume that fertility rates remain constant.

the experience of the new towns. Four new towns, Hemel Hempstead, Harlow, Crawley and Bracknell, were able to provide us with
estimates of the age-structure of the total immigrant population since
designation (including children born since arrival) for the year 1961.
At that time all the new towns in question were thirteen to fifteen
years old. As can be seen from Table IV/13, the age distribution of the
new populations within the new towns is very similar indeed.

TABLE IV/13

PERCENTAGE AGE DISTRIBUTION OF IMMIGRANT
POPULATION IN SELECTED NEW TOWNS, 1961

	Bracknell	Crawley	Hemel	Harlow	Average
0– 4	13·8	12·9	14·3	14·6	13·9
5– 9	10·3	12·7	15·5	13·0	12·9
10–14	7·8	9·6	9·3	9·6	9·1
	31·9	35·2	39·1	37·2	35·9
15–24	9·9	10·1	9·7	11·2	10·2
25–34	20·2	19·2	20·4	20·4	20·1
35–44	18·0	18·9	18·0	17·6	18·1
45–54	10·2	8·6	7·5	8·1	8·6
55–64	5·1	3·0	3·2	3·0	3·6
	63·4	59·8	58·8	60·3	60·6
65–69	1·8	0·9	1·3	1·0	1·3
Over 70	2·9	1·5	0·8	1·5	1·7
	4·7	2·4	2·1	2·5	2·9
Not known		2·6			0·6
	100·0	100·0	100·0	100·0	100·0

Using these distributions as a basis, we have made one or two
adjustments to allow, first, for a more rapid build-up of migration in
Basingstoke, and second, for a slightly more balanced age-structure
of incoming migrants throughout. As a result, our projected age
composition of the *new* population in Basingstoke in 1976 as shown in
Table IV/14 is slightly 'older' than that of the new towns in 1961
(Table IV/13). We also give our estimates of the age-structure of the
immigrant population in 1966, 1971 and 1981, which again are based
on certain assumptions concerning the age composition of incoming
migrants year by year. For example, in the early years of the town
development scheme it is likely the proportion of working to total
population (i.e. of the immigrants) will be as high as 65 or 66 per cent,
partly because of the small number of migrants over 65 and partly

because the proportion of children in the 5–14 age group is likely to be considerably lower than it will be in ten to fifteen years' time.[1] From the late '60s to the late '70s, when the proportion of school-children will be at its peak, the proportion of working population will fall slightly, but by 1981 the number of school leavers seeking work will have increased markedly and the proportion of the potential labour force will probably return to around 64 per cent. By now, the migration programme will have been completed and with those over 65 probably accounting for about 6·5 per cent of the new population (e.g. this group now includes all migrants who were 45 or over in 1961), those under 15 will fall back to 29–30 per cent of the population.

TABLE IV/14

ESTIMATED AGE COMPOSITION OF CUMULATIVE
IMMIGRANT POPULATION IN BASINGSTOKE, 1966/1981

	Percentage of Total			
	1966	1971	1976	1981
0– 4	14·0	13·5	11·5	10·5
5–14	16·5	19·5	20·5	19·0
15–64	66·0	63·0	63·0	64·0
65+	3·5	4·0	5·0	6·5
	100·0	100·0	100·0	100·0

Rounded to nearest 0·5 per cent.

When the percentage age compositions indicated in Table IV/14 are applied to the difference between total population forecast in the five projections (A–E) as shown in Table IV/5 and the existing, i.e. 1961, population of Basingstoke plus the expected natural increase (i.e. that calculated in Stage I of the present exercise)[2] we obtain the number of migrants (and their children) split up into their various age groups. If we then add to the numbers in each group those of the existing population, plus the expected natural increase, we obtain the data which is presented in Table IV/15. Such estimates as these are the

---

[1] Thus an examination of the age structure of migrants entering *in any one year* into the new towns shows that those aged between 5 and 14 have consistently accounted for between 16 and 18 per cent of the total. A look at the age structure of all immigrants after fifteen years shows the proportion of this age group has risen to 22–25 per cent. More generally, the age pattern of new immigrants in any one year for the (London) new towns since designation has been: Under 15, 32–34 per cent; 15–64, 63·5–66·5 per cent; 65 and over, 1·5–2·5 per cent.

[2] See also Appendix I, p. 75–6.

66

BASINGSTOKE: ESTIMATED AGE DISTRIBUTION OF POPULATION 1966/1981

Age Group		A	%	B	%	C	%	D	%	E	%
1966	0-4	3,653	10·8	3,834	10·9	3,834	10·9	4,018	11·0	4,196	11·1
	5-14	5,725	16·9	5,938	16·8	5,938	16·9	6,156	16·9	6,366	16·9
	15-64	21,539	63·5	22,391	63·6	22,391	63·6	23,260	63·7	24,101	63·8
	65+	2,993	8·8	3,038	8·6	3,038	8·6	3,084	8·5	3,129	8·3
		33,910	100·0	35,201	100·0	35,201	100·0	36,518	100·0	37,792	100·0
1971	0-4	5,369	11·2	5,730	11·3	5,905	11·4	6,269	11·5	6,625	11·6
	5-14	9,228	19·3	9,750	19·3	10,003	19·3	10,529	19·3	11,043	19·3
	15-64	29,444	61·5	31,128	61·6	31,945	61·6	33,646	61·7	35,305	61·8
	65+	3,826	8·0	3,933	7·8	3,985	7·7	4,093	7·5	4,198	7·3
		47,867	100·0	50,541	100·0	51,838	100·0	54,537	100·0	57,171	100·0
1976	0-4	7,091	10·5	7,425	10·6	7,586	10·6	8,222	10·6	8,551	10·7
	5-14	13,114	19·4	13,708	19·6	13,996	19·5	15,130	19·5	15,715	19·6
	15-64	42,089	62·4	43,914	62·4	44,798	62·4	48,285	62·3	50,083	62·5
	65+	5,209	7·7	5,354	7·6	5,424	7·6	5,701	7·6	5,843	7·3
		67,503	100·0	70,401	100·0	71,804	100·0	77,338	100·0	80,192	100·0
1981	0-4	8,494	9·9	8,685	9·9	8,435	9·9	8,514	9·9	8,837	9·9
	5-14	15,857	18·5	16,202	18·5	15,749	18·5	15,892	18·5	16,476	18·5
	15-64	54,261	63·3	55,425	63·3	53,899	63·3	54,381	63·3	56,349	63·3
	65+	7,084	8·3	7,202	8·2	7,046	8·3	7,096	8·3	7,295	8·2
		85,696	100·0	87,514	100·0	85,129	100·0	85,883	100·0	88,957	100·0

raw material for various aspects of planning, e.g. assessing educational, employment and housing requirements at various stages in the future. We will deal more fully with industry and employment in the following chapter. Here we very briefly touch upon the implications of the statistics in Table IV/15 from the viewpoint of educational and housing needs.

(a) *Education.* Taking the average and two extremes of our projections A–E, we see that the number of school places likely to be required for children over 5 and under 15 living in the borough of Basingstoke in the coming years is as indicated in Table IV/16:

TABLE IV/16

BASINGSTOKE: ESTIMATED NUMBER OF CHILDREN
OVER 5 AND UNDER 15, 1961/1981

	Average of projections	(1961= 100)	Low projection	(1961= 100)	High projection	(1961= 100)
1961 (estimated)	3,974	100·0	3,974	100·0	3,974	100·0
1966	6,025	151·6	5,725	144·1	6,366	160·2
1971	10,111	254·4	9,228	232·2	11,043	277·9
1976	14,333	360·7	13,114	330·0	15,715	395·5
1981	16,035	403·5	15,857	399·0	16,476	414·6

Thus, ignoring the requirements of children aged 15 or above and adult education–the needs of both groups at first are likely to rise more slowly than the rates indicated above, but in the latter years may grow very fast indeed–we see that the number of school places will need to be increased by between three and a half and four times their present level. At first the greater part of this increase will be concentrated in the primary schools, but gradually the pressure on secondary schools will be equally felt. The experience of the new and other expanded towns suggests that in any one year there are likely to be between one and three-quarters and twice as many child migrants in the 5–9 group as in the 10–14 group; on the other hand, after ten to fifteen years, of the total (i.e. cumulative) child migrant population, about 60 per cent will be aged 5–9 and 40 per cent 10–14. Even assuming there is little new migration after the fifteenth year, by 1981 the second generation of migrants' children will begin to need education. If we assume the trend towards a greater proportion of (migrant) children in the 10–14 age group will be spread evenly throughout the twenty-year period, and take into consideration the existing structure

68

of Basingstoke's child population and predicted trends of natural rates of increase and child deaths, the approximate breakdown of our child population (up to 15) between primary and secondary school age – based on the average of the above projections – would be as follows:

TABLE IV/17

BASINGSTOKE: ESTIMATED DISTRIBUTION OF CHILD POPULATION (5/14) 1961/1981

	5–9		10–14	
	Total	Increase	Total	Increase
1961 (estimated)	2,024		1,950	
1966	3,525	1,501	2,500	550
1971	5,567	2,062	4,554	2,142
1976	7,726	2,159	6,607	2,053
1981	8,232	506	8,003	1,396

It need hardly be stressed that the above figures must be considered as little more than the broadest of guides; nevertheless, they serve to indicate the likely orders of magnitude involved and to that extent (assuming our arguments are correct) provide a useful basis for forward planning.

(b) *Housing*. The residential development envisaged as necessary to satisfy the Town Development Scheme and other requirements of the next twenty years in the Basingstoke Town Map area is given as follows:

TABLE IV/18

BASINGSTOKE: ESTIMATED REQUIREMENTS OF DWELLING UNITS 1961/1981

Under Town Development Scheme (to be completed by 1976)	11,500
Private migration; second generation of migrants, etc.	4,850
Slum clearance, natural increase, replacement of prefabricated buildings, improvements in housing standards	2,170
Rehousing due to Town Development Scheme	750
	19,270

Source: Town Map, 1962.

On the basis of an estimated increase in population of 56,120[1] this would appear an adequate programme. However, for reasons already indicated, we consider that the population forecast for the Town Map

[1] Forecast population of 86,230 minus 1959 population of 30,130. In the new towns the average number of rooms per dwelling up to 31.3.62. was 4·3 and the number of occupants per dwelling 3·05.

69

Area of 86,330 by 1982 is probably low by about 7,000. If we are right in our analysis, it follows that to maintain the same housing standards a further 2,000 or so dwelling units will be needed. This is not a serious difference: in fact, most population or housing projections allow for a much greater margin of error than 10 per cent.

What is much more difficult, however, is to estimate the *type* of dwelling units required–and the type of dwelling unit will, to some extent, affect the persons-to-house ratio. Houses, flats and bungalows can hardly be considered homogeneous. A town with a high proportion of old people might require a large number of small bungalows; to cater for single people and young married couples without children, attention may have to be given to single- or two-bedroom flats. On the other hand, a family of husband, wife and three or more children will ideally need a four-bedroom house or bungalow. To some extent, these extremes of wants will even themselves out and in any event it is impossible to plan rigidly for a particular structure of housing so many years ahead. The experience of the new towns has been mixed: thus in Harlow bachelor flats have proved very popular (particularly with commuters to London) and now account for an eighth of all dwellings; in the other new towns they are not nearly so popular. Hemel has built an above-average proportion of old people's bungalows[1]–about 6 per cent of all dwellings–and, with Crawley, is finding an increasing demand by second-generation migrants for flats or two-bedroomed houses. While some Development Corporations have concentrated on providing dwellings for letting purposes, others have found a ready demand for houses built for sale. In general however, it may be that the average accommodation of four or three rooms per dwelling in the new towns will serve as a reasonable guide to Basingstoke.

Improvements in living standards are also very difficult to forecast. What is, however, very likely is that a very high proportion of houses –possibly three in four–will require garage accommodation, if not attached to the dwelling unit then easily accessible from it.

CONCLUSIONS

It may be argued that our analysis of the present and predicted trends of Basingstoke's population, its age composition and its implications for housing and education, has been too tentative and

[1] J. H. Nicholson, *New Communities in Britain*, National Council of Social Service, 1961, p. 64.

70

imprecise to be of any value. However, the point of our exercise has been not so much to forecast what actually will happen in the future as to examine the basis of the population projections underlying the Town Development Scheme and Town Map; to suggest what this involves regarding rates of growth over the next fifteen years; to give some idea of the likely effect of alternative rates of growth on the age-composition and finally to hint at some of the implications of the projections obtained for housing and education.

APPENDIX

Population and Age Structure Projections Basingstoke, 1961–1981.
Method of Calculation

A. The 1961 age structure of Basingstoke
Data available

1. The 1951 age distribution of Basingstoke taken from the Census of Population 1951.

0–4	1,474	55–64	1,880
5–9	1,198	65–69	758
10–14	1,018	70–74	567
15–24	2,066	75–79	368
25–34	2,483	80–84	190
35–44	2,537	85+	75
45–54	2,364		
			16,978

2. The actual number of births between 1951 and 1961.

1951–52	313	1955–56	336	1959–60	541
1952–53	295	1956–57	406	1960–61	585
1953–54	343	1957–58	427		
1954–55	334	1958–59	463		

(Source: Registrar General's mid-year estimates)

3. The death rates per 1,000 as a function of age group. These are annual rates averaged over the period.

0–4	5·7	35–44	2·2
5–9	0·5	45–54	5·9
10–14	0·3	55–64	16·3
15–19	0·7	65–74	41·6
20–24	0·7	75–84	102·8
25–34	0·9	85+	224·0

(Source: Registrar General's life tables: see Annual Abstract of Statistics 1961, Tables 34 and 35)

4. The population of Basingstoke—mid 1961.

25,940

The calculations

These fall into four parts.

I. By substracting deaths for the ten-year period from the 1951 distribution and assuming no migration, the old (1951) population was regrouped for 1961.

Age group	No. in 1951	Mortality Calculations		No. in 1961	Age group
0–4	1,474	$\dfrac{-5 \times 5{\cdot}7 \times 1,474}{1,000}$	$-\dfrac{5 \times 0{\cdot}5 \times 1,432}{1,000}$	= 1,428	10–14
5–14	2,216	$\dfrac{-10 \times 0{\cdot}55 \times 2,216}{1,000}$		= 2,204	15–24
15–24	2,066	$\dfrac{-10 \times 0{\cdot}8 \times 2,066}{1,000}$		= 2,049	25–34
25–34	2,483	$\dfrac{-10 \times 1{\cdot}5 \times 2,483}{1,000}$		= 2,446	35–44
35–44	2,537	$\dfrac{-5 \times 2{\cdot}2 \times 2,537}{1,000}$	$-\dfrac{5 \times 5{\cdot}9 \times 2,509}{1,000}$	= 2,435	45–54
45–54	2,364	$\dfrac{-5 \times 5{\cdot}9 \times 2,364}{1,000}$	$-\dfrac{5 \times 16{\cdot}3 \times 2,294}{1,000}$	= 2,107	55–64
55–64	1,880	$\dfrac{-5 \times 16{\cdot}3 \times 1,880}{1,000}$	$-\dfrac{5 \times 41{\cdot}6 \times 1,727}{1,000}$	= 1,376	65–74
65+	1,958	$\dfrac{-10 \times 57{\cdot}4(^1) \times 1,958}{1,000}$		= 833	75+
				14,878	

[1] Representing average of death rates per 1,000 (see A3) weighted by 1951 age distribution (see A1).

II. By subtracting child deaths for the ten years, the births during 1951–61 were changed into the two living groups in 1961 of 0–4 and 5–9.

Year	No. born	Mortality calculations	No. living in 1961	Age group
1951–52	313	5 years at 5·7 per 1,000 + 5 years 0·5 per 1,000	303	
1951–53	295	5 years at 5·7 per 1,000 + 4 years 0·5 per 1,000	286	
1953–54	343	5 years at 5·7 per 1,000 + 3 years 0·5 per 1,000	333	5–9
1954–55	334	5 years at 5·7 per 1,000 + 2 years 0·5 per 1,000	324	
1955–56	336	5 years at 5·7 per 1,000 + 1 year 0·5 per 1,000	326	
1956–57	406	5 years at 5·7 per 1,000	394	
1957–58	427	4 years at 5·7 per 1,000	417	
1958–59	463	3 years at 5·7 per 1,000	455	0–4
1959–60	541	2 years at 5·7 per 1,000	535	
1960–61	585	1 year at 5·7 per 1,000	582	
			3,955	

III. The totals derived in I and II were added together. The result was subtracted from the 1961 *total* population figure and the residue (which represents those who have entered the town between 1951/61 exclusive of their children born there) was divided into prescribed age group percentages.

Total I =	14,878	Total population in 1961	= 25,940
Total II =	3,955	∴ 1951/61 net immigration	= 7,107
	18,833		

The assumed age structure of the immigrants, having regard to the situation in the new and other expanding towns, was as follows:

0–4	6 per cent	403
5–14	14 per cent	999
15–24	11 per cent	784
25–34	27 per cent	1,925
35–44	20 per cent	1,426
45–54	10 per cent	713
55–64	8 per cent	571
65+	4 per cent	286
		7,107

IV. Add age group contributions from I, II and III to give total age distribution for 1961

0–4	2,811	
5–14	3,974	
15–24	2,988	
25–34	3,976	
35–44	3,870	
45–54	3,148	(See Table IV/12 in text)
55–64	2,678	
65+	2,495	
	25,940	

This distribution is the main basis for the projected distributions at 1966, 1971, 1976 and 1981.

B. Calculating the future age structure of Basingstoke

The method of calculating these estimates is basically the same as above though the treatment of the migrant population differs slightly. Stage I relies upon the projection above (1961) and a knowledge of death rates of the existing population. These latter were derived from the Registrar General's figures as given in A3.

Stage II requires a forecast of births as well as infant mortality rates. The birth rate was set using a knowledge of the births in the new towns.

Stage III requires a full total population projection again using expected immigration figures and an eye on other build-ups.

Stage IV is a matter of addition but sometimes involves the division of population group into two halves.

The total population projection required for Stage III was for our purposes made on the basis of five hypotheses, A, B, C, D and E; the main difference between them was the number of immigrants and the time of their entry.

The method adopted for calculating these projections is described in the text. (See pp. 50/61) First we multiply the population in any given year (P_1) by the national natural rate of increase in population (X); second, we add the assumed number of immigrants (Y) and multiply the result by the deviation of the Basingstoke natural increase from the national natural increase (Z). The result will be the population in the following year (P_2). For example, reading from the text (p. 60)

Pop. 1961 $25,940 \times 1,00445 + 1,500 \times 1 \cdot 007 = 27,750$ Pop. 1962

or $\quad P_1 \quad \times \quad X \quad + \quad Y \quad \times \quad Z \quad = P_2$

N.B. The loading figure was applied to people in Basingstoke *after* the addition of immigrants.

The following tables provided values for X, Y, Z, for each year.

(*a*) *Value of X for each hypothesis A–E*

X	1·00445	1·00454	1·00452	1·11450	1·00448	1·00456	1·00454	1·00463
Years	1961	1962	1963	1964	1965	1966	1967	1968
X	1·00477	1·00500	1·00508	1·00516	1·00513	1·00490	1·00494	1·00491
Years	1969	1970	1971	1972	1973	1974	1975	1976
X	1·00489	1·00487	1·00484	1·00484	1·00484			
Years	1977	1978	1979	1980	1981			

These figures were taken from the Registrar General's returns but rounded-off to give a continuous function.

(*b*) *Value of Y according to hypothesis A–E*

A

Y	+ 1,500	+ 2,500	+ 3,000	+ 2,250	+ 250
Years	1962–65	66–70	71–75	76–80	81–82

B

Y	+ 1,500	+2,750	+ 3,500	+ 1,500	+ 250
Years	1962–65	66–70	71–75	76–80	81–82

C

Y	+ 1,250	+ 2,250	+ 3,000	+ 2,500	+ 250
Years	1962–65	66–70	71–75	76–80	81–82

D

Y	+ 1,750	+ 3,000	+ 3,500	+ 500	+ 250
Years	1962–65	66–70	71–76	76–80	81–82

E

Y	+ 2,000	+ 3,250	+ 3,500	+ 500	+ 250
Years	1962–65	66–70	71–75	76–80	81–82

(*c*) *Value of Z for each hypothesis A–E*

Z	1·007	1·0085	1·011	1·010
Years	1962–65	1966–70	1971–75	1976–80

The Calculations

The essential population projections for this part of the calculation are given below.

	1961	1966	1971	1976	1981
A	25,940	35,201	50,541	70,401	87,514
B	25,940	35,201	51,838	71,804	85,129
C	25,940	33,910	47,867	67,503	85,696
D	25,940	36,518	54,537	77,338	85,883
E	25,940	37,792	57,171	80,192	88,957

Stage I attempts to assess the age distribution of the *existing* population in Basingstoke in 1966, 1971, 1976 and 1981 assuming no migration into or out of town. Death rates are calculated from life tables published by the Registrar General (see *A*3 and *Annual Abstract of Statistics* 1961, Tables 34 and 35). Thus for 1966 we are able to estimate the age distribution of population from five years and above, for 1971 from 10, for 1976 from 15 and for 1981 from 20 years and above.

Stage II involves estimating the number of births in each year of the existing population from 1962 to 1981, deducting deaths in each year and classifying in age groups. 0–4 for 1966, 0–4 and 5–9 for 1971, 0–4, 5–9 and 10–14 in 1976 and 0–4, 5–9, 10–14 and 15–19 in 1981. The following assumptions were made in this calculation:

(i) that the excess of births over deaths in Basingstoke was at the rate of 12 per 1,000 in 1962–1966, 11 per 1,000 in 1967–1971, 10 per 1,000 in 1972–76 and 9 per 1,000 in 1977–81.

(ii) that the birth rate in the town was 21 per 1,000 in 1962–66, 20 per 1,000 in 1967–71, 19 per 1,000 in 1972–76 and 18 per 1,000 in 1977–81.

For example, births in 1962–3 $= \left(25,940 \times \dfrac{1,012}{1,000} \right) \times \dfrac{21}{1,000} = 551$.

For subsequent years ending June 30th and starting in 1964, births worked out at 558, 564, 571, 557, 563, 569, 575, 582, 558, 564, 569, 575, 581, 555, 560, 565, 570.

When deaths are deducted and Stage I taken into consideration, the broad pattern of the age distribution of Basingstoke's existing population *plus* excess of births over deaths for the years 1966, 1971, 1976, and 1981 is obtained. This is set out below:

	1961	1966	1971	1976	1981
0–4	2,811	2,743	2,794	2,800	2,783
5–14	3,974	4,653	5,509	5,464	5,522
15–64	16,660	17,249	17,427	18,579	19,450
65+	2,495	2,765	3,063	3,343	3,548
	25,940	27,410	28,793	30,186	31,303

Stages III and IV. By deducting the estimated population obtained in Stages I and II in 1966 and onwards from the projections of *total* population (see Table IV/V in text) an estimate of the number of migrants (and their children) still living in Basingstoke in those years can be obtained. By means of the assumed age distribution of such immigrants given in Table IV/14 we can calculate for each projection the total numbers of people in the four age groups 0–4, 5–14, 15–64 and 65 and over. These are set out in Table IV/15.

CHAPTER V

INDUSTRY AND EMPLOYMENT

BACKGROUND

ALTHOUGH an important proportion of Basingstoke's wealth still
arises from its situation as an agricultural and market centre for the
surrounding downland areas, the role of manufacturing industry has
become increasingly significant in recent years. The only published
figures of the occupational distribution of the labour force within the
Borough are those contained in the decennial Censuses of Population.
In 1921, of a total working population in the Borough of 5,800, some
2,400 or 41·4 per cent were engaged in manufacturing or secondary
industry, 3,070 or 52·9 per cent in the service trades (including those
ancillary to manufacturing), and 330 or 5·7 per cent in agriculture.
Thirty years later, by which time the total labour force had reached
9,425, the proportion employed in industry had risen to 45·1 per cent,
while that in the remaining two categories had fallen to 50·8 per cent
and 4·1 per cent respectively.[1] In the last ten years the share of secondary industry has continued to increase. On the basis of information
provided by individual firms in Basingstoke, the number employed in
manufacturing at the end of 1961 was 6,300. Assuming the numbers
employed in agriculture and the proportion of the working labour
force to the total population in the 15–64 age group have remained
fairly constant,[2] this would mean that about 47 per cent of Basingstoke's labour force is now engaged in manufacturing compared with
50 per cent in the services trades and 3 per cent in agriculture.

More regular and detailed statistics are published by the Ministry
of Labour for the Basingstoke Employment Area, which incorporates
not only the Borough but certain outlying areas as well.[3] As may be

The 1951 Census records three people engaged in mining and quarrying,
which, for the purposes of classification, are included in agriculture.

[2] For basis of this latter assumption, see p. 89. Allowance has had to be made
or the 1,000 or so workers living in Basingstoke but working at Aldermaston.

[3] viz. Basing, Basingstoke M.B., Bradley, Bramley, Candovers, Cliddesden,
Dean Dummer with Kempshott, Ellisfield, Farleigh Wallop, Greywell, Hartley
Wespall, Heckfield, Herriard, Hook, Laverstoke, Long Sutton, Mapledurwell and
Up Nately, Mattingley, Monk Sherborne, Newnham, North Waltham, Nutley,
Oakley, Odiham, Overton, Popham, Preston Candover, Rotherwick, Sherborne,
St. John, Sherfield on Loddon, South Warnborough, Steventon, Stratfield Saye,
Stratfield Turgis, Tunworth, Upton Grey, Weston Corbett, Weston Patrick,
Winslade, Woodmancott, Wootton St. Lawrence and that part of Kingsclere
south of a line from the village of Sydmonton to the village of Wolverton.

expected, the number employed in agricultural pursuits[1] in these latter areas is considerably higher than in the Borough, but the expanding role of secondary industry is no less in evidence, as shown in Table V/1 and Figure V/1.[2]

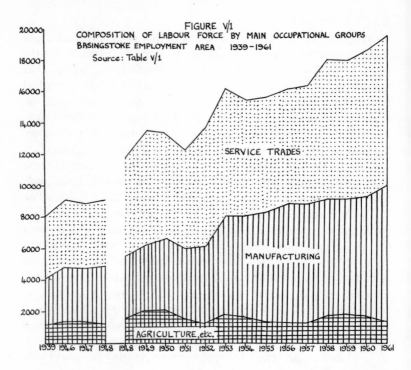

FIGURE V/1
COMPOSITION OF LABOUR FORCE BY MAIN OCCUPATIONAL GROUPS
BASINGSTOKE EMPLOYMENT AREA 1939-1961
Source: Table V/1

SERVICE TRADES

MANUFACTURING

AGRICULTURE, etc.

The comparative rates of growth in the numbers employed in the three main occupational groups (which correspond broadly to the economist's classification of primary, secondary and tertiary industries) are shown even more clearly in Table V/2, which presents indices of growth related to a base year of 1948. Since in that year important changes in the coverage and composition of employment statistics took place, there is an unavoidable break in the series, but a linkage is possible as both 'old' and 'new' coverage statistics for 1948 are available.

[1] Only nine people were employed in mining and quarrying activities in 1960.
[2] Although there have been certain changes in the size and composition of the Employment Area since the war, these have not been important enough to affect the general trend.

TABLE V/1

ESTIMATED NUMBER OF EMPLOYEES IN THE MAIN OCCUPATIONAL GROUPS: BASINGSTOKE EMPLOYMENT AREA, 1939/1961

	Agriculture, etc.	%	Manufacturing	%	Service Trades	%	Total
1939	1,260	15·4	2,839	34·9	4,061	49·7	8,160
1946	1,440	15·9	3,350	37·0	4,285	47·1	9,075
1947	1,405	15·8	3,313	37·2	4,192	47·0	8,910
1948 (old)	1,380	15·2	3,483	38·2	4,282	46·6	9,145
1948 (new)	1,633	13·8	3,983	33·7	6,237	52·5	11,853
1949	2,285	16·7	4,210	30·9	7,159	52·4	13,654
1950	2,287	16·9	4,503	33·1	6,788	50·0	13,578
1951	1,525	12·2	4,607	37·0	6,351	50·8	12,483
1952	1,375	9·9	5,012	36·1	7,509	54·0	13,896
1953	1,800	11·0	6,349	38·7	8,256	50·3	16,405
1954	1,622	10·5	6,565	42·3	7,357	47·2	15,544
1955	1,541	9·8	6,854	43·6	7,346	46·6	15,741
1956	1,552	9·6	7,186	43·8	7,632	46·6	16,370
1957	1,547	9·4	7,245	44·0	7,724	46·6	16,516
1958	1,741	9·5	7,384	40·8	9,054	49·7	18,179
1959	1,757	9·7	7,354	40·6	9,015	49·7	18,126
1960	1,667	8·9	7,643	41·0	9,361	50·1	18,671
1961	1,579	8·1	8,431	43·1	9,571	48·8	19,581

Source: Data kindly provided by Ministry of Labour. For details of the coverage of 'old' and 'new' industrial classification see *Ministry of Labour Gazette*, August 1948, p. 260.

TABLE V/2

RELATIVE GROWTH OF NUMBERS EMPLOYED IN THE MAIN OCCUPATIONAL GROUPS: BASINGSTOKE EMPLOYMENT AREA, 1939/1961

1948 = 100

	Agriculture	Manufacturing	Services	All industry
1939	91	81	95	89
1946	104	96	100	99
1947	102	95	98	97
1948 (old)	100	100	100	100
1948 (new)	100	100	100	100
1949	140	106	115	115
1950	140	113	109	114
1951	93	117	102	105
1952	84	126	121	117
1953	111	159	133	138
1954	99	165	118	131
1955	95	172	118	133
1956	95	180	122	138
1957	95	182	124	140
1958	107	186	146	154
1959	108	185	145	153
1960	102	191	151	157
1961	97	212	154	165

Source: Table V/1.

It is likely that the post-war growth of manufacturing in Basingstoke would have been even more pronounced had it not been for Government policy encouraging industrial development in other parts of the country. At the same time, at least part of the recent expansion in the tertiary industries is itself due to the growth of manufacturing, either directly through the provision of ancillary service facilities, e.g. repair shops, storage depots, distributing outlets, etc., or indirectly through the increased spending power made available by the workers employed in new secondary industry.

Within manufacturing industry, the expansion in the labour force has been particularly noticeable in three trades–engineering, vehicles and paper and printing. Since 1952 the numbers employed in these groups, both in the Borough and Employment Area, have more than doubled. Figures for 1961 indicate that they account for 56·3 per cent of the total manufacturing workers in the Borough and 78·4 per cent in the Employment Area. The chemical industry has also increased its share of the total labour force, while that of the food, textiles and clothing, non-metalliferous mining products, and leather trades has fallen. Further particulars of the distribution of manufacturing workers for various years since before the war and the comparative rates of expansion in the Basingstoke Employment Area are given in Tables V/3 and V/4. If anything, it would appear that the industrial structure of the town is more specialised than in 1939: however, within the engineering industry there are a wide variety of diverse trades serving different markets and with their own particular labour requirements. The products made by such firms in Basingstoke range from heavy lorries to aeronautical instruments and from road rollers to lawn mowers. It would, however, be true to say that most industries in Basingstoke are of a kind requiring light or medium rather than heavy machinery and equipment, are market rather than material orientated, and employ skilled or semi-skilled rather than unskilled labour.[1] Outside manufacturing, the most pronounced expansion has been in the professional and scientific services, the numbers in which have risen from 684 in 1948 to 2,111 in 1960: most of this increase is accounted for by educational and medical services. Further details of the changing pattern of the labour force in the tertiary industries for the Employment Area are given in Table V/5. The distributive trades, most miscellaneous services (including domestic

[1] See also Appendix I, p. 120 and R. C. Estall and R. Ogilvie Buchanan, *Industrial Activity and Economic Geography*, Hutchinson University Library, 1961, Chapter 2.

TABLE V/3

DISTRIBUTION OF EMPLOYMENT IN MANUFACTURING INDUSTRY: BASINGSTOKE EMPLOYMENT AREA, 1939/1960

	1939		1948 (old)		1948 (new)		1952		1955		1960	
Bricks, pottery, glass, etc.	81	2·8	27	0·7	34	0·8	46	0·9	32	0·5	35	0·5
Chemicals	45	1·6	163	4·7	210	5·3	249	5·0	288	4·2	504	6·6
Engineering and electrical	178	6·3	623	17·9	678	17·0	1174	23·4	1033	15·1	1859	24·3
Vehicles	1387	49·0	1583	45·4	1669	42·0	2305	46·0	2490	36·3	2784	36·4
Textiles and clothing	395	14·0	421	12·1	507	12·7	479	9·6	585	8·5	438	5·7
Leather	67	2·4	75	2·2	217	5·5	115	2·3	111	1·6	107	1·4
Food, drink and tobacco	81	2·8	131	3·8	156	3·9	99	2·0	91	1·3	14	0·2
Timber, furniture, etc.	106	3·7	90	2·6	99	2·5	40	0·8	97	1·4	70	0·9
Paper and printing	34	1·2	66	1·9	82	2·1	98	2·0	1368	20·0	1396	18·3
Other industries	465	16·2	304	8·7	331	8·3	407	8·1	759	11·1	436	5·6
All industries	2839	100·0	3483	100·0	3983	100·0	5012	100·0	6854	100·0	7643	100·0

TABLE V/4
RELATIVE GROWTH OF MANUFACTURING INDUSTRIES: BASINGSTOKE EMPLOYMENT AREA, 1939/1960

	1939	1948 (old)	1948 (new)	1952	1955	1960
Bricks, pottery and glass	300	100	100	135	94	103
Chemicals	27	100	100	118	137	240
Engineering and electrical	29	100	100	173	152	274
Vehicles	87	100	100	138	149	167
Textiles and clothing	94	100	100	95	115	85
Leather, etc.	89	100	100	53	51	49
Food, drink and tobacco	62	100	100	63	58	9
Timber, furniture	118	100	100	40	98	71
Paper and printing	52	100	100	120	1,670	1,702
Other	153	100	100	123	229	132
All Industries	81	100	100	126	172	191

services) and constructional activities are now relatively less important than before the war; the public utilities and transport undertakings occupy about the same position; only insurance, banking and finance and the professional services have substantially increased their share of the labour force.

Thus, over the years, the economy of Basingstoke may be said to have evolved around a dual personality of a light industrial town and an agricultural-cum-commercial centre, with the former aspect gradually becoming the more important. Even without the projected expansion scheme, it is likely that the share of manufacturing industry would have increased in the next decade. Firms currently producing in the Basingstoke Employment Area already have plans for expansion which would increase the size of their labour force by between 25 and 30 per cent by 1965. A good deal of speculative building is being undertaken to the south of the town in anticipation of the migration of industry to Basingstoke, and providing the Board of Trade is willing to grant the necessary industrial development certificates,[1] there is no reason to suppose there will be any difficulty in finding tenants for these premises.[2]

In recent years, Basingstoke has attracted firms partly because of

[1] Under the Local Employment Act, 1960, industrial development certificates are required for all industrial building of 5,000 sq. ft. or more. The Board of Trade has power to refuse such certificates to industrialists if it considers that the development in question would not be in the best interests of the area concerned or would be better still (from the social viewpoint) in another area.

[2] Already since the new agreement there have been enquiries from firms wishing to produce the following goods in Basingstoke – precision instruments, cigarettes, paper tags, plastics, medical equipment, machine tools, display material, electronic equipment, printing and publishing.

DISTRIBUTION OF EMPLOYMENT IN SERVICE INDUSTRIES:*
BASINGSTOKE EMPLOYMENT AREA, 1939/1960

	1939	%	1948	%	1948	%	1952	%	1955	%	1960	%
Construction	1157	28·4	898	21·0	964	15·5	1191	15·9	1110	15·1	1449	15·5
Gas, electricity and water	89	2·2	95	2·2	111	1·8	213	2·8	213	2·9	233	2·5
Transport and communications	254	6·3	490	11·5	676	10·8	885	11·8	861	11·7	759	8·1
Distributive trades	1019	25·1	998	23·3	1146	18·4	1198	16·0	1215	16·5	1813	19·4
Insurance, banking and finance	28	0·7	37	0·9	173	2·8	272	3·6	208	2·8	221	2·4
Professional and scientific services	1514	37·3	256	6·0	684	11·0	1169	15·6	1140	15·5	2111	22·5
Miscellaneous services			467	10·9	1279	20·5	1301	17·4	1390	19·0	1460	15·6
Public administration			1034	24·2	1195	19·2	1264	16·9	1209	16·5	1315	14·0
All service industries	4061	100·0	4275	100·0	6228	100·0	7493	100·0	7346	100·0	9361	100·0

* Excluding Ex-Service personnel not classified by industry.

83

the general drift of industry southwards, itself a reflection of changing locational forces, e.g. the substitution of electricity for raw coal as motive power, and partly on account of the growing pull of the market (as opposed to the source of raw material supplies) for most of the newer industries. In this latter connection, Basingstoke is particularly well situated midway between London and Southampton, with good rail and road communications and easy access to the Midlands and North. The initial reasons which attracted firms already producing in Basingstoke were varied and some would no longer hold good today. These ranged from the cheapness and availability of labour and the excellent transport facilities to the desire of a particular managing director to set up a factory midway between the parent plant and his family home on the South Coast. In general, most firms appear very satisfied with their existing location. One company which has its main production units in London is shortly to transfer these lock, stock and barrel to Basingstoke. Since the war there has been a steady flow of applications from industrialists for Industrial Development Certificates to set up plants in or around Basingstoke but, until recently, far more have been refused than granted. The town has, nevertheless, remained highly prosperous: unemployment has been generally lower than the national average since 1945 and, in spite of the rapid growth in the working labour force, vacancies have far exceeded the available labour supply for most of the last six years.[1]

THE FUTURE PATTERN OF ECONOMIC ACTIVITY

Although there is nothing specific in the Town Development Act, 1952, concerning the structure of economic activity in expanded towns, in so far as an essential purpose of post-war legislation dealing with overcrowded cities has been to promote the dispersal of jobs as well as people, and that manufacturing industry is the most mobile type of job-provider, it may be assumed that the greater proportion of the new labour force in Basingstoke will find work in secondary

[1] BASINGSTOKE EMPLOYMENT AREA

	Vacancies for adults	Unemployment of adults	% of total labour force	% of U.K. labour force unemployed
1956	408	70	0·4	1·2
1957	391	134	0·8	1·4
1958	221	225	1·2	2·1
1959	268	227	1·3	2·2
1960	409	243	1·3	1·6
1961	571	136	0·7	1·5
1962	379	214	1·2	2·1

industry. This would also appear to be the assumption of those who drafted the new Town Map, as a very substantial area of land has been set aside 'primarily for the re-location of industry from London'.[1] Nevertheless, it might be helpful, at this stage, to look at one or two general principles governing the economic structure of towns. We will then seek to apply those principles to the particular situation now facing Basingstoke and, in so doing, draw upon the experience of other towns in Britain comparable in size and general character to that envisaged for Basingstoke at the end of its expansion scheme.

More particularly we shall seek to use such data in an attempt to answer three questions:

1. What is the total size of the working labour force likely to be in Basingstoke fifteen to twenty years hence?
2. What is likely to be the broad distribution of this labour force as between the main income-earning activities, i.e. agriculture, manufacturing industry and the service trades?
3. What guidance, if any, can be given regarding the 'best' distribution of the labour force within manufacturing industry?

CLASSIFICATION OF ECONOMIC ACTIVITY IN TOWNS

For the purpose of the first part of our argument it might be helpful to distinguish between three main groups of job-providers usually common to all towns:

(1) The 'exporting' industries, i.e. those trades supplying products or services which are sold mainly outside the immediate area of production. Most agricultural, mining and manufacturing activities fall into this category and, apart from firms supplying highly perishable products, e.g. bread and milk, the market served is a national or international rather than a local one. Some service industries, e.g. insurance, civil engineering, the provision of scientific and technical services, non-local transport, etc., may also be of this kind. Such trades are not solely or even mainly dependent upon local demand for their prosperity. Instead, for good or bad, they are subject to national or international economic forces over which little regional control is possible. As far as the local area is concerned, they are *investment* or employment-stimulating industries, since spending power is created without the corresponding quantity of consumer goods and services being produced. Most of these latter are 'imported' into the area and sold through local distributive and service trades.

[1] Draft Amendment Map, 1962, p. 5.

The great majority of industrial towns rely upon the 'exporting' industries for their wealth and prosperity, and such trades usually account for between one-third and one-half of the total labour force.

(2) The service trades ancillary and directly (if not wholly) linked to (1) for their prosperity. These may be suppliers of specialised services or products, such as the various public utilities or distributing agents of one kind or another, i.e. dealers in industrial machinery or materials, storage depots, (local) transport undertakings, wholesale outlets, etc. Part of the services of local bankers, solicitors, builders, etc., also fall into this category, although in these cases the economic interdependence may be less important.

(3) Those trades catering mainly for local expenditure, i.e. the *consumption* or the 'importing' and home supplying industries and service trades. These also are of two kinds:

(i) those selling wholly or partly to shoppers living outside the town in question, e.g. hotels, motor distributors and filling stations, agricultural merchants, certain types of wholesaler, and retail distributors; these are in a sense 're-export' trades and often play a vital role in the economy of a market centre which caters for a large catchment area;

(ii) those selling and servicing, or otherwise dependent upon the income and wealth of the town's own residents, e.g. most distributive trades and local constructional activities.

In so far as free mobility of both resources and goods exists across regional boundaries, there need be no necessary relationship between the numbers employed in these various groups of industries. Thus, in some towns, e.g. those in close proximity to larger cities, economic activity is confined to one or two local industries, a certain amount of constructional activity, the provision of local authority services, the distributive trades and certain professional and miscellaneous services. In such instances most of the labour force earn their living by working outside the boundaries of the town, but spend the greater part of their money within them. By contrast, other towns, e.g. the City of London and certain specialised industrial conurbations, are little more than workplaces where income is created but comparatively little is spent. If we assume, however, that the aim of Basingstoke is basically similar to that of the new towns, viz. 'to build up self-contained and balanced communities for work and leisure',[1] and that any commuting

[1] *New Towns*, Committee Final Report, p. 30; 1945–6 Cmd. 6876, xiv.

to work outside the town is more or less balanced by a movement in the reverse direction, then it follows that the economic future of the town must lie in the satisfactory development of the exporting or the re-exporting industries, i.e. groups (1) and (2) above.

THE SIZE OF THE WORKING LABOUR FORCE

Before discussing further the possible distribution of employment between the *investment* and *consumption* industries (as defined above), let us look further into the question of the likely size and composition of the future labour force in Basingstoke.

Given the size of the population at any given time, the primary consideration governing the supply of labour will be the age-distribution of that population. In addition there are the various secondary factors–particularly as regards the provision of married female labour, such as the income of the husband, the size and age composition of the family and the available opportunities for work. Nevertheless, according to a recent study made of the economic structure of towns,[1] there appears to be a close statistical association between the size of the working labour force and the total number of people in the 15–64 age group. Based on data derived from the Census of Population for 1951, it was found that for 153 towns with a population of 50,000 or over this ratio varied between 85 and 102 per cent, being lowest in the London suburbs and highest in the Lancashire cotton towns. This did not mean, of course, that such a proportion of those aged between 15 and 64 were actually employed; a number outside this age group also worked. But apart from one or two unusual cases, the proportion was fairly consistent, and applying a figure of 90 per cent in the case of Basingstoke, it would mean that by 1976 the town would have a labour force of between 37,880 on our 'low' population projection and 45,075 on our 'high' projection:[2] the corresponding figures for 1981 would be 48,835 and 50,714.

However, while in this particular instance national averages might suffice as a basis for estimating future trends in Basingstoke, other kinds of data relating to the economic structure of towns, e.g. *commuting* and *job* ratios[3] and the occupational distribution of the labour

[1] *British Towns*, by C. A. Moser and W. Scott, Oliver and Boyd, 1961.
[2] See Chapter III, p. 50.
[3] *Commuting* ratio is defined as the number of persons either living in the area but working outside it, or working in the area but living outside it per 100 occupied resident population; *job* ratio is defined as the population working in the area per 100 resident occupied population. Moser and Scott, op. cit., p. 103.

force, differ so widely between different types of towns that general comparisons are of little value. Moreover, there are one or two features of Basingstoke's likely development, e.g. the marked changes likely to occur in the age structure of the population over the next fifteen to twenty years (of particular significance from the viewpoint of the female and juvenile labour force) which, apart possibly from Dagenham, had not been experienced by any town of 50,000 or above prior to 1951. In consequence, in an attempt to obtain useful comparative data, we made a careful analysis of the 153 towns examined by Messrs. Moser and Scott, eliminating those which for one reason or another had (in 1951) an economic structure substantially different from that envisaged for Basingstoke by 1981. For example, we considered only towns with a population between 65 and 120,000–taking an upper limit rather higher than Basingstoke's immediate target of 80,000 to allow for subsequent (natural) growth of the town after 1980. Second, for reasons already stated, we considered only towns with a *job*-ratio lower than 80 and those with a *commuting* ratio higher than 50. This eliminated almost all the London suburbs and some of the industrial work-towns of the North. Third, in assuming that the main economic wealth of Basingstoke in fifteen to twenty years' time will be derived (as it is at present) from the *investment* rather than the *consumption* industries (in the sense defined above), we discounted from consideration all towns which were so obviously different in their make-up that they offered no useful basis for comparison at all. There were quite a large number of these, most of which were highly specialised in their functions, e.g. seaside resorts, ports, spas, administrative centres and so on.

This left us with only thirty-two towns,[1] to which we added the eight London new towns and two expanding under the Town

[1] viz.

Barrow	Norwich
Blackburn	Oxford
Burnley	Peterborough
Carlisle	Preston
Chesterfield	Reading
Darlington	Slough
Gloucester	South Shields
Grays	St. Helens
Grimsby	Swindon
Halifax	Treorchy
Ipswich	Walsall
Lincoln	Warrington
Merthyr	West Hartlepool
Newport	Wigan
Northampton	Worcester
North Shields (Tynemouth)	York

Development Act,[1] Aylesbury and Bletchley, taking the population and economic structure of these last ten as of June 1961.

Returning to the relation of the size of the labour force to the total population in the age group 15–64, in all the above towns except the Lancashire cotton and Welsh mining centres it ranged from 85·5–95·6 per cent. However, on the basis of the experiences of the new towns, the proportion of the female labour force seeking work is likely to rise slightly as the average age of the newly settled immigrants increases. Taking into consideration this fact, and also the present percentage of those working in Basingstoke to the total population between 15 and 64 (85 per cent in 1961), we have assumed a slightly rising labour force proportion from 87 per cent in 1966 to 91 per cent in 1981. In terms of our five population projections (see p. 50) and for the years 1966, 1971, 1976 and 1981, we obtain the following estimates of the size of the labour force in the Town Map Area.[2]

TABLE V/6

ESTIMATED SIZE OF WORKING LABOUR FORCE IN
BASINGSTOKE 1966/1981

Population Projection

	A	B	C	D	E
1966	18,739	19,480	19,480	20,236	20,968
1971	25,911	27,393	28,112	29,609	31,068
1976	37,459	39,084	39,870	42,974	44,574
1981	49,378	50,437	49,048	49,487	51,278

THE DISTRIBUTION OF ECONOMIC ACTIVITY

In their examination of the economic structure of towns in Britain, Messrs. Moser and Scott found there were marked variations in the relative significance of the *primary* and *secondary* goods industries, i.e. agriculture, extractive industry and manufacturing, and the *tertiary*, i.e. service trades, as employers of labour. For example, the proportion of the working population engaged in the former kind of economic activity (i.e. the *investment* or 'export' industries) ranged from only 10·3 per cent in Gillingham (Kent) to 78·4 per cent in Oldham (Lancs.), while those employed in the service trades (i.e. the *consumption* industries) (excluding building and contracting) ranged from 14·4

---

[1] i.e. apart from Swindon which is included in the first group.
[2] Less the existing working population in the new intake area and any increase in that labour force resulting from the natural increase in population.

per cent in Oldham to 75·4 per cent in Gillingham.[1] With such a wide dispersion between towns, average (i.e. mean) figures of 41·6 per cent for agriculture and industry and 37·4 per cent for services are almost meaningless.

However, when we look at the distribution of the labour force in the forty-two towns selected for comparison we find the following picture:

| | Primary and Secondary Industry % of total labour force | | Tertiary Industry* % of total labour force | |
	All towns	New towns	All towns	New towns
Average†	49·8	59·3	33·4	27·5
High	69·7	69·7	49·9	40·8
Low	27·6	41·4	19·3	19·3

* Excluding those employed in building and construction.
† Arithmetic mean.

The range between high and low in each case is still fairly considerable, but there is a considerable bunching around the average. The median percentage (i.e. that half-way between highest and lowest) is 49·3 per cent for the goods industries and 33·6 per cent for the tertiary trades.

On the basis of the mean figure for all towns, and making allowances for a small increase in the numbers employed in agriculture,[2] Basingstoke would have a working labour force engaged in manufacturing of between 18,250 and 21,750 in 1976 and between 24,000 and 25,000 in 1981 according to the population projections (A–E) made. However, apart from the new and expanded towns, these ratios are based on 1951 statistics. Between that date and 1959 the numbers employed in manufacturing industry rose by 5·1 per cent, while those in the service trades increased by 8·3 per cent, (in spite of a decline in those engaged in road and sea transport and national government service): during these years, the proportion of the civilian labour force engaged in secondary industry fell slightly from 38·5 per cent to 36·5 per cent.[3]

The reason for this trend is not hard to find. On the one hand, a rising standard of living has meant a relatively higher expenditure on

[1] Gillingham is a good example of a residential town of an exceptionally high commuting ratio; Oldham is a Lancashire cotton workplace but is within six miles of Manchester which supplies the main services of the town's residents.
[2] i.e. in the Borough. Once again, it is emphasised that we have excluded the working population of the new intake area from our calculations.
[3] Calculated from data contained in the *Annual Abstract of Statistics 1961*. Information is not readily available for 1960 and 1961 due to changes in industrial classification.

most kinds of services, and in particular on education, medical and dental services and retail distribution; on the other, automation and other labour-saving methods of production and increases in productivity have proceeded faster in manufacturing industry than in the service trades. While it seems likely that Britain will remain predominantly a manufacturing nation throughout the next twenty years, it may well be that the movement of the last decade towards a more service-based economy will continue. Studies of the comparative situation in other advanced industrial countries are of limited value, though it is interesting to note that over the past decade the numbers employed in commerce in European countries have risen twice to three times as fast as those in manufacturing industry. There is also quite a close positive correlation between the proportion of a nation's labour force engaged in the service trades and living standards. In the United States and Canada (the wealthiest countries in the world), for example, this proportion is appreciably higher than in Britain and Continental Europe, but as the manufacturing methods of these latter countries become more capital-intensive and living standards rise, the role of commerce and distribution in their economies is likely to become even more significant.[1]

At the same time there are other factors more specific to Basingstoke and to other towns similarly placed which must be taken into consideration. We have already indicated that in addition to selling the products of secondary industry a town may earn its livelihood by exporting various services or by developing as a commercial and/or administrative centre for the surrounding region. To deal briefly with this second point first, while it is probable that Basingstoke will develop a considerably larger hinterland or catchment area as it grows and improves its shopping facilities, professional services and recreational amenities, it is doubtful (though not impossible) that the town (assuming a population of 80,000–100,000) will ever become as important a commercial outlet as Reading or Southampton (both of which serve regions more highly populated than that around Basingstoke) or an administrative centre as Winchester (unless the seat of the Hampshire County Council were removed from there to Basingstoke).

On the other hand, there is a distinct possibility that Basingstoke could develop as an important supplier of certain types of 'export' services, e.g. insurance and research. For example, some planners feel

[1] See Employment in Distribution, *Treasury Bulletin for Industry*, May 1962.

that too much emphasis has been placed by the new towns on the provision of factory work for manual workers. This is only partly true as, in comparison with the other towns analysed earlier, the new towns employ above the average proportion of professional and scientific workers (9·7 per cent compared with 6·7 per cent.) Government research and scientific establishments currently employing over 2,000 people have been set up in Harlow, Bracknell and East Kilbride. In addition some 334 new commercial offices employing 6,636 people and occupying 937,643 sq. ft. had been established in the new towns by 31 December 1961 and a further twenty-three offices of some 464,648 sq. ft. were under construction at that date.[1] Moreover, it is a little premature to argue that such towns are unduly dependent on manufacturing industry as a source of employment. For one thing, most service industries naturally tend to develop after the goods industries; for another, many new towns have embarked upon a programme of providing shopping centres more labour saving in character than those now generally in existence.

Much, of course, depends upon the particular aspect from which development is being discussed. Compared with other light industrial towns, the new towns, having due regard to their stage of development, would not seem to be underprovided with tertiary industry. However, from the viewpoint of decentralising people and work from London, it may reasonably be argued that too little attention has been paid to the dispersal of new office building and too much to secondary industry. In spite of the contributions made by the new (and expanded) towns in this respect, the proportion of office workers to all workers in the London area has almost certainly risen very substantially in the last ten years. This has inevitably created an increasing shortage of office accommodation and has been accompanied by steeply rising rentals and building costs. This being so and bearing in mind the need to attract and retain as balanced an occupational and social structure as possible in Basingstoke, it might merit the particular attention of the Joint Development Committee to induce offices to expand or transfer their London premises to Basingstoke. Much may well depend on future Government policy. At present there are few controls over the building of office blocks. If, as has been suggested by various bodies, an office development certificate, similar in principle to the industrial development certificate in secondary industry will eventually be required for all new office building in London

[1] *Town and Country Planning*, January 1962, p. 32.

92

(and/or other congested areas), Basingstoke could well benefit from the decentralisation which would almost certainly follow.[1]

At the same time, from a town-planning aspect, office accommodation is much more economical in its use of space than most manufacturing industry. Not only is it possible to concentrate a greater floor area on a given site for offices than for factories but an acre of office space usually accommodates many more people than a similar area of factory floor space.[2] Basingstoke is particularly suitable for such development for topographical reasons, the area between the existing town centre and the railway having natural height differences formed by the Loddon valley; in addition, more multi-storey office building would result in a more compact town and less encroachment on the surrounding agricultural land or countryside. Finally, in connection with social and occupational balance, it is essential to present as wide a range of employment opportunities as possible to cater for the varying aptitudes and abilities of the population, not only of the initial new-comers but of the children of immigrants, so that they may not simply become commuters or drift back to London whence their parents were induced to migrate several years earlier.

The Town and Country Planning Association has frequently drawn attention to this problem[3] and in one of its most recent publications *The Paper Metropolis* (1961) specifically suggested the development of new regional centres up and down the country to act as counter-magnets to London.

. . . they should be made into places which could attract branch or divisional offices of London firms, and also attract firms which would otherwise seek to establish themselves in London.[4]

A study of relative costs also published in the booklet, based on questionnaires completed by firms which have moved out of London to new towns in recent years, revealed that as much as £140 a year per employee can be saved on office rent and office workers' salaries alone. Another estimate, by an insurance company, put the annual

[1] Since this chapter was written the Government has taken further steps to make planning control over office building in the London area more effective. These include the setting up of a Location of Offices Bureau to encourage the dispersal of offices from central London to other areas. See *London. Employment: Housing: Land*; 1962–63, Cmnd. 1952.

[2] For example, in the new towns at the end of 1961 there were 214 people employed per acre of manufacturing industry and 309 employed per acre of commercial offices.

[3] See, for example, *Office location in the London region*, Report of Conference held in July 1958, Town and Country Planning Association, 1958.

[4] ibid. p. 60.

saving of dispersal of this kind at $7\frac{1}{2}$–10 per cent of total costs. In most cases, it appears that increased communication and administrative costs associated with decentralisation are more than outweighed by easier recruitment of staff, lower rentals and an improvement in productivity resulting from the reduced 'wear and tear' of the journey to work.[1]

To some extent this difficulty of achieving the right occupational balance can be solved by attracting manufacturing firms with an above-average proportion of administrative, technical and scientific workers, but this is a poor substitute for more office building in other respects. It may be that economic pressures will eventually drive more firms out from the centre, but it is unlikely to be as effective as a direct embargo on office building in London. In either case Basingstoke stands in an extremely good position to share in the expansion which seems bound to come.

For all these reasons an average figure of 50–55 per cent as derived from the table on p. 89 may well prove too high a proportion when estimating the labour force likely to be engaged in secondary industry in Basingstoke in 1981. A yardstick of 45–50 per cent would perhaps give a better balance (although in the early years it might be somewhat higher) and may well prove to be more in line with the likely course of events. It may be argued that differences in forecasts of this magnitude are unimportant in any case and that as long as the planned programme of expansion is sufficiently flexible, marginal adjustments in the substitution of office for factory workers (or vice versa) can be made as the needs arise. This is perfectly true, save that, in so far as office accommodation may have to be sited in the centre of the town (and is thus not directly competitive with factory space), basic policy towards the (eventual) build-up of offices needs to be formulated at the planning stage. On the other hand, it may be that we have underestimated the trend towards automation in retail distribution. If the numbers employed here, per unit of sales, fall significantly below the national average, this could mean that proportionately fewer rather than more people would be engaged in the service trades in the years to come. How far this may prove to be so will depend at least in part on the type of retail outlets envisaged and the importance of the numbers employed in this particular service industry to all tertiary industry.

[1] J. V. Aucott, 'Dispersal of offices from London', *Town Planning Review*, Vol. XXXI, 1960/61, p. 37. *The Paper Metropolis*, pp. 33 ff.

WHAT KIND OF SECONDARY INDUSTRY?

With the above considerations in mind, our estimate of the supply of workers for manufacturing industry in the Basingstoke Town Map Area[1] (percentage of the total labour force assumed to be so employed given in brackets) for the period of the Town Map is as follows:

TABLE V/7

ESTIMATED SIZE OF MANUFACTURING LABOUR FORCE IN BASINGSTOKE 1966/1981

Population Projection

		A	B	C	D	E
1966	(49)	9,183	9,545	9,545	9,906	10,274
1971	(48)	12,437	13,149	13,494	14,212	14,913
1976	(47)	17,606	18,370	18,739	20,198	20,950
1981	(46)	22,714	23,201	22,563	22,764	23,588

The question next arises – what kinds of secondary industry are (a) best likely to suit the needs of Basingstoke and (b) most likely to want to locate their production units in Basingstoke?

Again, it should be stressed that our purpose here is not so much to predict what the pattern of industry will be in fifteen to twenty years' time, but rather to examine some of the general principles on which it is possible to assess whether a particular distribution of firms is in the town's best interests or not. Obviously it would be foolish to envisage a particular structure of industry as desirable from the start and to adhere rigidly to this conception come what may. To begin with, factors influencing the choice of firms (assuming that a choice is possible) in five, ten or fifteen years' time may well be quite different from those of today. Nevertheless, whenever more firms or industries wish to set up factories in Basingstoke than the town can accommodate, decisions as to the attractiveness of one by comparison with the other will have to be made, and it is clearly better that they should not simply be taken on an *ad hoc* basis, with only the needs of the particular situation in mind, but in the light of some overall (albeit flexible) plan which adheres to certain principles underlying any satisfactory balance of industry.

What are these principles and what is meant by a satisfactory balance of industry?

The twin aims of national economic policy – full employment without inflation and an adequate and stable rate of growth – are well

[1] Less population in new intake area.

known, but these are no less desired or sought after by particular sections of the economy. Thus a prosperous town might be thought of as one in which there are sufficient and continuous employment opportunities for its labour force and an improving standard of living for its inhabitants. Partly, of course, such prosperity will be dependent on circumstances outside the control of the local policy, e.g. trends in national (or international) demand, technological developments, Government policy and so on, but the extent to which a town, or the regional economy of which it is part, can cushion any depression, or benefit from a boom, may be closely related to the pattern of its industrial composition which, to a certain extent at least, is (or can be) of its own making.

At one time, a town or region could not control its destiny to the same extent as it can today. Even now, of course, some locations are unavoidably highly specialised. One particular town, for example, may be responsible for supplying the total, or a large part of the total national output of a particular raw material which, for one reason or another, is also its sole resource of any real value. Another may attract a nucleus of firms in a particular industry which act as a magnet to other firms in similar businesses, and soon a highly specialised industrial structure is evolved. Again, some industries are more or less tied in their choice of location for technological reasons such as the cotton textile industry, centred in Lancashire and Scotland because of climatic suitability, or the shipbuilding industry necessarily sited on the estuaries of important rivers. A town with fortunes based on one industry or group of related industries is always liable to suffer more in times of depressed or fluctuating demand for its products than towns with a more diverse structure. This was perhaps the main reason why certain geographical areas of Britain, e.g. South Wales, mid-Scotland and north-east England, were so much more affected by the economic depression of the inter-war years than other parts of the country, and why the need for some kind of overall industrial location policy became so acute.

Changing locational forces, and in particular the development of ubiquitous sources of secondary energy, more efficient transport and the growth of new industries, market-oriented and using a wide range of raw or semi-finished materials, none of which exercises a strong locational pull,[1] have tended to shift the regional distribution of

[1] See R. C. Estall and R. Ogilvie Buchanan, *Industrial Activity and Economic Geography*, H.U.L., 1961, Chapters 2 and 3.

industry and employment towards the Midlands and the South of England, at the same time creating a new class of town and a new congeries of industry. Such industries, drawn to one location rather than another on economic rather than purely technical grounds, are mainly of the 'mobile' or 'footloose' character which, as the names suggest, are industries capable of producing quite satisfactorily in various places, although cost-wise there is almost always an optimum or best location (if this can be found). Similarly, a town or region may be capable of attracting a wide range of these 'footloose' or 'mobile' industries but, in theory at least, there is, for any given moment of time, one particular pattern of industry which is preferable to any other. The difficulty inherent in either kind of choice is that over time not only may demand and supply conditions for different industrial products change, and thus the desirability of alternative patterns of industry for a region's economy, but so might the locational attractions of the region itself *vis-à-vis* other regions, due, for example, to a marked reduction in transfer costs. What is the best structure of industry and the best location today may be only second best tomorrow. In planning for the future, then, a degree of 'built-in' flexibility is essential for any industrial programme envisaged.

To some extent, of course, since the industrial growth of Basingstoke is bound to be gradual, any movement away from the type of industrial structure thought desirable can be corrected as and when it occurs. Nevertheless, there are a number of principles of economic planning which might well serve as a useful basis for decision-taking from the start. These must necessarily be related to expected national and international trends on the one hand and Basingstoke's particular pattern of resources on the other. Fortunately, there is nothing in the present industrial composition of the town to suggest that it possesses any inherent location disadvantages which makes the continuing achievement of balance an impossible or even a difficult task.

THE CONCEPT OF INDUSTRIAL BALANCE

Though frequently applauded as a desirable aim, the concept of industrial 'balance' is a nebulous one and one which can be defined differently for different purposes. Usually it refers to the diversity (or otherwise) of products manufactured by a particular group or cluster of firms; equally, however, it might be related to the character of the markets serviced, the type and variety of skills demanded, the proportion of male to female and adult to juvenile labour employed and the size dispersion of the manufacturing units. Clearly, to achieve one's target in one direction does not necessarily ensure success in another. However, in the light of the aims of economic stability and growth mentioned earlier, we might define a well-balanced pattern of industry as one which incorporates

> a range of manufactures, which offer favourable growth prospects, a minimum risk of employment instability, a diversity of occupational opportunities, and a proper balance of work for men, women and juveniles.

To some extent the attainment of even this end involves a compromise. For, on the one hand, it is desired to achieve a sufficient degree of industrial diversification to avoid the dangers inherent in overspecialisation and to provide opportunities for different skills and aptitudes of the working population. On the other hand, in order to derive full benefit from the natural or acquired advantages of a location (e.g. a pool of skilled labour) and other external economies of scale, some concentration of industries and skills is essential. One way of achieving this compromise is to aim at attracting groups of different industries with comparable labour requirements or, in other words, encourage an occupational structure which allows for industrial mobility or flexibility. Such a pattern of industry would guarantee both a choice of employer and a basic group of employees capable of changing from one job to another. It also means that in a town of 80,000–100,000 inhabitants it is possible for there to be a certain degree of specialisation while maintaining the necessary viability to counteract any sectional slump. The engineering industry, for example, contains a fairly wide group of separate trades requiring similar skills.

The question of the distribution of skills is also related to that of technological education. Most of the new towns, for example, are constructing technical colleges which will serve a wider region than the designated area, but will, nevertheless, relate their curricula to the local industrial needs. If there is some specialisation of skills it may be possible to provide a balanced curriculum particularly in the more advanced stages. Thus, for example, Hatfield is likely to cater for aircraft engineering, Corby for metallurgy and general engineering, Harlow for electrical engineering and woodwork.

With these somewhat conflicting aims in view, let us now examine the types of industry most likely to meet the requirements of a town seeking a 'balanced' industrial structure, i.e. in this instance Basingstoke, ignoring for the moment the town's relative attractions or disadvantages to particular firms as a locational site. In our analysis, we first look at the factors influencing industrial balance (or absence of it) from the viewpoint of the *demand* for the products being supplied and second from that of the *supply* of factor resources available and required.

(i) *The demand side*–the nature of goods produced. From past experience it is clear that instability of regional employment and sometimes also a lack of economic growth may arise from an undue specialisation of that region's resources on the production of three types of products:

(*a*) Products the demand for which is particularly sensitive to changes in the economic climate. These are mainly to be found in the investment or producer goods' industries. In times of recession or fluctuating demand (or even of rates of change in demand) it is these industries which not only are usually affected the first, and thus initiate a chain of repercussions throughout the local economy, but are also affected the most, if for no other reason than it is easier to postpone buying capital goods than consumer goods. Even if one examines the post-war period, fluctuations in the rate of growth of the industries supplying producer goods, e.g. iron and steel, machine tools, most kinds of plant and equipment, have been more pronounced than that of those supplying consumer goods.[1] While it is unlikely that Basingstoke will prove an attractive location for the main (industrial) materials processing industries, e.g. manufactured fuel, metal smelting, heavy chemicals (which are usually material oriented), where the main fluctuations usually occur, it may well be suitable for producer-good industries a little farther down the chain, e.g. agricultural machinery, mechanical handling equipment, industrial electronic apparatus, which are still vulnerable to economic instability above the average. If anything, Basingstoke may be a little overspecialised in this respect now. Of the numbers employed in manufacturing, between 75 and 80 per cent are engaged in producer-good industries and nearly one-half in the supplying of commercial transport equipment and mechanical handling machinery: nationally the proportion of the total manufacturing labour force employed in capital goods is about three-fifths.

[1] See Appendix I to this chapter and particularly Col. 2 on the table.

(b) Products which cater for a specialised or limited market. One of the main reasons why the basic industries were so severely affected by the inter-war depression was that they exported an unduly high proportion of their output to declining markets. While it is possible by one means or another to stimulate home spending, it is less easy to exercise control over overseas demand. On the other hand, a firm which sells only in the home market has no outlets to fall back on in times of domestic recession. Firms or industries supplying the Government (or other firms) with defence equipment are vulnerable for similar reasons (illustrated by the closure in 1962 of part of the English Electric Company's factory at Stevenage due to the cancellation of a Government contract), as are companies whose whole output consists of specialist components or parts for one firm or small groups of firms. How far such 'single-marketness' is a (potential) disadvantage depends on whether the market supplied is a stable and growing one and if the capacity of the industry in question is sufficiently flexible to allow a speedy change to the supply of other kinds of products.

(c) Products the demand for which is growing less than the national average or is even falling. Again, while *prima facie* it is improbable that the less prosperous industries will wish to set up plants in Basingstoke, it must be remembered that applications for factory space will come from firms seeking removal from other parts of the country, especially London, and it by no means follows that these will be within the growing industries. Important in this context is the likely impact of Britain's proposed entry into the European Community, of which further mention will be made later in this chapter.[1] Guides as to which are potentially the fastest and most consistent 'growth' industries can be obtained partly from examining recent output trends in this country and other advanced industrial nations, and partly from a study of Britain's performance in export markets where she competes on equal terms with other exporters. In terms of the first criterion the industries with the best record over the last decade are pharmaceuticals, office machinery, industrial instruments, mechanical handling equipment, most types of electrical apparatus, plastics and man-made fibres and on the second, chemicals, motor cycles, electrical equipment, aircraft manufacturing, rubber, hosiery and clothing. The

[1] Since this chapter was written negotiations for Britain's entry have been suspended.

trades with apparently the least prospects for expansion on the former basis include cotton textiles, leather and non-ferrous metals and on the latter, cotton fabrics, china and glassware, watch and clock and shipbuilding or repairing.

The above account presents the economic case for planned industrial selection from the demand side. Similar arguments can be put forward on planning and sociological grounds. From a planner's viewpoint (and assuming the economising of space is one of the crucial issues in successful town expansion) industries which require above the average floor space or below the average density of workers per acre may well be less welcome than those more labour or capital intensive in character. The amended Town Map sets aside 350 acres for manufacturing industry.[1] In the new towns and larger trading estates, factory floor space as a proportion of industrial acreage in 1960 varied from 28 per cent to 40 per cent, but the median is around 35 per cent. Nationally, the factory/site proportion is nearer 50 per cent. However, most firms have bought or rented land with expansion in view and it is unlikely that any modern light industrial zoning scheme would envisage as high a ratio, as only a small proportion of buildings are likely to be multi-storey in character. Similarly the average density of workers per acre ranged from 58 to 77. Assuming that the manufacturing labour force will reach 23,000 by 1981, then the density assumed for Basingstoke would work out at 66 per acre, which is around the average.

In terms of individual industries, space requirements vary considerably. The *average number of workers per acre of site*, for all industries suitable for location in residential areas, was 233 in 1951. However, the figure ranged from 75 in the case of cigarette, metal furniture or shop fitting manufacture to 500 for telegraph and telephones, biscuits and precious metals. Similarly, the *average floor space per worker* required varied from 75 sq. ft. in the production of electrical equipment, hosiery and clothing, to 175 sq. ft. in the production of photographic papers, most metal goods, biscuits and cigarettes.[2]

In part, of course, the latter figure will be influenced by the capital/output ratio of the industry in question. Those industries highly mechanised or automated, with a high productivity per head, e.g. those producing such goods as cigarettes, electrical lamps, etc., will

[1] For further details see Appendix II to this chapter, p. 125.
[2] See G. Logie, *Industry in Towns*, Allen and Unwin, 1952, pp. 106 ff.

101

need a larger floor area per employee than those with a low capital/ output ratio. It is worth bearing in mind that over the years the average floor space per employee has been increasing in British industry and that the most dynamic and productive industries are among the most prodigious users. This means that, compared with 1951, a town planning its economic structure will either require more land to employ a given number of workers in manufacturing industry or will have to engage a higher proportion of workers in non-manufacturing undertakings. To some extent, then, the type of industries likely to be attractive to a town will be influenced by the acreage set aside for industrial uses – itself a reflection of the general structure of economic activity.

Finally, from the viewpoint of social balance in the community, it is desirable that the pattern of industry evolved should be such as to require as wide a selection of types or grades of labour as possible. For example, it is probable that a good proportion of firms wishing to establish in Basingstoke will take the form of branch units of a parent plant located elsewhere in the country. A predominance of such businesses in any industrial conurbation creates various problems, one of which is the possible lack of employment opportunities for administrative and higher-grade technical and scientific staff. Even if firms transfer their activities lock, stock and barrel, a proper balance between operative and non-operative labour needs to be achieved.

In their awareness of this problem, the new towns have made special efforts to attract science-based industries with research and development facilities. Fortunately, as a recent F.B.I. survey has shown, there is a close connection between growth and expenditure on research.[1] Thus, if the growth industries are attracted to Basingstoke, and a good proportion are transfers or new firms, this particular problem should not present any great difficulties.

(ii) *The supply side*

(a) *Types of employment*

The above points all relate to the industrial pattern of a community considered from the demand aspect. Equally important, however, is the need to achieve industrial balance from the viewpoint of the local resources available. There are several features of

---

[1] *Industrial Research in Manufacturing Industry 1959/60*, Federation of British Industries, 1961.

this problem, the first and perhaps the most important being the distribution of employment as between the sexes and between adult and juvenile labour. If the experience of the new towns is anything to go by, for the first ten years or more of the Town Development Scheme, the ratio of the (incoming) supply of female to male labour is likely to be considerably less than the national average, due partly to the high proportion of married women with children of ten years or under, and partly to the low proportion of juveniles and older spinsters in the population.[1] In due course, however, the situation will drastically change and the numbers of women looking for work will rise substantially. Although a case may be made out on non-economic grounds against married women working (particularly those with young children) there is little doubt that adequate employment opportunities for women greatly add to the stability of family incomes. The lack of such opportunities in the depressed areas during the inter-war years was one of the reasons why such areas suffered so much more than the rest of the country. Though it is to be hoped that governments now have sufficient knowledge to prevent the occurrence of severe and prolonged periods of unemployment, adequate employment openings for women are not only an insurance against the loss of earnings of male workers but also add to the economic flexibility of the region as a whole, as the type of industry which employs a high proportion of female labour is usually of a different kind from that employing predominantly male labour. According to national statistics the industries employing the highest proportion of female to male workers are tobacco, confectionery, pharmaceuticals, watches and clocks and clothing, and those employing the lowest proportion, metal manufacture, most kinds of heavy or medium engineering, vehicles and the timber trades. It is perhaps significant that the lighter and more 'footloose' industries are of the type likely to be considerably more attractive to female labour than those strongly oriented either to a particular market or raw material.

At present, compared with a national average of 46 women employed per 100 men in manufacturing industry, the number in Basingstoke Employment Area is 47. This ratio might well fall in the next decade or so, due to the age composition of the immigrants, to around 45 in 1966 and 43 in 1970. It is then likely to rise as the number of school-leavers begins to increase rapidly, and by 1976 will probably

---

[1] Both of these are reflections of the age-structure of incoming migrants. See Chapter IV, p. 65.

be back to the current level of 47, and by 1981 to 50 or more. At this latter date it would mean between 11,280 and 11,800 jobs would be required for women according to the population projection accepted.[1]

The appropriate balance between juvenile and adult labour is even more difficult to assess and will also fluctuate considerably over time. The proportion of school-leavers to total working population is unlikely to be significant for several years, but for a period from about 1975–1980 it will rise very rapidly indeed. For example, on the average of our age projections the number of those between 15 and 20 is likely to rise from an estimated 1,400 in 1961 to 5,000 in 1976 and 6,500 in 1981. The number of school-leavers in Basingstoke in 1961/1 was 534, a 36 per cent increase compared with 1955/6. Comparative figures for the London new towns show that in 1960 the proportion of juvenile to total workers was 6·3 per cent: for the country as a whole it was 6·4 per cent. But the average number of school-leavers in these latter towns was substantially higher than in 1956 and is still increasing.

A study of national statistics reveals that the industries with the highest ratio of juvenile to adult labour are woodworking, paper and printing, precision instruments, clothing, cans and metal boxes, hosiery, leather goods and sports equipment, and those with the lowest, metal manufacture, motor vehicles, man-made fibres, and chemicals and dyes.

However, it may be that, with further expansion of higher education, the demands for work from boys will be somewhat less than that for girls. Distribution and commerce usually cater for much of the juvenile female labour, but these avenues of employment will most certainly need to be supplemented by additional facilities for office and industrial work.

(b) Types of firms

There are two other senses in which the word balance can be interpreted in connection with the structure of industry. These refer less to the type of employment required than to the type of production unit. One relates to the size of firm both absolutely and relatively to other firms demanding similar kinds of labour in the surrounding area; the other to the administrative structure of the firm, i.e. whether it is the

[1] This, of course, assumes there is no marked change in the attitude or willingness of married women to work in the next fifteen to twenty years. It should be noted that one is here concerned only with manufacturing industry. However, basically the same trend is likely to occur for service industries as well, except that the proportion of female to male labour will be higher throughout.

sole production unit or a branch plant, or whether it is simply a manufacturing unit or also engages in administration, research and development and so on.

With regard to the first question, it is difficult to lay down any general principles concerning the 'best' size of a firm for a particular region. So much depends on the range of products supplied, their price and income elasticity of demand, and the type of markets served. There are, however, undoubted dangers in a single firm or group of firms becoming so large as to dominate the economy. Thus the dependence of such towns as Eastleigh on the railway works, Hatfield and East Kilbride on aircraft engineering, Corby on steel,[1] and so on, while of no apparent disadvantage in times of prosperity, can be a serious drawback if and when the particular firm or industry in question suffers from falling or fluctuating markets.

Though Basingstoke (unlike Corby) is not an isolated town in that there are alternative opportunities for work within reasonable distance, e.g. in Aldershot, Reading and Farnborough, its economy is still largely self-contained and likely to be even more so in future years. Taking Mr. W. Luttrell's view, that no one establishment in a particular geographical area should employ more than 10 per cent of the industrial labour force, unless its range of products is highly diversified,[2] and assuming a manufacturing employment of 23,000 in 1981, this would mean that no one firm ought to employ more than 2,300–2,500 people. This, however, is probably a somewhat conservative estimate, particularly as the 'optimum' size of business units in many industries is growing all the time; the principle is nevertheless valid. For example, the employment of 10,000 people by any one firm in Basingstoke (unless it is highly diversified in product-structure) would inevitably bring with it risks (however dormant they may appear in times of prosperity) to the stability of the local economy. Naturally, firms attracted to Basingstoke will vary in size and growth potential; some will be industries where the average size of production unit is necessarily large (space and/or labour wise); others will reach their least cost position at a much smaller output. The point is, however, that care should be taken not to encourage a particular cluster of firms, each of which, by the nature of their production processes,

[1] For example, no less than 75 per cent of the male labour force in Corby is employed by Stewart & Lloyds, while De Havillands account for about the same proportion of workers living in Hatfield.
[2] W. F. Luttrell, 'Industry and Small Town Expansion', *Official Architectural and Planning Review*, Vol. 18, August 1955.

must necessarily employ a larger number of people than is thought desirable for the balance of the town. Size dispersion is important. It may also be considered worth while to set aside a certain area of industrial land for the establishment of workshops or 'nursery' factories of between 1,500 and 5,000 sq. ft. to encourage the development of new business talent. Such factories have proved both successful and popular in the new towns.

It is, of course, almost inevitable that in the early stages of industrial growth a certain amount of ill balance in size dispersion will occur. There are certain advantages in attracting one or two large firms of national or international repute in growth industries rather than a host of small and little-known companies. The fact that a well-known enterprise has expressed its faith in the town's future in this way often stimulates the setting up of other concerns. But looking at a particular expansion scheme as a whole, any advantages of the economic domination by particular firms should be carefully weighed against the inevitable risks involved.[1]

A final problem of balance concerns the administrative structure of firms. It is likely that the great majority of firms setting up in Basingstoke will be one of two kinds:

(1) Removals, lock, stock and barrel from London (due, for example, to antiquated premises, cramped accommodation or requisitioning orders): these will probably be mainly small companies employing 50–150 people at the time of removal.

(2) Extensions or branches of existing companies, mostly, but not entirely, those with headquarters in London.[2]

It is this second group of firms (which in the new towns, expanded towns and development areas are usually much larger in size than the first group) which might create special problems. The first of these is more a question of occupational and social balance referred to earlier; the second relates to the economic and administrative position of the branch in relation to the parent plant. Studies in the costs of alternative locations have frequently underlined the dangers of a particular

---

[1] In 1958 of the 271 firms in the new towns engaged in manufacturing industry, 227 employed less than 250, 38 employed between 250 and 1,000, and 6 employed more than 1,000. They provided, respectively, for 33·4 per cent, 38·5 per cent and 28·1 per cent of total employment.

[2] Of the same 271 firms quoted above, 188 represented plant removals from other parts of the country and 69 branch units with their headquarters elsewhere. The important point to notice, however, is that these latter accounted for 44 per cent of the total labour force.

geographical area becoming too dependent on the activities of branch plants for their industrial prosperity. There is always the possibility, borne out by the experiences of radio firms in the South Wales Development Area in 1951–2, and more recently in Basingstoke itself with the closure of Associated Transistors Ltd., that should a recession develop, or other economic difficulties arise, it is the branch plant which is the first to curtail its output, particularly where it is simply duplicating the products manufactured by its parent company. It is thus preferable that, wherever possible, branch plants should be complementary rather than competitive with their parent companies in their product structure and not simply be 'overflow' units. In the new towns the great majority of plants manufacture a different range of products from their parent companies; the same would appear to be true of the Basingstoke firms which are subsidiaries of other manufacturing units. It is, of course, possible that, due to the locational advantages of the town, a branch might find it more profitable to expand (or not to curtail) production there than in its alternative location.

These, then, are some of the aspects of 'balance' which warrant consideration in any attempt to assess the 'ideal' pattern of industry for a particular town. As has been said, while the principles outlined may be of lasting validity, their relative significance and value may change over time. Usually, too, a compromise is called for in that it is rarely the case that one industry or group of industries satisfies *all* the criteria of a satisfactory balance between diversification and specialisation; in the last resort some kind of value judgment has to be made as to which of the qualities of industrial balance are considered the most desirable. Thus, if a choice has to be made, is growth to be regarded more highly than stability; diversification of output more than an ideal balance between juvenile and/or adult labour; a proper occupational balance more than a particular administrative structure, and so on? Moreover, such expansion as may take place in Basingstoke must always be viewed in the light of national trends which may not always accommodate themselves to local requirements. Thus, at present, in those industrial communities set up since 1945, due to the emphasis of the national economy on the expansion of capital goods industries, some four-fifths of the manufacturing labour force is engaged in such industries, while at the same time, and for much the same reason, the export performance of others is well above average.

For similar reasons most of Basingstoke's new industry since 1945 has been directed to the manufacture of producer goods.

BASINGSTOKE AS A CHOICE OF LOCATION

So much for the broad issues facing Basingstoke when deciding on the pattern of industry it would like to attract. But what of the value of Basingstoke as a choice of location for firms wishing to set up factories in the next fifteen to twenty years? This will depend on first, (*a*) the general prospects of particular firms and industries and (*b*) the locational value of Basingstoke itself.

Let us briefly discuss each of these considerations in turn:

(*a*) Forecasting what type of industries are likely to develop most rapidly for more than two or three years ahead is little better than crystal-gazing. The variables are far too many and complex to allow one to gain any more than the broadest of impressions. As mentioned earlier in this chapter, the main guideposts which we have are (i) the immediate past performance of individual industries and, in particular, whether the *rate* of change in their output and/or employment is increasing more or less than the average for industry as a whole and (ii) comparative trends in other industrial countries and especially those more advanced than our own. To these two general pointers a third may be added which is of particular significance to Britain in her bid for overseas markets viz. the competitiveness of U.K. exports. Let us examine this latter factor in a little more detail.

If and when this country enters the Common Market,[1] the forces making for further concentration of population and manufacturing industry in Southern England might well become overwhelming. Obviously such a movement would be likely to have vital repercussions on the economy of Basingstoke and any data on the possible effects of the entry of this country into the E.E.C. on industry and commerce needs to be studied carefully. Even if Britain does not join the E.E.C., the impact of this economic bloc on world trade and production is bound to affect the pattern of industrial output in this country. Such considerations as these must be taken into account when attempting to assess what types of industry and what firms within particular types of industry it is desirable to attract to Basingstoke.

Implicit in the whole concept of the Common Market are the advantages of specialisation, i.e. the maximum advantage is obtain-

[1] The current round of negotiations was broken off in January 1963

108

able when each country concentrates its production on those goods it is best fitted to produce and exchanges its surplus commodities for the products of other countries. This in turn implies that if this country joins the E.E.C. some of our existing industries will decline, and indeed, may well disappear, while others will expand. As a result, then, of our entry into the Common Market and of the removal of tariffs and other impediments to the free movement of the factors of production, this country will tend to increase the production of those goods in which she has a comparative cost advantage and will tend to reduce her output of those goods in which she has a comparative cost disadvantage. In other words, the entry into the E.E.C. will induce a movement of resources from the relatively less efficient to the relatively more efficient industries. In so far as it is within Basingstoke's power, then, it would be to her advantage to attract firms in those industries which enjoy a comparative cost advantage *vis-à-vis* the rest of the world and in particular Continental Europe, in preference to firms in industries which produce at a comparative cost disadvantage.

The difficulty arises in trying to determine which U.K. industries are most likely to prosper and expand and which industries are likely to encounter difficulties and contract, if the U.K. joins the E.E.C. Such research as has been conducted on this question has been designed mainly to study the performance of a range of industries in recent years in world markets where competitive conditions are more or less equal for all exporters. A rising British share suggests this country will benefit considerably from the opening up of new markets or the reduction of tariffs; a falling share suggests Britain might fare badly. From studies of this kind carried out by H. Liesner and the Economist Intelligence Unit,[1] the industries which appear most likely to gain from Britain entering the E.E.C. are tractors, cosmetics, plastics, medicinal and pharmaceutical products, agricultural machinery and implements, commercial vehicles, sports cars, electric domestic appliances, television and radio and electronic equipment. Those which are likely to lose (or at least gain the least) from Britain's entry (usually industries which up to now have been protected by a high tariff or import control) include watches and clocks, wood and cork manufactures, glass and glassware, furniture and fixtures, organic chemicals, leather manufactures, scientific instruments and footwear.

[1] H. H. Liesner, 'The European Common Market and British Industry,' *Economic Journal*, Vol. LXVIII, June 1958, pp. 302 ff. *Britain and Europe*, Economist Intelligence Unit, 1957.

These conclusions are, of course, only tentative: there may be special reasons to explain Britain's poor or good performance in a particular export market. Moreover, it by no means follows that all firms in an expanding industry are equally efficient or firms in a contracting industry equally inefficient. In this connection the particular authority dealing with applications of firms wishing to come to Basingstoke might do well to contemplate making (or arranging to have made) as close an examination as possible of their financial standing, managerial policies and recent production performances. The reputation of such firms as employers of labour and their attitude towards their civic responsibilities need also to be taken into account. Apart from cases where the applying firm is starting manufacturing for the first time, consideration should be given to visiting the existing factory premises and conducting as detailed a discussion as possible on the firm's future plans and prospects. In some instances it might prove useful to seek independent expert advice.

(b) Basingstoke's main locational advantage lies in its good geographical position midway between London and Southampton and its convenient situation for the Midlands and North. All firms at present located in the town appear satisfied with the road and rail facilities into and out of the town. Basingstoke is particularly fortunate in its proximity both to the London docks and the main South Coast ports. As the trade of both London and Southampton may be expected to expand considerably if Britain joins the E.E.C. – and, in fact, according to the Rochdale Committee report, this is likely to happen in any event[1] – Basingstoke stands in a good position to develop as an important centre supplying the new market. Labour need not be a serious problem, except for specialised grades, as the housing accommodation which is initially tied to jobs will be an important inducement for people to move to the town. Although most materials and parts have to be brought a distance from other areas of the country and the major sales outlets are in London and the Midlands, Basingstoke is nevertheless well sited as a transport-oriented centre in that the costs of procurement and distribution are no higher, and, in some cases are considerably lower, than for other towns in South or South-East England. As has already been mentioned, it is a feature of the lighter industries that they purchase from many sources and, though the market influence is an important one, high rents, service costs and

[1] *Major Ports of Great Britain*, Committee Report; 1961–62 Cmnd. 1824.

labour costs in the city centres are gradually pushing firms out to the suburbs.[1]

Surveys on comparative locational costs are few and far between. There is nothing written, for example, concerning the relative attractions of the individual expanding towns–or even of the new towns. Some four years ago, however, a qualitative survey was conducted on the manufacturing conditions in new towns, compared with those in the London area and the Development Areas. As the production facilities offered by Basingstoke are likely to be very similar to those currently prevailing in the new towns, it might be helpful to reproduce the main conclusions of the survey and summary table.[2]

'The following are some impressions and observations of cost and production conditions in the new towns, as related to those in the Development Areas and the London region. First, are there any factors which normally influence costs which in this particular case may be treated as constant? Broadly speaking, in each of the three areas, providing he can obtain the necessary permits, there is no reason why an industrialist should not build an identical factory and equip it exactly the same way. However, the cost of land and building operations will vary, being highest in London and lowest in the Development Areas. Alternatively, a firm might prefer to rent an existing or a newly-built factory, in which case his choice, in effect, is restricted to the Development Areas or the new towns. In our experience, the accommodation offered in both cases is first class, and, except perhaps for a little more willingness to cater for individual requirements by the Development Corporation, there is little to choose between the two locales. And, certainly, irrespective of his choice of location, there is nothing to stop the industrialist employing virtually the same production methods (the only variable factor being the marginal difference in the cost ratios of the productive services).

'The differences are more marked than the similarities. In favour of the Development Areas is the fact that factory rents are generally only about one-third of those in the new towns[3] and about one-quarter of those in the London area. From a sample of rates and public tariffs studied, these, too, would appear to be slightly cheaper in the Development Areas than in either of the other two regions.

'Labour costs are a function of both the wages paid and the productivity of the work. As regards the former, in the new towns these are, in practice,

[1] R. C. Estall and R. O. Buchanan, op. cit.

[2] 'Manufacturing Industry in the New Towns', by J. H. Dunning, *Manchester School of Economic and Social Studies*, May 1960. We are indebted to the editor of this journal for his permission to use material previously published there.

[3] In the new towns these average 5s. 6d. to 7s. sq. ft. according to location, size and type of factory; in the Development Areas the corresponding range is between 1s. 6d. and 2s. 6d. per sq. ft. For a 50,000 sq. ft. factory of comparable design and constructional quality the difference would be around £10/12,000 a year. In the London area, the range of rents is even wider—as a guide, between 7s. 6d. and 10s. 6d. for the type of accommodation in question.

comparable to London rates (and hence higher than in the Development Areas) partly because employees who move out from the capital expect to maintain their money incomes (and this means the newly established firms have to offer similar amounts), partly because living costs are generally as high, and in some cases higher, in the new towns. Unskilled labour is generally much less plentiful than in the Development Areas – particularly female labour, which is often at a premium in the new towns and can command more than the minimum union rates. In all three areas skilled labour is difficult to come by and variations in rates mainly reflect cost of living differences.

'Productivity is notoriously difficult to measure. Most of the main productive services, e.g. factory accommodation, capital equipment, entrepreneurship, etc., are not, in this instance, influenced by locational considerations. The main variables are, in fact, the quality of labour and its attitude to work. Here the important difference between conditions in the new towns and those in the Development Areas is that the average wage-earner in the former is usually far more experienced in his work, either because factories who remove from the London area bring most of their workers with them (the usual proportion is 75–80 per cent) or because skilled labour is not difficult to recruit owing to the attraction afforded by a new house. Productivity is thus higher in the new towns, at least for an initial period, while suitable workers are being trained in the Development Areas. Some firms in the new towns, however, have found that shift work is less readily accepted by their employees than previously, mainly, it seems, because of its domestic unpopularity now wives are without the company of their relatives. It is generally thought that the response of the labour to incentives is also more pronounced in these areas as early commitments of new town dwellers are usually very heavy with large sums of money being spent on purchasing furniture, etc., for the new homes (and high rents). This has lessened absenteeism to below the level previously experienced and that usually experienced in the Development Areas. Compared, too, with their previous location, firms are finding that the improved living conditions of their staff and the shorter distances they have to travel from home to work make for more sustained labour effort. The lower proportion of women at work in the new towns has usually meant a lower rate of absenteeism, except where relatives are far removed and mother is ill, in which case father has to stay away to look after the children.

'Labour turnover is influenced by many factors. Usually factories on industrial estates tend to find removals high. Some firms in the new towns have concluded 'gentlemen's agreements' not to poach on each other's labour. Nevertheless there is often quite a high turnover to begin with – particularly in the case of branch units and mainly in the lower grades of skill, where a worker has joined the firm primarily to get a house rather than because he is interested in the work. It is perhaps unfortunate that at one period when new industrial developments were moving particularly rapidly in the new towns, that a large firm adjacently sited – Vauxhall Ltd. of Luton – was demanding large quantities of labour (and paying 25 per cent above the average wage rates) and that there were insufficient houses to enable local recruitment. Employers in Stevenage and Hemel Hempstead were particularly embarrassed by this situation at the time. Where female labour is employed, turnover is particularly high for pregnancy reasons – one firm noting its turnover in this respect in 1957 was 22 per cent compared with the national average of 4 per cent. Nevertheless the record of the new towns, as a whole, is encouraging. Crawley, for example, reports its labour turnover as less than half the national average in 1957.

'What of administration costs? To begin with, industrialists setting up in the new towns have often had to incur a number of additional costs as a result of imperfect synchronisation between their demand for labour on the one hand and the provision of houses on the other. Houses are rarely completed in large batches, whereas a production programme is, to some extent at least, indivisible and a firm usually wishes to engage numbers of workers at a time. In spite of every effort made to stagger production, this has meant that personnel have often had to travel daily from London to the new towns or *vice versa* for a limited period of time – anything up to eighteen months – with the employer bearing the differential cost of fares. Then there is the question of industrial site layout. In these days, when it is the aim of the planner to make an industrial estate look anything but what it is, the costs (e.g. of camouflaging chimneys and ugly buildings, screening of yards, gardens round buildings, etc.) are more important than hitherto, and this is probably more so in the new towns than in the Development Areas. Furthermore, due to the comparative lack of social amenities in the new towns, some of the larger firms have had to provide additional canteens and recreational facilities for their workers, whereas the unusual age structure of the female labour force has increased the need for nursery and playground facilities for young children.

'Firms moving from London are naturally finding their communication costs considerably higher.[1] In most cases the telephone bill has been multiplied several times: what was previously a 3d. call (e.g. to suppliers or customers in the London area) is now 1s. 3d. or more; in addition, there is a certain loss of flexibility and speed in meeting urgent production requirements. At the same time, where possible, most factories in the new towns are deliberately sited as near to their original place of production or main works as possible since it appears that the extra communication costs associated with the operation of a branch increase substantially once this distance exceeds 30 miles, the new towns gain considerably on the Development Areas in this respect. Some new factories are better provided with transport facilities than hitherto; e.g. several firms in the Temple Fields industrial estate at Harlow have their own railway siding. Communication and managerial costs naturally vary according to the type of branch operated: whether, for example, it is a separate entity functioning independently of its parent unit employing its own executive staff or whether it is solely a production unit. The fact that most branches are manufacturing quite different products from their parent concerns means that a separate administration is often required. On the other hand, a number of firms

[1] The exception is where one factory has replaced a group of dispersed units in London.

113

who have transferred their manufacturing facilities *in toto* have found it convenient to retain a London sales office and/or service depot.

'Lastly, transport costs. Compared with their original location, transport costs inwards and outwards differ very little (e.g. it is cheaper to buy from and market to the Midlands and North than previously, but dearer to Central London); the main exception is the case of imported raw materials (e.g. timber, copper, etc.) which have to be brought from the London docks, and end products which are destined for the Metropolis or exported from there. Even here, however, the basic costs of loading and unloading are constant and as we have seen, there is not more than thirty miles in it either way.

'Here, also, the new towns gain on most of the Development Areas, except where these latter are satisfying a localised market. The differential cost is the greatest where branch plants are supplying the main works with components and parts, or where the markets for the products in question are in the South or Midlands.

'The following Table summarises these broad impressions. Thus it can be seen that while the prices of most factor services in the new towns are somewhat higher than those in the Development Areas, there are equally important savings in the form of higher labour productivity, and lower communication and transport costs. Compared with the original place of manufacturing (or location of the main works), the new town industrialists score slightly on their lower labour costs and rents, which after the initial period of 'settling-in', are sufficient to outweigh the higher transport and communication costs. The Development Areas, if anything, come off worse in these comparisons, with overall costs working out at 1-2 per cent of turnover above those in the London Area,[1] and similarly in excess of those in the new towns.'

In an earlier study of the industries which might be suitable for dispersion to rural areas,[2] Professor Sargant Florence listed several trades, e.g. electrical apparatus, cabinet and furniture making, printing and bookbinding, food, drink and tobacco, clothing and toys and games, etc., some of which are now well represented in the new towns. Florence used a negative criterion of locational desirability, in that, while recognising that rural areas have few positive advantages to offer new factories, there are those industries–designated as 'footloose'–whose production costs would seem to be little affected by their actual siting. These are in contrast to those trades whose location is more or less tied to their markets or raw materials, or those which

[1] For further details see D. C. Hague and P. Newman, *Costs in Alternative Locations: the Clothing Industry*, paper XV of N.I.E.S.R., C.U.P., 1952 and D.C. Hague and J. H. Dunning,'Costs in Alternative Locations: the Radio Industry,' *Review of Economic Studies*, 1954/5, XXII, No. 59.
[2] 'The Selection of Industries for Dispersal into Rural Areas', *Journal of the Royal Statistical Society*, 1944, Vol. CVII, Pt. II.

IMPRESSIONS OF COSTS IN ALTERNATIVE LOCATIONS

	(a) Costs in New Towns compared with London Area*	(b) Costs in Development Areas compared with London Area	(c) Costs in New Towns compared with Development Areas
(1) Labour costs			
(a) Price of labour			
(i) Unskilled	About the same or a little lower	Considerably lower	Higher
(ii) Skilled	About the same or a little lower	A little lower	Slightly higher
(b) Productivity of labour†			
(i) Technical efficiency	Probably very slightly higher due to reduced travelling fatigue	Considerably lower	Generally higher particularly in early years of production
(ii) Incentive to work	More pronounced particularly in early years of production	Varies, but probably a little less pronounced apart from Scottish Development Areas	Generally more pronounced
(iii) Labour turnover	Given similar circumstances, probably a little greater	Slightly higher	About the same
(iv) Absenteeism	Generally a little lower	Higher	Generally less
(2) Rents and building costs‡	Somewhat lower	Very much lower	Appreciably higher
(3) Service costs	Much the same	Somewhat lower	Slightly higher
(4) Administrative and communication costs			
(a) Telephone and postage	Considerably higher	Very much higher	Lower
(b) Extra travelling by executives, etc.	Higher	Very much higher	Considerably lower
(c) Other managerial costs	Slightly higher	Very much higher	Considerably lower
(5) Transport costs	Slightly higher	Considerably higher	Higher
(6) Total costs	Slightly lower	Slightly higher	Lower, particularly in the early years of production

*Assuming an extension of, or an adjacent factory to, existing works.

†Assuming a given plant layout and capital equipment.

‡Assuming present scales of subsidies, etc.

are linked, e.g. by way of manufacturing processes, with other industries or firms. In the case of the new towns, however, where there is a deliberate policy to evolve an industrial conurbation, it is important to take a long-term viewpoint, since certain economies of manufacturing will only become apparent once the estates are well on their way to completion. Present production costs are thus not always a good guide to ultimate production costs.

What of the impression of firms already producing in Basingstoke of the locational value of the town? All the companies engaged in manufacturing activities were contacted and it is on the basis of interviews and replies to questionnaires that we have formed the following impressions.

Compared with other possible locations for the industries in question, Basingstoke comes out well. Apart from its generally good access to suppliers and markets (though not all firms felt this), the advantages most cited by firms were the price of labour, and favourable service costs.

Advantages of Basingstoke as a location	No. of firms in 10 noting advantage	Disadvantages of Basingstoke as a location	No. of firms in 10 noting disadvantage
Good access to suppliers and markets	6	Shortage of skilled labour	5
Lower labour costs	6	Shortage of unskilled labour	2
Lower rent and building costs	4	Lack of planning by local authority	1
Lower service costs	2	Lower productivity of labour	2
Lower administrative costs	5	Higher transport costs	2
		Higher administrative costs	2

There was no unanimity about the possible disadvantages, save the general complaint of the shortage of special types of skilled labour. One or two of the firms found it difficult to get other kinds of labour, e.g. juvenile and seasonal, but surprisingly few firms mentioned any other problems specific to their location which they were experiencing.

Concerning the reasons originally chosen for coming to Basingstoke the answers might be summarised as follows:

Good road and rail centre	5
Good local markets	1
Board of Trade direction	1
Availability of labour	1
Availability of factory accommodation at right time	1
Amenities and personal reasons	4
Availability of water power	1

116

Only three of the firms operated factories in other parts of the country. Each thought Basingstoke compared favourably as a production centre.

The materials of firms were bought from all parts of the country but mainly in the Midlands and the South. Markets were mainly concentrated in the South. Some firms were important exporters using London docks.

TABLE V/8

DISTRIBUTION OF MANUFACTURING EMPLOYMENT
IN THE LONDON NEW TOWNS (1960)
AND SLOUGH INDUSTRIAL AREA (1961)

	New Town Employment Areas		Slough Employment Area	
	Nos. employed	% of total employment	Nos. employed	% of total employment
Food, drink and tobacco	3,727	3·7	5,037	13·0
Chemicals and allied trades	7,327	7·2	5,620	14·5
Metal manufacture	1,632	1·6	3,351	8·7
Engineering and electrical goods	35,157	34·9	12,377	32·0
Vehicles	22,728	22·5	3,125	8·1
Metal goods not elsewhere specified	4,514	4·5	1,982	5·1
Textiles, leather, clothing, etc.	3,203	3·2	2,256	5·8
Bricks, pottery, glass, etc.	3,371	3·3	568	1·5
Timber and furniture	4,060	4·0	312	0·8
Paper, printing and publishing	7,797	7·7	1,297	3·4
Other manufacturing	7,390	7·3	2,770	7·2
Total	100,906	100·0	38,695	100·0

Data kindly provided by the Ministry of Labour.

Of the total labour force employed about 85 per cent came from inside the Basingstoke Employment Area.

Finally, for comparative purposes, we illustrate the present-day pattern of industry in the London new towns and the Slough industrial area: either structure might well suit Basingstoke today, but forecasting fifteen to twenty years ahead is another matter. However, as a summary to the argument of this chapter we incorporate, as Appendix I, a list of manufacturing industries, taken from the Standard Industrial Classification, which potentially at least are capable of being sited at Basingstoke. In the columns beside each we have given a numerical assessment of 1 to 5 which indicates the likely contribution of such industries towards a particular type of industrial structure, not necessarily acceptable to Basingstoke, but for the most part so. For example, in Column 1, which deals with rate of growth

117

over the last decade, a figure of 1 would suggest that the industry in question has grown considerably faster than the national average (further details of what constitutes 'considerably faster' are given in an explanatory note to the Appendix). A figure 5 in the Column 5 would reveal that the ratio of juvenile to adult labour employed in the industry in question is considerably lower than the average for all industry; a figure 3 in Column 8 points to the industry in question having about the same space needs as industry in general. We have examined nine main indicators of industrial suitability, viz. rate of growth 1951–8, stability of employment 1951–9, possible success in E.E.C., productivity, ratio of female/male labour, ratio of juvenile/adult labour, density of workers per acre of factory space, ratio of non-operative/operative labour, and potential aptness of Basingstoke as a choice of location. This table thus provides a guide as to how far a particular industry fulfils the requirements of a particular conception of the most desired pattern of industry. Thus, if the aim is to attract industries which are expanding above the national average, are stable in their employment requirements, employ above the average ratio of female/male labour, around the average juvenile/ adult labour and above the average non-operative/operative labour, and for which Basingstoke may prove suitable as a location, then our table tells us that (among others) the radio and electronics industry would (on present indications) be an extremely good choice. If there is a particular need to provide employment for juvenile labour, then the paper, clothing, scientific instruments, toys, games and sports equipment trades would be ideal; if space economy is all important, then industries most likely to appeal would be the cigarette, office machinery, radio, furniture and hosiery.

More important, this table reveals that there is a reasonably close relationship between those industries likely to be attracted to Basingstoke and those which the town is likely to *want* to attract. As mentioned before, however, the requirements of Basingstoke will differ according to the particular period of the fifteen-year plan being considered and data given in respect of industries tells us nothing about the efficiency of individual firms.

With these reservations in mind, we cite the ten industries which, in our opinion, come nearest to meeting the requirements of Basingstoke for the years 1962–9, 1970–5, 1976–81 respectively. The choice has involved some compromise in that certain assumed aims (e.g. a good rate of growth) have been valued more highly than others. With a slight revaluation of the priority of objectives, the lists presented might be read quite differently.

Up to 1969	1970/5	1976/81
1. Agricultural machinery	Fruit and vegetable products	Tobacco
2. Office machinery	Pharmaceuticals and cosmetics	Pharmaceuticals and cosmetics
3. Scientific and photo-graphic instruments	Biscuits	Clothing
4. Domestic electrical appliances	Chocolate and sugar confectionery	Toys and games
5. Motor vehicles (and accessories)	Office machinery	Shop and office fittings
6. Printing and publishing	Radio and electronics	Scientific and photo-graphic instruments
7. Plastics moulding	Scientific and photo-graphic instruments	Made-up textiles
8. Radio and electronics	Cans and metal boxes	Furniture and upholstery
9. Pharmaceuticals and cosmetics	Clothing	Bread, flour confectionery
10. Mechanical handling equipment, bacon curing, meat and fish products	Stationers' goods	Woodworking

Finally, it should be repeated that many of the existing firms in Basingstoke have impressive expansion plans for the future. Of the fifteen firms giving information on this point eight noted they intended to expand their labour force. This will mean between 1,500 and 2,000 jobs will be created by 1964 and 2,500 by 1967–9. It is also understood that the Joint Development Committee has already allocated sites to new firms who plan to employ nearly 2,000 people.

We have not dealt with questions of administrative procedure and site allocation in this chapter: Appendix II summarises three of these– (1) the main steps of negotiation; (2) the provisions of the industrial selection scheme as it affects labour recruitment and (3) the proposed areas to be reserved for industry as scheduled in the Town Map.

APPENDIX I

GUIDE TO SUITABILITY OF INDIVIDUAL MANUFACTURING INDUSTRIES FOR LOCATION IN PARTICULAR GEOGRAPHICAL AREAS

(For definitions see explanatory notes at end of table)

Standard Industrial Classification	THE DEMAND SIDE (a) Rate of Growth 1951–8	(b) Stability of Employment 1951–9	(c) Productivity (net-output per head 1958)	(d) Possible Impact of E.E.C.	THE SUPPLY SIDE (a) Female/Male Employment	(b) Juvenile/Adult Employment	(c) Non-operative/operative Employment	(d) Space Requirements	Potential suitability of Basingstoke as a Location
1. Food, drink and tobacco									
(a) Grain milling	2	2	2	3	2	3	4	n.a.	3
(b) Bread, flour confectionery	3	2	1	No details for individual industries. Much depends on the particular agricultural agreements made between the U.K. and the Six.	4	4	2	4	3
(c) Biscuits	2	2	4		2	2	5	2	3
(d) Bacon curing, meat, fish products	1	2	3		1	3	4	2	3
(e) Milk products	1	2	2		2	4	2	3	3
(f) Sugar	5	1	3		3	4	4	4	2
(g) Cocoa, chocolate, sugar confectionery	1	3	3		4	3	4	4	3
(h) Fruit and vegetable products	2	1	3		1	4	3	4	2
(i) Animal and poultry foods	1	n.a.	1		4	4	1	4	3
(j) Food industries (other)	2	1	2		2	3	2	n.a.	2
(k) Brewing and malting	3	3	1		4	4	3	3	5
(l) Other drink industries	2	2	1		3	2	3	3	3
(m) Tobacco	4	1	1		1	2	3	2	3
2. Chemical and allied industries									
(a) Coke ovens, manufactured fuel	1	2	1	2	4	4	1	n.a.	3
(b) Lubricating oils, greases	1	2	2	n.a.	5	5	4	n.a.	2
(c) Chemicals and dyes	2	2	1	4	4	4	1	5	1
(d) Pharmaceutical, toilet preparations	2	2	1	3	1	5	1	3	4
(e) Explosives and fireworks	3	2	3	2	3	2	4	1	1
(f) Paint and printing ink	4	2	2	n.a.	3	4	1	3	5
(g) Vegetable and animal oils, fats, etc.	1	3	1	1	3	4	1	1	2
(h) Synthetic resins, plastics materials	1	n.a.	1	3	4	5	1	3	4
(i) Polishes, gelatine, adhesives, etc.	4	n.a.	2	2	3	5	1	n.a.	3
3. Metal manufacture									
(a) Iron and steel (general)	3	5	2	3	5	4	4		5
(b) Steel tubes	2	4	2	2	5	4	4		5
(c) Iron castings, etc.	2	5	2	4	5	4	3	4/5	5
(d) Light metals	5	n.a.	3	4	4	5	5		3
(e) Copper, brass, other base metals	4	3	3	3	4	4	3		4

120

Table (row categories with numeric ratings across columns; column headers not printed on page):

Category										
4. Engineering and electrical goods										
(a) Agricultural machinery (except tractors)	2	2	n.a.	n.a.	2	2	2	3	3	2
(b) Metal working machine tools	2	4	4	n.a.	2	3	3	3	3	3
(c) Engineers' small tools and gauges	3	3	4	3	3	3	3	5	3	3
(d) Industrial engines	2	4	5	n.a.	1	4	4	3	n.a.	4
(e) Textile machinery and accessories	4	4	n.a.	n.a.	1	4	3	3	3	5
(f) Contractor's plant, quarrying machinery	3	3	n.a.	4	1	4	3	3	3	3
(g) Mechanical handling equipment	1	3	n.a.	3	2	4	3	5	n.a.	3
(h) Office machinery	2	3	n.a.	3	2	5	3	4	3	1
(i) Other machinery	3	2	1	3	2	4	3	5	n.a.	3
(j) Industrial plant, steelwork	4	1	1	2	1	5	2	4	4	4
(k) Ordnance and small arms	3	1	n.a.	3	3	4	3	5	2	3
(l) Other mechanical engineering	2	n.a.	1	3	1	3	3	3	2	1
(m) Scientific, surgical, photographic instruments, etc.	4	2	4	3	3	4	2	1	5	2
(n) Watches and clocks	4	2	4	3	1	3	2	4	4	4
(o) Electrical machinery	2	2	4	3	1	4	3	3	2	2
(p) Insulated wires and cables	4	2	2	3	1	4	4	2	2	1
(q) Telegraph and telephone apparatus	1	4	3	3	1	4	4	2	4	2
(r) Radio, other electronic apparatus	1	2	{	3	1	5	3	3	1	1
(s) Domestic electric appliances	2	3	} 3	3	3	4	3	2	4	2
(t) Other electrical goods	2	3		3	3	3	3	3	3	2
5. Vehicles										
(a) Motor vehicle manufacturing	2	2	2	2	4	5	4	3	2	3
(b) Motor cycle, three-wheel vehicle, cycle manufacturing	n.a.	3	2	1	4	5	4	3	2	2
(c) Aircraft manufacturing and repairing	1	3	1	3	1	5	5	4	5	3
(d) Locomotives, railway track equipment	5	4	2	4	4	4	5	5	5	4
(e) Railway carriages, wagons, trams	4	3	3	4	5	4	4	5	4	4
(f) Perambulators, hand-trucks, etc.	5	1	1	4	4	1	5	5	4	3
6. Metal goods not elsewhere specified										
(a) Tools and implements	4	4	3	3	4	3	3	3	4	4
(b) Cutlery	4	5	3	3	4	3	2	3	5	5
(c) Bolts, nuts, screws, rivets, etc.	5	5	2	4	5	1	2	4	4	5
(d) Wire and wire manufactures	n.a.	5	3	4	5	2	2	4	4	5
(e) Cans and metal boxes	3	3	3	4	5	4	4	3	3	5
(f) Jewellery, plate, refining precious metals	n.a.	3	3	3	5	1	3	4	2	3
(g) Metal industries (other)	4	4	n.a.	3	3	2	2	4	3	2
7. Textiles	n.a.	3	3	3	3	3	3	3	3	4
(a) Production of man-made fibres	5	5	4	4	5	4	2	5	4	5

121

APPENDIX I continued

	THE DEMAND SIDE				THE SUPPLY SIDE				
	(a) Rate of Growth 1951–8	(b) Stability of Employment 1951–9	(c) Productivity (net-output per head 1958)	(d) Possible Impact of E.E.C.	(a) Female/Male Employment	(b) Juvenile/Adult Employment	(c) Non-operative/operative Employment	(d) Space Requirements	Potential suitability of Basingstoke as a Location
(b) Spinning, etc., of cotton, flax, man-made fibres	n.a.	5	5	4	1	3	5	5	5
(c) Weaving of cotton, linen, man-made fibres	n.a.	5	4	4	1	3	5	5	5
(d) Woollen and worsted	5	4	4	1	1	3	5	5	5
(e) Jute	4	4	4	3	2	4	5	4	5
(f) Rope, twine and net	5	3	4	n.a.	1	2	5	5	5
(g) Hosiery, other knitted goods	4	5	4	1	1	1	5	1	5
(h) Lace	5	5	4	n.a.	1	3	4	3	5
(i) Carpets	5	4	3	n.a.	1	2	4	n.a.	5
(j) Narrow fabrics	5	5	4	} 4	2	1	4	n.a.	3
(k) Made-up textiles	3	2	4	}	4	3	4	n.a.	3
(l) Textile finishing	5	5	4	}	3	3	4	3	3
(m) Other textile industries	4	4	3	n.a.	3	3	2	3	3
8. Leather, leather goods and fur									
(a) Leather (tanning and dressing), fellmongery	5	4	4	3	2	2	4	4	3
(b) Leather goods	4	4	3	2	4	3	4	4	4
(c) Fur	5	3	4	5	1	1	4	4	2
9. Clothing and footwear									
(a) Weatherproof outerwear	4	2	5	} 2	1	1	5	n.a.	3
(b) Men's and boys' tailored outerwear	4	} n.a.	5	}	1	1	5	1	4
(c) Women's and girls' tailored outerwear	4	}	4	}	1	1	5	1	1
(d) Overalls, men's shirts, underwear, etc.	4	1	5	2	1	1	5	4	3
(e) Dresses, lingerie, infants' wear, etc.	4	4	5	} 4	1	1	5	4	2
(f) Hats, caps, millinery	5	3	4	}	1	4	5	n.a.	1
(g) Dress industries (other)	4	5	4	}	1	1	4	n.a.	2
(h) Footwear	4	3	4	4	1	1	5	1	4
10. Bricks, pottery, glass, cement, etc.									
(a) Bricks, fireclay, refractory goods	4	4	3	3	4	3	4	4	3
(b) Pottery	5	3	3	3	5	3	5	4	4
(c) Glass	3	5	3	4	1	3	4	3	5
(d) Cement	3	3	1	4	4	4	3	5	4
(e) Abrasives, building materials, etc. (other)	3	2	3	n.a.	5	4	3	4	3

122

11. Timber, furniture, etc.									
(a) Timber	4	3	4	4	4	2	4	n.a.	2
(b) Furniture and upholstery	4	2	3	n.a.	5	1	4	5	2
(c) Bedding, etc.	4	4	4	4	4	2	4	1	1
(d) Shop and office fittings	2	4	4	n.a.	2	3	4	3	3
(e) Wooden containers, baskets	5	1	3	n.a.	5	3	5	1	2
(f) Miscellaneous wood and cork manufactures	n.a.	3	4	5	4	1	4	2	2
12. Paper, printing and publishing									
(a) Paper and board	4	3	3	4	3	2	3	3	2
(b) Carboard boxes, cartons, etc.	3	3	2	5	4	2	4	4	2
(c) Manufactures of paper and board (other)	4	4	3	4	1	1	4	2	3
(d) Printing, publishing of newspapers, periodicals	2	1	2	n.a.	4	4	1	3	1
(e) Other printing, publishing, etc.	3	3	3	n.a.	2	1	3	3	2
13. Other manufacturing industries									
(a) Rubber	3	3	3	n.a.	2	3	3	n.a.	3
(b) Linoleum, leather cloth, etc.	4	3	3	1	3	4	3	2	3
(c) Brushes and brooms	4	4	3	n.a.	4	4	2	n.a.	5
(d) Toys, games, sports equipment	3	2	4	5	1	2	4	n.a.	2
(e) Miscellaneous stationers' goods	3	3	4	n.a.	1	1	2	n.a.	2
(f) Plastics moulding and fabricating	1	3	3	1	2	3	4	n.a.	1
(g) Miscellaneous manufacturing industries	3	2	3	n.a.	2	2	4	n.a.	2
All manufacturing industries	3	3	3	3	3	3	3	3	3

n.a. = not available

123

EXPLANATORY NOTE TO APPENDIX 1

1. *The demand side*

   (a) *Rate of growth 1951/58.* We have adopted the composite of two measures (i) the change in net output, i.e. value added, and (ii) the change in labour force, both data derived from Censuses of Production 1951 and 1958. The average increase in net output for all manufacturing industry over this period was 54 per cent and the average increase in labour force was 4·8 per cent. Where output rose by 90 per cent and the labour force rose by 15 per cent we allocated a figure of 1; similarly an increase of output of 65–89 per cent and employment of 7·5–14·9 per cent was designated 2; 45–64 per cent and 2·5–7·4 per cent, 3, 20–44 per cent and –5·0– + 2·4 per cent 4, and under 20 per cent and under 5 per cent, 5.

   The two sets of figures (1 to 5) were then averaged and the resulting index listed in Column 1.

   (b) *Stability in Employment 1951/59.* Source: *Ministry of Labour.* End May 1951 was representative of a period of high prosperity in the U.K.: end May 1953 and 1959 were months of recession. (Figures are only available for the end of May each year.) The proportional change in the *rate* of unemployment in each industry was calculated as between 1951 and 1953 and 1951 and 1959. The result was averaged. For all manufacturing industry the average increase in unemployment rose by 130 per cent. A figure 1 indicates an increase of unemployment of under 85 per cent; 2, 85–119 per cent, 3, 120–144 per cent; 4, 145–174 per cent and 5, 175 per cent and over.

   (c) *Labour productivity or net output per head.* Net output (= value added) divided by total numbers employed. Source: 1958, *Census of Production, Summary Tables.* The average net output per head for all manufacturing industry was £1,009. A figure 1 denotes a value of £1,600 or above; 2, £1,150–1,599; 3, £850–1,149; 4, £600–849 and 5, below £600.

   (d) *Possible impact of E.E.C.* This represents the assessment of H. Liesner and the E.I.U. which is constructed on the basis of Britain's performance in world export markets and estimated trends in Europe's production facilities and world demand, (see pp. 108 to 110 and footnote 1 on p. 109). A figure 1 denotes an estimated gain (i.e. in terms of production and employment) if Britain joins the E.E.C.; 2, a possible gain; 3, a negligible impact; 4, a possible loss and 5, a definite loss.

2. *The supply side*

   (a) *Female/Male Employment.* As at end May 1961. Source: *Ministry of Labour Gazette,* February 1962. For all manufacturing industry there are 48·3 female workers to every 100 male workers. A figure 1 denotes a ratio of 100 or more; 2, 60–99; 3, 40–59; 4, 20–39 and 5, under 20.

(b) *Juvenile/total employment.* As at end May 1961. Source: *Ministry of Labour Gazette*, February 1962. For all manufacturing industry the proportion of juvenile to all workers employed was 6·95 per cent. A figure 1 denotes a proportion of 30 or more; 2, 25–29; 3, 20–24; 4, 15–19 and 5, under 15.

(c) *Space requirements.* Data here derived from Table 7, p. 107 of G. Logie's *Industry in Towns.* A figure 1 denotes a floor space[1] of under 150 sq. ft. per worker is required; 2, 150–249 sq. ft.; 3, 250–449 sq. ft.; 4, 450–699 sq. ft.; 5, 700 sq. ft. or over. In certain instances where data were not available in this form, an indication was derived from similar data in respect of workers employed per acre of site.

3. *Potential suitability of Basingstoke as a choice of location*

As an approximate guide to this we have taken the proportion of insured workers employed in various industries in the London and South Eastern, and Eastern and Southern regions as a proportion of workers employed in all areas in Great Britain. Source: *Ministry of Labour Gazette.* The combined share of these regions of all workers employed in manufacturing industry in 1961 was 31 per cent. A figure 1 denotes a proportion of all employees of 55 and above (this indicating a well above average concentration of industry in the areas concerned); 2, 40–54 per cent; 3, 25–39 per cent; 4, 15–24 per cent; and 5, 14 or under.

APPENDIX II

(a) *Industrial development procedure*

Both the Basingstoke Borough Council and private owners have land which is available for the establishment of factories.

1. Firms wishing to move from the London area first approach the London County Council which has an information centre designed to provide industrialists with information about sites and facilities in towns such as Basingstoke.

2. If a firm is interested in coming to Basingstoke, it sends its representatives to the town. They, accompanied by London County Council officers, have discussions with the Town Clerk and visit the sites available.

3. If, as a result of the visit to the town, the firm is interested in making a firm offer for a site, it asks for a form of application. The completed form is submitted to a committee of the Borough Council, which either recommends or rejects the application.

4. The firm next approaches the Board of Trade (for an Industrial Development Certificate).

5. It then draws up plans and submits them to the Planning Authority.

[1] Defined as 'space used for manufacturing, packing, warehousing and office, laboratories, drawing offices, etc., where these are ancillary to manufacture'. Logie, op. cit., p. 107.

6. Under present arrangements, on the Borough Council's industrial estate, the firm builds its own factory.

7. Under this procedure, the intention is that there shall be co-ordinated construction of factories and houses.

N.B.

(1) Private owners of land scheduled for industry can enter into negotiations with any firm which wishes to establish itself on a new site. A number of firms have already set up factories on the private industrial estate situated to the south west of the town

(2) Firms may come from places other than London, but only if they are so authorised by the Board of Trade.

(3) The Borough Council may enter into an agreement with another 'overspill' authority – e.g. Birmingham – but only with the consent of the Minister of Housing and Local Government.

(b) Industrial selection scheme

In the new and expanded towns, houses and factories or other work places are being built so far as possible *pari passu*. To assist such synchronisation and the dispersal of people in the London region most in need of accommodation, the Ministry of Housing and Local Government has arranged with the Ministry of Labour a scheme by which workers required by firms moving to new and expanding towns, or already established there, will be recruited, so far as practicable from lists of persons prepared by a group of London and Greater London boroughs and urban districts. These 'exporting authorities' so called are linked with authorities in a number of reception areas of which Basingstoke is now one, and from the names of those seeking a move to the area (or more generally to one of the new or expanded towns) the Ministry of Labour makes known to industrialists and other employers of labour in the reception area those available for work in different occupational groups. The official procedure of the scheme is given below as recommended by the Ministry of Housing and Local Government.[1]

Registers

Exporting authorities should get into touch with those people in their area who have an urgent housing need, including families occupying requisitioned properties, to find out which of them are willing to move to a new or expanded town if work is available there. They should also, at their own discretion, approach council tenants in the same way, since a council tenant who moves out will give up accommodation which can be used to relieve urgent housing needs. Those who are willing to move out should state what type of work they could take. From the replies, the authority will be able to compile an up-to-date register showing both the skills and the housing need of people willing to move. Employment Exchange managers will give any guidance they can to authorities in compiling and maintaining their registers, and the authority's responsible officer should

[1] Quoted from Circular No. 29/53, 'Movement of Population to New and Expanded Towns'.

discuss the matter with the local Exchange manager at the outset. When replies come in Exchange managers will be prepared to examine them as to the adequacy of the industrial information given, and to classify them by occupation and classification number in accordance with the Ministry of Labour's system. This will facilitate the subsequent selection procedure.

Selection of workers

Firms requiring workers, whether they are firms already established in the receiving area or are about to go there, will notify their requirements in the usual way to the Ministry of Labour, and the Ministry will then approach local authorities for names from their registers. Registration will not, of course, ensure any automatic right to selection, but other things being equal, preference will be given to people on the register.

Allocation of houses

Anyone selected for employment by this means will be nominated by the employer engaging him to the development corporation or the receiving authority for a tenancy of one of the houses being made available under the agreed programmes. The exporting authorities will understand that no promise can be given to any applicant of an early offer of a house in a receiving area. The rate at which accommodation will become available will depend on many factors, including the proportion of existing employees who move out with their firms, to whom housing priority must be given, and the types of employment which firms moving into the receiving areas are able to offer. Nevertheless it should be remembered that an opening may occur at any time for a particular class of worker, and if authorities want to get the maximum relief out of this arrangement they should keep always available a pool of applicants, classified according to the kinds of work they can do, on which the Ministry of Labour can draw without delay. People interested in moving to new or expanded towns will want to have some information about rents and conditions generally: development corporations and (later) authorities of expanded towns will no doubt give exporting authorities as much information as they can from time to time, when asked.

(c) Site allocation for industry[1]

The approved Town Map provided 150 acres of additional land for industrial development. This comprised two areas, one of 112 acres in the Houndmills area north of the Waterloo-Southampton railway, and the other of 28 acres on the north side of Road A.30.[2] At Houndmills, 63 acres have been developed and the remaining 49 acres are committed to various industrial concerns. North of road A.30 11 acres have been developed, 3 acres are in the course of development, leaving 14 acres available.

[1] Quoted from Basingstoke Town Map Amendment Map, 1961, Report of Survey and Conclusions, Part I, Hampshire County Council, 1962, p. 8.
[2] Eleven acres of additional industrial land were allocated on the approved Town Map adjoining Thornycrofts' works. This is now increased by a further 21 acres which will include a playing-field area.

As part of the town development scheme, new industry (intended primarily for the relocation of industry from London) will be located in Basingstoke in the following principal areas:

56 acres north of Houndmills

8 acres in the south west corner of Houndmills

96 acres to the south of road A.30 and north of the new motorway

51 acres south of the railway and north of the new central area spine road and to the east of the present Town Yard (primarily for warehousing and Central Area service industry)

53 acres lying between Reading road and Reading railway to the north of the diverted road A.339

86 acres east of the Reading railway line and north of the Waterloo-Southampton railway. This site is at present occupied by the existing sewage farm and will not be available for development until about 1970, and the completion of the eastern principal town road.

These areas will be owned by the Borough Council.

In addition, it is proposed to locate some 10 acres of industrial land in sites of up to 2 or 3 acres each adjacent to residential areas and in the Central Area. Five acres of land allocated for industry at Worting are at present being used for this purpose.

The total industrial land thus allocated can provide accommodation for all new manufacturing industry and certain categories of new service employment (e.g. maintenance, storage and service depots) required for the increased population.

The amount of land provided for industry includes:

(a) individual factory sites including buildings, roads, car parking, storage and room for expansion;

(b) sufficient land for roads capable of eventual adaptation to carry estimated future car, commercial and industrial vehicle traffic;

(c) open space for tree planting and recreational areas for the use of factory employees. Tree belts will be planted between industry and other adjoining uses where required.

128

THE SERVICE TRADES AND RETAIL DISTRIBUTION

ON the basis of the population projections made in Chapter IV and the structure of economic activity postulated in Chapter V, the numbers employed in the various service industries in the Basingstoke Town Map Area are likely to rise from a figure of 9,571 in 1961, to between 25,505 and 26,665 by 1981.[1] Estimates for the intermediate years are given in Table VI/1. The figures in brackets indicate the proportion of the town labour force estimated to be employed in tertiary industry.

TABLE VI/1

BASINGSTOKE: ESTIMATED EMPLOYMENT
IN SERVICE TRADES 1966/1981
Population Projection

	A	B	C	D	E
1966 (49%)	9,182	9,545	9,545	9,916	10,274
1971 (50%)	12,956	13,697	14,056	14,806	15,534
1976 (51%)	19,104	19,933	20,334	21,917	22,733
1981 (52%)	25,677	26,227	25,505	25,733	26,665

As previously indicated, the accuracy of these projections will depend upon the extent to which:

(*a*) automation or other labour replacing production methods are introduced in manufacturing industry relative to the service trades;

(*b*) offices and other 'exporting' services (i.e. substitute, as opposed to complementary, employment for secondary industry) can be attracted to Basingstoke;

(*c*) increases in consumers' real income are spent on services rather than goods;

(*d*) innovations in the service industries (and particularly retail distribution) are labour saving or labour creating.

[1] See also Table V/7. We have assumed that 2 per cent of the labour force will be engaged in agriculture throughout the projections.

None of these developments can be easily forecast and, once again, it is essential that flexibility should be 'built in' to any overall plan for labour deployment. Unforeseen technological or economic changes might well swing the proportion of those engaged in the service trades in the opposite direction to that thought likely at present.

Similar difficulties are involved in trying to forecast the distribution of the labour force *within* the various service trades. Once again, some indication can be gained by studying the pattern of employment in towns of similar size and economic structure to that envisaged for Basingstoke by 1981, but the inherent difficulty still remains that, at best, one is only able to compare a pattern of tertiary industry which was essentially determined by events of many years ago (even though, subsequently, marginal adjustments to that structure may have been made) with one which is intended to suit a town's need in fifteen to twenty years' time. Even assuming it was possible to know what was the 'ideal' balance of service industries for a particular town, it is highly improbable that it is the one which the town actually has. For various reasons the forces of supply cannot always fully adjust themselves to changes in demand. Hence, the actual distribution of employment among the service industries in Reading in 1962, for example, may be quite different from that which might arise if the town were to create an economic structure for a population of 120,000 *ab initio* today – and even this may be unsuitable for the situation in twenty years' time.

Indeed, it may be true to say that there are as important developments now taking place within the service trades as there are within manufacturing. Let us first compare the pattern of employment in the Basingstoke Employment Area in 1952 and 1960 with that in the U.K. for those years. Table VI/2 illustrates these figures. The final two columns indicate the *rate* of growth in these trades over the period. From this table it can be seen that both from a national and a local viewpoint, employment in the distributive trades and professional services has risen faster than in the other groups over the last eight years (in Basingstoke these industries employed 40 per cent of the labour force in the tertiary industries in 1960 compared with 32 per cent in 1952), while the public utility, transport and public administration services have lost ground. In Basingstoke the numbers employed in the building and contracting industry have risen while those in the country as a whole have fallen slightly; the reverse has occurred in the insurance, banking and finance sector.

TABLE VI/2

BASINGSTOKE AND UNITED KINGDOM:
NUMBERS EMPLOYED IN THE MAIN SERVICE TRADES*
1952/1960

	1952				1960				1960 1952 = 100	
	Basingstoke	%	U.K. (000)	%	Basingstoke	%	U.K. (000)	%	Basingstoke	U.K.
Building and contracting	1,191	15·9	1,359	12·6	1,449	15·5	1,505	12·2	122	90
Gas, electricity and water	213	2·8	381	3·5	233	2·5	381	3·1	110	100
Transport and communications	885	11·8	1,761	16·3	759	8·1	1,667	13·5	85	94
Distributive trades	1,198	16·0	2,218	20·6	1,813	19·4	2,870	23·3	150	130
Insurance, banking and finance	272	3·6	448	4·2	221	2·4	550	4·5	81	122
Professional and scientific services	1,169	15·6	1,540	14·3	2,111	22·5	2,025	16·4	180	132
Public administration	1,264	16·9	1,390	12·9	1,315	14·0	1,294	10·5	104	93
Miscellaneous services	1,301	17·4	1,677	15·6	1,460	15·6	2,036	16·5	112	121
Total	7,493	100·0	10,774	100·0	9,361	100·0	12,328	100·0	125	114

*1952 figures based on 1948 Standard Industrial Classification: 1960 figures on the 1958 classification. These classifications are broadly the same, the main difference being that in 1960 motor repairers and garages are included in the distributive trades, but were not in 1952.

If we look at those same towns whose economic structure we took as the basis for estimating the distribution of employment between the main occupational groups in Basingstoke,[1] but widen the Census classification to include the employment area of those towns and update the statistics to 1960, we obtain the following pattern of tertiary activity.[2] We record data for the new towns separately.

TABLE VI/3

DISTRIBUTION OF NUMBERS EMPLOYED IN SERVICE
TRADES IN THIRTY-TWO SELECTED TOWNS, LONDON
NEW TOWNS AND BASINGSTOKE, 1960

	Selected Towns %	New Towns %	Basingstoke %
Building and contracting	13·5	18·7	15·5
Public utilities	3·8	2·6	2·5
Transport and communications	13·7	7·6	8·1
Distributive trades	22·9	19·9	19·4
Insurance, banking and finance	2·9	2·5	2·4
Professional and scientific services	19·1	25·0	22·5
Public administration	8·8	6·3	14·0
Miscellaneous services	15·3	17·4	15·6
	100·0	100·0	100·0

What may we conclude from these figures? Apart from an unusually high proportion of Basingstoke workers employed in public administration and a low proportion in the transport industry, the structure of service industry in Basingstoke and the selected towns is broadly comparable. It is also very similar to that in the new towns save in the latter, for obvious reasons, the building and contracting industry assumes a greater significance, while the numbers employed in public administration are relatively much smaller. The high proportion of workers engaged in the professional and scientific services reflects some of the success which the Development Corporations have had in attracting Government research and scientific establishments in recent years.

Using the above data, we have estimated the possible distribution of Basingstoke labour force in the service trades for 1981, on the basis of (a) an average of our population projections and (b) for three separate hypotheses:

[1] See p. 88.
[2] Although it does not necessarily follow that towns which have much the same structure of economic activity as between manufacturing and service undertakings should have a similar distribution *within* these groups, the coefficient of correlation between the numbers employed in the tertiary industry and the individual service trades is very close.

(1) assuming a distribution of labour as in the 32 selected towns in 1960;

(2) assuming a distribution as in the London new towns in 1960;

(3) assuming a distribution which is a synthesis of (1) and (2) adjusted to allow for

    (i) Basingstoke's existing pattern of employment,

    (ii) national trends within the service trades 1952–60 (as shown in Table VI/2), and

    (iii) related trends in other advanced countries.

RETAIL DISTRIBUTION

The conclusions of this chapter have so far been highly speculative. An alternative approach is to try to estimate the numbers likely to be employed in the main tertiary groups and assume the balance as a residual.

Of the service industries, the distributive trades occupy a pre-eminent position—not only as a complement to other forms of economic activity but as 're-export' services in their own right. Nationally, about one-half the output of manufacturing industry is bought by the public through wholesale or retail outlets which, in 1961, gave employment to 23·3 per cent of those occupied in the service industries; in Basingstoke the corresponding proportion was 19·4 per cent. According to the projection in Table VI/4 the numbers likely to be employed by 1981 will rise to 5,323. However, since the adequacy of shopping facilities must, in the last resort, be judged on their ability to satisfy consumer needs rather than (as in manufacturing industry)

TABLE VI/4

BASINGSTOKE: ALTERNATIVE PATTERNS OF TERTIARY INDUSTRY, 1981
(Estimated numbers employed)

	1960	1981 (1) Selected Towns	(2) New Towns	(3) Author's estimate	%*
Building and contracting	1,449	3,505	4,855	3,634	14·0
Public utilities	233	987	675	649	2·5
Transport and communications	759	3,557	1,973	2,336	9·0
Distributive trades	1,813	5,944	5,166	5,323	20·5
Insurance, banking and finance	221	753	649	1,038	4·0
Professional and scientific services	2,111	4,957	6,489	6,360	24·5
Public administration	1,315	2,285	1,636	2,596	10·0
Miscellaneous services	1,460	3,972	4,517	4,024	15·5
	9,361	25,960	25,960	25,960	100·0

*To nearest 0·5%.

as a means of employment for the local labour force, it is necessary to adopt other criteria of estimated requirements.

Now, until recently, studies of shopping needs have usually based their calculations of the number of retail outlets required in a particular town or region on the size of the population to be served. This was partly because of the unavailability of more reliable data, e.g. consumers' expenditure, classified by different types of shops, from which estimates might be formulated, and partly because the shop or establishment (like a dwelling unit) is a convenient planning unit. Thus the Reith Committee recommended that provision should be made in the new towns for one shop per 100–150 persons[1]–a calculation based on the belief that the number of shops in established towns of one to 40–50 persons was too high–while the corresponding ratio in the newer housing estates of one shop to 150–325 persons was too low. Both Birmingham and the London County Council work on this recommended standard. For residential neighbourhoods, an average of one shop to every 200 persons was recommended by the authors of the Middlesborough Survey and Plan and one shop to between 208 and 277 persons by the Retailers' Advisory Committee on Town Planning. In its provision for shops in redevelopment areas Coventry is allowing one shop per 250 persons in neighbourhood centres and one shop per 250 persons in 'scattered' shopping areas.[2]

This method of estimating shopping requirements is, however, being increasingly questioned. Quite apart from the fact that global shopping requirements cannot be divorced from the structure of retail facilities which itself is dependent on the size and character of the town in question and its proximity to other towns–and that population size and consumers' spending are not always closely correlated, much of the usefulness of this particular criterion of measuring shopping needs has disappeared as the heterogenity of retail outlets–in terms of size, composition and efficiency–has increased over the years. Thus a shopping unit or selling point[3] may mean anything from a back-street tobacconist to a centrally situated multiple self-service store; from a boot and shoe repairer to an auto-

---

[1] *New Towns* Committee, Final Report, para. 136; 1945–6 Cmd. 6876, xiv.
[2] Quoted in, *Cumbernauld: Report on the Central Area*, 1960, p. 71.
[3] The difference between selling point and shopping unit is often an important one. Thus in the Census of Distribution for 1950 the unit for enumeration was the trading premises or single establishment. Establishments with a number of separate activities are treated as one unit; thus no attempt is made to calculate the number of selling points which obviously is very much greater than the number of shops and increases as the structure of retailing becomes more multi-product in character.

matic vending machine. Even the optimum or most efficient size of a particular type of shop is difficult to calculate, for shops supply not only different kinds of goods but different kinds of services. The 'best' size of a high-class bespoke tailor may be quite different from that of a multiple clothing outlet; that of a specialist bookshop quite different from the run-of-the-mill bookseller-cum-stationer. The type of hardware or confectionery outlet most acceptable to a neighbourhood shopping area may be quite unsuited to the town centre.

In general, recent trends in retailing and, in particular, the growth of branding, standardization and pre-packaging of goods have tended to make for larger shopping units. Between 1950 and 1961, for example, while the number of retail establishments in Great Britain fell slightly, the (real) sales turnover rose by 36 per cent and the numbers employed by 5·3 per cent per establishment.[1] The increase in sales per shop was particularly noticeable in the grocery trade, where the introduction of self-service has been most in evidence. It is estimated that, for all trades, there were more than 7,300 self-service stores in 1960–twelve times the number in 1950: of these, some 594 were supermarkets in the food trade–each with a sales area of 2,000 sq. ft. or more, and accounting in mid-1961 for about 4 per cent of total food shops. It has been further calculated that to be profitable a supermarket must obtain a turnover equivalent to that of sixty grocer's shops.[2]

With the change in the optimum size of existing outlets, and as new shopping techniques and town layouts are evolved, it is clearly hazardous to predict future requirements from any existing ratio of shops to population. However, if we reject the size of population as a demand indicator for shopping needs and the number of shops as a supply indicator, what should take their place? Now, it is clear that the immediate and principal factor influencing the demand for shopping facilities is consumers' expenditure or, more specifically, the total outlay by consumers on various types of goods and services bought at retail outlets per period of time. From the supply side, whereas it has been found that there need be no (or little) correlation between consumers' expenditure on retail goods and the *number* of shops, there would appear to be a fairly close relationship between

[1] Census of Distribution and Other Services, 1957, H.M.S.O., 1959. First results of the 1961 Census of Distribution. *Board of Trade Journal*, 8.2.1963.

[2] *The Observer*, 21.1.62, quoted by D. R. Diamond and E. B. Gibb, 'Development of new Shopping centres: area estimates', *Scottish Journal of Political Economy*, Vol. IX, p. 132. See also W. G. McLelland, 'Economics of the Supermarket', *Economic Journal*, Vol. LXXII, No. 285, March 1962.

the former variable and the *area* of shopping space required. We shall return to this point later in our discussion but, accepting for the moment that variations in shopping space do reasonably measure variations in the supply of retail services, and that it is desired to translate the shopping needs of Basingstoke's population in the future into terms which can be used as a basis for planning, four questions need to be answered:

1. What area of land in the town needs to be set aside for shops in 1981?
2. How is this area to be distributed between different kinds of shops?
3. What policy is to be adopted as regards the competitive structure of shops? and
4. How is this area to be distributed geographically in the town, e.g. as between central and neighbourhood shopping centres?

We shall attempt to deal with each of these questions in turn, some more cursorily than others, as there is only a limited amount which the economist *qua* economist can say.

First we return to the demand side of the shopping equation, i.e. consumers' expenditure.

RETAIL SALES AND POPULATION

Any estimate of the likely total consumers' expenditure on shopping provisions in Basingstoke in fifteen to twenty years' time involves separate calculations of the two main elements of such expenditure: (i) size of shopping population and (ii) expenditure per head. As regards the first variable, although the population of the Town Map area is scheduled to rise to 86,000 by 1981 (93,000-94,000 on our calculation), if one extends Basingstoke's catchment area to include the rest of the population in the rural district, the total figure rises to between 110,000 and 120,000. Naturally, the effective hinterland for retail trade is likely to be larger for certain types of shopping than others, while its exact size is difficult to calculate due to the proximity of two larger shopping centres – Reading to the north and Southampton to the south. Assuming, however, that Basingstoke will gain from these towns as many customers as it loses,[1] and making certain allowances for shopping done in the rural districts outside the town, it might be

---

· [1] Since, on the one hand, there are certain facilities of retail distribution, e.g. certain types of specialist shops, which improve as a town enlarges in size and, on the other, the ultra-modern features planned for Basingstoke's central shopping precinct, will undoubtedly prove a major attraction to shoppers at present drawn to Southampton (and to a lesser extent Reading).

fair to assume that shopping facilities in the Town Map area will be required for a population of at least 120,000–if not by the late 1970s then by the middle of the following decade.

The question next arises–how far can retail expenditure be related to the size of town in terms of population? Such studies as have been conducted on this subject, based on the data provided by the Censuses of Distribution 1950 and 1957, suggest that a close positive correlation exists between the total volume of retail sales in a particular area and the population of towns (plus their hinterlands) in that area;[1] this means that the average retail sales per head in any geographical region, multiplied by the population of a particular town, is likely to give a reasonably accurate picture of the likely expenditure on all goods sold by retail outlets in that town. The coefficient of correlation between population and expenditure on *particular types* of retail goods appears to be more erratic: in some cases it works out very high–in others quite low.[2] It also varies according to the coverage of the analysis; thus, other things being equal, a pattern of retail expenditure is most likely to be the same where towns of similar economic character and geographical locality are being compared. Cumbernauld, for example, is a new town situated only twenty miles from the city of Glasgow and is scheduled to grow to a population of 70,000. In order to obtain some idea of the likely retail expenditure of the town when complete, correlation coefficients between population and expenditure were obtained for 100 towns in Great Britain of similar size and geographical situation. For all retail trade the coefficient worked out at 0·84 per cent; for individual outlets, as Table VI/5 shows, it varied from 0·92 per cent in the case of chemists to 0·42 per cent for booksellers. Since however, the first five trades–in which the coefficient is seen to be very high, currently account for three-quarters of national retail sales, it might be rewarding to apply this type of exercise to our present case study.

[1] Two of these studies may be briefly noted. The first is that of J. B. Fleming in an article 'An analysis of shops and service trades in Scottish towns' published in the *Scottish Geographical Magazine*, December 1954. He correlated the population of all Scottish towns plus their hinterland against sales and obtained a coefficient of 0·98. The second study is that of the Cumbernauld Development Corporation which undertook a similar exercise in respect of 100 towns whose populations were between 25,000 and 100,000 and situated within a 20 mile radius of one of the nine largest cities in Great Britain. The coefficient of correlation between retail sales and population in this instance worked out at 0·84.

[2] For example, taking three towns of equal size – Reading, Oxford and Luton – the annual amount spent in grocery shops in 1950 was approximately the same, viz. £2,716, £3,243 and £2,785,000 respectively; on the other hand, sales of booksellers varied a great deal–£241, £295 and £127,000.

For Basingstoke the situation is rather different in that the town is not situated within easy reach of any shopping centre substantially larger than its own projected size twenty years hence, and is thus likely to be much more self-contained in its shopping habits than Cumbernauld. No doubt, as we have said, Southampton and Reading will continue to exercise some pull, particularly for certain types of durable goods,[1] but this is likely to become less important as the central area is developed and the (effective) catchment area of Basingstoke is enlarged. Thus, *ceteris paribus*, one would expect a rather larger proportion of the population in outlying areas to shop in the town than previously. Only empirical investigation would reveal (and this only approximately) the likely proportion of shopping expenditure in Basingstoke undertaken by people who live at different distances from the town. For this study however, we must rely on published information of the Census of Distribution, updated by statistics obtained from the National Income and Expenditure 'Blue Book'.

TABLE VI/5

CORRELATION COEFFICIENTS BETWEEN POPULATION AND EXPENDITURE ON RETAIL AND SERVICE TRADES, 1950

Kind of Trade	Cumbernauld (100 Town Sample)	Basingstoke (36 Town Sample)
1. Grocery	+ 0·89	+ 0·92
2. Other food retailers	+ 0·87	+ 0·82
3. Confectioners, newsagents and tobacconists	+ 0·85	+ 0·83
4. Clothing	+ 0·78	+ 0·78
5. Chemists and photographic goods	+ 0·92	+ 0·73
6. Hardware	+ 0·59	+ 0·64
7. Furniture	+ 0·69	+ 0·47
8. Jewellery, leather and sports goods	+ 0·75	+ 0·73
9. Booksellers	+ 0·42	+ 0·53
Total retail trade	+ 0·84	+ 0·87
10. Catering	+ 0·67	+ 0·76
11. Hairdressers	+ 0·65	+ 0·69
12. Motor vehicle repairs	+ 0·60	+ 0·60

Sources: *Cumbernauld – Report on the Central Area*, 1960. Author's researches.

[1] A sample survey undertaken by the Sociology Section of the Basingstoke study group has revealed that about one-third of the population living in recently established housing estates in the town shop for clothing at least once every three months in Reading and about one-quarter buy clothing in Southampton. Some 15 per cent appear to purchase furnishings in Reading and another 15 per cent in Southampton, although in both cases about one-half visit the two towns oftener than every three months. For foodstuffs, 'large items' in general and special items, the proportion of purchases undertaken outside Basingstoke was very small.

The methodology we have adopted is as follows:

(1) From the Census of Distribution 1950 we obtained the average retail sales per head of population in that year in various kinds of trades and services, in thirty-six towns in England and Wales of approximately similar market size and hinterland, economic structure and situation to that envisaged for Basingstoke in 1981.[1] Correlation coefficients between population and retail sales in these towns, are given in Col. 2 of Table VI/5 above: if anything, they suggest an even closer statistical association than in the case of the towns chosen as the basis for the Cumbernauld survey.

(2) We next updated the figures so obtained to take account of changes in the price level and living standards between 1950 and 1960, deflating the result by the percentage increase in population. This information was obtained from the Index of Retail Sales taken from the Annual Abstract of Statistics, although certain adjustments had to be made to render the statistics comparable.[2] Multiplied by the expected population in Basingstoke, which we have assumed to be 120,000, we arrive at the anticipated *total* sales of the various groups of retail outlets *at current, i.e. 1960, prices and living standards.*

Side by side with the sales figures obtained from the sample towns, which are presented in Table VI/6, we also give the equivalent figures for Basingstoke in 1950 based on an estimated population of 20,000. It is interesting to note that apart from the furniture and, to a lesser extent the chemist, motor vehicles and hardware trades, the figures in the two columns show a remarkable similarity.

Column 4 of Table VI/6 gives alternative figures computed on the basis of sales in Great Britain as a whole in 1960. As there is no reason to suppose that the average income of Basingstoke citizens is greatly different from the national average, no adjustment has to be made to

[1]

Barnsley	Colchester	Lincoln	Reading
Barrow	Darlington	Luton	Romford
Bedford	Doncaster	Maidstone	Scunthorpe
Blackburn	Exeter	Newport	Slough
Burnley	Gloucester	Northampton	Swindon
Cambridge	Grimsby	Norwich	Warrington
Carlisle	Halifax	Oxford	West Hartlepool
Cheltenham	Ipswich	Peterborough	Worcester
Chesterfield	Lancaster	Preston	York

[2] Two exceptions to this procedure were the motor vehicles and catering trades. In the former case we calculated the average sales per person employed by motor repairers, distributors, garages and filling stations in the United Kingdom in 1960 multiplied this figure by the numbers engaged in this trade in Basingstoke in that year and divided the result by the population in the Borough. In the case of catering trades we assumed the increase in average sales over the period to be equivalent to that of food sales in general.

TABLE VI/6

RETAIL SALES FOR SELECTED AREAS 1950/1960

	(a) Basingstoke 1950 (est.)	Sales per Head (£) (b) 36 Town Sample		(c) G.B. 1960 (est.)	Total Sales (£000s) Assuming population of 120,000	
		1950	1960 (est.)		Col. 3 basis	Col. 4 basis
Grocery	28·9	27·0	46·0	43·1	5,520	5,172
Other food retailers	22·1	22·3	38·0	33·9	4,570	4,068
Clothing	29·9	30·6	41·3	25·0	4,956	3,100
Hardware	5·4	6·1	9·8	9·2	1,176	1,104
Chemists and photographic goods	6·0	4·6	7·5	6·4	900	768
Furniture	5·6	9·7	16·4	10·7	1,968	1,284
Jewellery	2·6	2·6	4·4	3·7	528	444
Confectioners, tobacconists, newsagents	9·6	10·1 }	16·5	18·6	1,980	2,232
Booksellers	n.a.	2·1 }				
Department and variety stores	n.a.	n.a.	15·0	16·8	1,800	2,016
Motoring	14·9	18·0	30·0	20·7	3,600	2,484
Catering	6·3	6·5	10·5	8·4	1,260	1,008
Miscellaneous group*	n.a.	n.a.	5·0	2·5	600	300
					28,858	23,980

*Including hairdressing, funeral furnishers and repair trades.
Source: Census of Distribution 1950. Author's researches.

140

these figures. However, though incomes might be comparable, patterns of expenditure differ considerably according to region and size of town. Thus, in 1950 the average expenditure on food of those living in Wales and the North-West of England was consistently lower than the expenditure of those living in London and the North Midlands.[1] As a whole, the total retail sales for a town of 120,000, calculated by taking national figures, gives an expenditure of 17 per cent less than the thirty-six town sample. The difference in the individual categories varies from 12 per cent more in the case of booksellers to 33 per cent less in the case of clothing.

(3) In planning for the future, some allowance must be made for the developing pattern of consumers' expenditure. Overall consumers' expenditure is comparatively easy to calculate, given assumptions concerning the rate of increase in income. If, for example, one assumes that the increase in average real income and the relationship between income and retail sales over the next two decades is of the same order as that of the last ten years, then total (real) expenditure by 1980 should be half as much again as in 1960.

Unfortunately, any model which might be constructed to estimate trends in the expenditure on particular types of retail goods would be based on highly conjectural assumptions. Some commodities bought today may well be non-existent in 1980: others, not even thought of today, may be in common use. While certain pointers might be obtained partly from U.S. or Canadian experience (where living standards are the highest in the world) and partly from trends of consumption in this country in the last decade or so, it is impossible to make any accurate predictions. For example, even discounting relative price changes, it can be seen from Table VI/6 that the sales of motor vehicles, furniture and hardware have increased well above the average since 1950, while the proportion of expenditure on clothing has fallen. The volume of business done by radio and electrical shops, photographic dealers and chemists, hairdressers, gas and electricity showrooms and special forms of trading, e.g. mail order houses and the automatic machine trades, has also risen rapidly.

The developing pattern of consumers' expenditure as a whole over the last decade *assuming constant prices*, is illustrated in Table VI/7. Col. III indicates the proportionate change in expenditure since 1950 and Cols. IV and V calculate the percentage change by 1980 on two

[1] See M. Hall *et al.*, *Distribution in Britain and North America*, Oxford University Press, 1961.

bases: (a) the eleven-year increase 1950–60 and (b) extrapolation of the existing trend. In actual fact, the projections are very similar: we have taken the mid-point of the two for our purposes, discounted slightly to allow for (national) population changes. If one turns to the U.S., one sees a similar pattern of rise in retail expenditure, save that the proportion allocated to food and clothing is declining faster and that allocated to motor vehicles, hardware, hairdressing and catering services is rising faster than in the U.K.

TABLE VI/7

UNITED KINGDOM: TRENDS IN CONSUMERS' EXPEN-
DITURE ON SELECTED ITEMS: 1950/1960 AND ESTIMATES
FOR 1980 (1954 prices)

	Expenditure (£m)			Estimated Expenditure		Assumed
	1950	1960	% increase	1980(a)	1980(b)	% increase
Food	3,602	4,248	17·9	5,422	5,920	33·0
Tobacco	783	981	25·3	1,341	1,350	37·6
Durable goods	562	1,269	125·8	2,554	2,800	113·0
Other household goods	366	487	33·1	707	805	54·0
Clothing	1,227	1,537	25·3	2,101	2,340	43·0
Books, newspapers, etc.	176	186	5·7	204	205	10·2
Miscellaneous recreational goods	117	202	72·7	357	366	81·0
Chemists' goods	145	214	47·6	339	355	64·0
Other goods	134	230	71·6	405	424	83·0
Motoring	145	409	182·1	900	1,250	200·0[1]

Source: Annual Abstract of Statistics
Author's estimates

[1] See footnote 1 p. 143.

The above data is given in terms of commodity groups, but these can be converted into retail trade categories, thus giving a broad picture of the possible pattern of sales per head in Basingstoke[1] in 1970 and 1980, again assuming constant prices and an even rate of change. These are given side by side with the 1960 pattern in Table VI/8. Among the assumptions made in re-classifying the figures in Table VI/7 are that the sales of groceries and other foods (the retail trade breakdown of the 'food' category) rise at the same rate as do those of hardware and furniture – calculated as a weighted average of household and durable goods. Tobacco and books, newspapers and magazines have been grouped together and, in addition, some allowance has been made for confectionery sales on the basis of consumption since 1950. The same applies to photographic equipment, which has

[1] Or more correctly that of the 36 towns used as a basis for estimating Basingstoke's requirements.

been taken into account in the chemist group. The *number* of newly registered vehicles and expenditure on motoring have been taken as reasonable indicators of motor vehicle sales and expenditure at motor repairers and garages.[1] No separate estimate has been possible for jewellery, sports equipment, etc. Instead, a new group–other goods purchased–is included. Sales in this group and also in department stores are assumed to increase by the average for all trades. When these *per capita* sales figures are multiplied by the population estimated for Basingstoke in 1970 and 1980 (which will naturally vary according to the projection chosen), we obtain the anticipated total expenditure at various retail outlets. For the purposes of this exercise we have assumed an *average* population served by the retail outlets listed as 40,000 in 1960, 70,000 in 1970 and 120,000 in 1980.[2]

TABLE VI/8

BASINGSTOKE: ESTIMATED RETAIL SALES PER HEAD
AND PROJECTED TOTAL SALES 1960/1980
(Based chiefly on 36 town sample)

	Sales per head (£)			Total sales (£000s)		
	1960	1970	1980	1960 pop. 40,000	1970 pop. 70,000	1980 pop. 120,000
Grocery	46·0	53·6	61·2	1,840	3,752	7,344
Other food	38·0	44·3	50·6	1,520	3,101	6,072
Clothing	41·3	50·2	59·1	1,652	3,514	7,092
Hardware	9·8	14·6	19·3	392	1,022	2,316
Chemists' and photographic goods	7·5	9·9	12·2	300	693	1,464
Furniture	16·4	24·4	32·7	656	1,708	3,724
Confectioners, tobacconists, newsagents and booksellers	16·9	18·1	20·8	676	1,267	2,496
Jewellery	4·4	6·3	8·1	176	441	972
Department and variety stores	15·0	19·5	24·0	600	1,365	2,880
Motoring	30·0	60·0	90·0	1,200	4,200	10,800
Catering	10·5	12·2	13·9	420	854	1,668
Miscellaneous*	5·0	6·5	8·0	200	455	960
Total	240·8	319·6	399·9	9,632	22,372	47,788

*Including boot and shoe repairing, hairdressers, builders' materials, corn merchants, portrait photographers, funeral furnishers, etc.

[1] The assumed increase of 200 per cent in Table VI/7 takes into account new car sales, which on the basis of the trend of the period 1952/60 will rise by between 175 per cent and 250 per cent by 1980 (i.e. on 1960 sales).
[2] As mentioned before, the size of the market served will clearly vary according to the type of commodity in question. Thus, while people in the rural areas outside Basingstoke will probably buy their groceries locally, they may shop for clothes and durable household goods in the town.

So much for the calculation of our demand indicator of shopping requirements. The reader may choose whether he bases the supply indicator on the 1960, 1970 or 1980 assumed sales per head, and also the population projection he thinks most likely.

AREA OF SHOPPING SPACE NEEDED

To convert the expenditure estimates into area terms, data on the space requirements and trends in space requirements of different kinds of shopping outlets are needed. These are usually expressed in terms of square feet. The very different nature of commodities, their bulk and the rate of sales turnover of each unit per year give rise to markedly different sales per square foot. Two recent investigations have elicited some information on this matter. The first was that carried out by the Cumbernauld Development Corporation in 1960 by means of extensive enquiries of trade associations, chambers of commerce and individual businesses to assess the appropriate conversion factors for an 'efficient' firm in each of the retail and service trade categories; the results of this survey are summarised in Column 1 of Table VI/9.[1] The second investigation, undertaken by the Norfolk County Planning Department in respect of King's Lynn Borough[2] was conducted on a rather different basis; first, the sales turnover for each of the main retail trades was computed for 1961 by projecting forward known expenditure in 1950, using national consumption expenditure trends, and second, individual retailers in King's Lynn were asked for details of their floor areas (used for sales and not storage) in 1961. The relationship between sales and floor area was then calculated at 1954 values of the £. We have recalculated these sales in terms of 1959 values (to correspond with those used in the Cumbernauld survey) and we give the result in column 2 of Table VI/9. It can be seen that there is a reasonably close correlation between the two sets of data in spite of some differences in methods of calculation; the main exception is that of motor vehicles, where the estimated sales turnover is quite different, mainly it would seem because the Norfolk survey confined its analysis to motor-car showrooms, while the Cumbernauld survey also included garage and filling stations with or without service and repair facilities.

[1] See R. J. Green and R. M. Beaumont, 'The size of shopping centres', *Journal of the Town Planning Institute*, Vol. XLVIII, No. 10, Dec. 1962.
[2] For further particulars see Cumbernauld Development Corporation *Report on the Central Area* 1960 and D. R. Diamond and E. B. Gibb, 'Development of New Shopping Centres: Area estimation', *Scottish Journal of Political Economy* Vol. IX, pp. 131ff.

In looking to the future, consideration must be given to trends in retail efficiency and the changing nature of consumers' demand. The latter shows itself in the expected volume of sales; as regards the former, the Norfolk survey revealed that the efficiency of retail outlets in King's Lynn Borough, as measured in terms of the (real) sales per sq. ft., increased by an average of 55 per cent between 1950 and 1961. The average increase per annum in terms of sales per square foot is given in column 3 of Table VI/9. If one assumes that this rate of increase will continue in the future, then by 1981 the average sales per sq. ft. would increase to £55 per sq. ft. (at 1959 prices). This, so far as is known, is the only estimate which has been made of recent and anticipated trends in retail efficiency.

Since comparable data are not available for Basingstoke, we use the above information as the basis for our own projections. However, when calculating the likely future shopping requirements for Basingstoke, account must also be taken of the town's *existing* provisions. Unfortunately the latest detailed information available is that contained in the Census of Distribution for 1950, but, assuming that the average expenditure per person in Basingstoke on all retail services increased by the same proportion as that in the country as a whole in the following decade, then total sales expenditure in the town in 1960 would have been between £8,000,000 and £9,000,000. Since, at present, there appear to be at least 20 acres of shopping facilities in the central area,[1] this means that the average turnover per sq. ft. of Basingstoke shops is in the order of £12 per sq. ft. There is, however, no information available on the distribution of retail accommodation amongst the various kinds of shops. But the overall figure would strongly suggest that the conversion factors assumed for Cumbernauld and King's Lynn in 1961 would prove too high as a starting-point in the case of Basingstoke.

The data at our disposal for estimating Basingstoke's future shopping needs are, then, as follows. First, a very approximate estimate of the present (overall) ratio of sales per sq. ft. in the town; second, an estimate of the sales per sq. ft. of 'efficient' shopping outlets in various retail categories; and third, an estimate of trends in the retail efficiency of shopping groups in one town over the past ten years. We also know from the Town Map that while part of the existing shopping facilities at Basingstoke are to be redeveloped, *some* of the present retail outlets

[1] Derived from a figure of 23 acres given in the Town Map for the Comprehensive Development Area. This, however, probably includes storage area.

will exist side by side with new shops in 1980. This being so, even assuming a radical improvement of sales per sq. ft. efficiency of existing shops, it is doubtful whether such efficiency will reach the average of newly built (and assumed more efficient) shops by that time.

Because of the above factors, we consider that the estimates of the sales per sq. ft. for King's Lynn in 1981 are too high to be applicable to Basingstoke; on the other hand, the estimates of the Norfolk survey for the *increase* in shopping efficiency between 1950 and 1961 may well be used as a basis for forecasting. We therefore made the following calculation. First, we have taken a composite of the King's Lynn and Cumbernauld sales per sq. ft. figures for 1961 and multiplied these by the increase of the estimated sales per sq. ft. in King's Lynn in 1980 over those of 1961.[1] We have then deflated these figures by an arbitrary 15 per cent to allow for the existing shopping services in Basingstoke where it may be difficult to increase sales per head to the figure anticipated for the new facilities. Using these factors and the total estimated sales for 1980 for each retail and service trade in Table VI/8, the final column of Table VI/9 gives the estimated area of shopping space required by this date. If the retail use is distributed over more than one storey the site area will be reduced correspondingly.

How does this estimate of shopping area required by 1980 compare with that projected in the Town Development Plan for Basingstoke? Seemingly quite well; that our estimate is slightly lower than that envisaged in the Plan is probably due to our rather optimistic forecast of retail efficiency. According to the Town Map, between 30 and 40 acres are to be set aside for shopping purposes in the Comprehensive Development Area.[2] At present there are some 23 acres of shopping accommodation, but part of this is subject to redevelopment proposals. An additional 19 acres is to be allocated to meet the demands of the new population. This means that although the population of

[1] The least satisfactory of these estimates concerns that for motor vehicles, as no satisfactory breakdown is available between showrooms where the sales per sq. ft. are very high and increase much faster than for other retail groups and garages, other repair facilities and filling stations, which are more prodigal of space and where sales per sq. ft. are below the average. However, with national expenditure figures on 'new motor vehicles' and 'motoring' in mind, we have 'guestimated' an overall figure of £40 per sq. ft. for 1961, and assuming a *slightly reduced rate of increase* in retailing efficiency in the next twenty years (there was considerable 'surplus capacity' in garages and showrooms in 1950) have postulated a figure of £75 per sq. ft. in 1980.

[2] Basingstoke Comprehensive Development Area Map. Report of Survey and Conclusions, Part II, p. 8.

TABLE VI/9

CALCULATION OF SHOPPING AREA ESTIMATED FOR BASINGSTOKE 1980

(£ per sq. ft. at 1959 prices)

Type of Retail Outlet	Cumbernauld (1959)	King's Lynn (1961)	King's Lynn Increase per annum (£) (1950/61)	Assumed 'Efficient' Outlet (1960)	Assumed 'Efficient' Outlet (1980)	Basingstoke Estimated shopping area* (acres) (1980)
Grocery	32 }	35	0·53	33	45	8·1
Other food	30 }					5·3
Clothing	26	20	0·41	25	37	2·3
Hardware	15	15	0·63	15	29	0·9
Chemists' and photographic goods	23	28	0·70	25	42	3·7
Furniture	17	18	0·41	18	27	3·1
Confectioners, newsagents, tobacconists and booksellers	n.a.	18	0·10	18	22	
Jewellery	20	14	0·25	18	27	1·0
Department stores	n.a.		0·90	17	30	2·6
Motor vehicles	13	116	6·50	40	75	4·0
Catering	n.a.	15	0·00	15	15	3·1
Miscellaneous	n.a.	10	0·73	10	12	2·3
All types	n.a.	32	0·90	n.a.	n.a.	36·4

*Col. 6 of Table VI/8
 Col. 5 of Table VI/9

147

Basingstoke and its hinterland is scheduled to increase threefold, the area reserved for shops is to be less than doubled. Since, then, the average turnover per sq. ft. of retail space in Basingstoke is about £12, it follows that a considerable increase in the efficiency of existing retail establishments is envisaged by the planners, and indeed may be considered a necessary condition for the achievement of a retail turnover of £48,000,000 by 1980.

Data in respect of shopping facilities now available in the new towns is of only limited guidance to the future needs of Basingstoke. This is because, in most cases, retail development has lagged behind population and industrial growth and only Hemel, Crawley and Harlow have completed (or nearly completed) their town centres. However, for illustrative purposes, we compare (*a*) the relationship between the increased population in the London new towns since their designation and the 31st December 1961, and the area of new shops completed, with (*b*) the (estimated) population of the *Basingstoke Town Map area (plus hinterland)*[1] in 1961 and 1981 and the area in (or planned for) retail use, and with (*c*) the proposed future population of Cumbernauld and the area estimated to be required for their shopping facilities.

TABLE VI/10

SHOPPING AREAS AND PERSONS SERVED IN SELECTED TOWNS

(a) The new towns	Increase in population up to 31.12.61.	Increase in shopping area (acres)	Persons served per acre
Basildon	31,000	12·5	2,480
Bracknell	16,421	3·4	4,830
Crawley	46,000	8·2	5,610
Harlow	52,200	9·1	5,736
Hatfield	13,000	2·5	5,200
Hemel Hempstead	35,300	8·8	4,001
Stevenage	37,000	7·9	4,684
Welwyn	17,500	4·5	3,889
Total	248,421	56·9	4,366

(b) Basingstoke	Population served (a) in 1961 (b) 1981	Shopping area	
	(a) 40,000	23·0	1,739
	(b) 120,000	40·0	3,000
(c) Cumbernauld	Projected population 70,000	Shopping area 14·2	4,930

[1] Unlike the new towns, except possibly for Crawley, Basingstoke is a market centre for the surrounding area and therefore serves a wider population than that contained in its boundaries.

This table serves to confirm our earlier estimate that the area set aside for retail distribution in Basingstoke is fully adequate for the needs of the town on completion of the Town Development Scheme, and may indeed be sufficient to accommodate a further expansion of the buying population of the town and hinterland up to 150,000, which may well occur due to natural increase in the last two decades of the century.

TYPES AND STRUCTURE OF SHOPS

So far our exercise has been concerned with estimating areas of shopping space, both for retail trade in general and for different classes of goods. The method by which space is allocated, as between the numbers and type of shopping units and the geographical siting of outlets is a further problem facing the planners. As we have seen, previous projections of shopping requirements have been expressed in terms of numbers of retail outlets. Table VI/11 shows that in 1950 there were 285 such establishments in Basingstoke, which, on average,

TABLE VI/11

BASINGSTOKE: NUMBER OF RETAIL AND SERVICE ESTABLISHMENTS, SALES TURNOVER AND TOTAL EMPLOYMENT 1950

	Establish-ments	Sales Turnover (£m.)	Total employment
Total Retail	217	2,507	1,103
Grocery group	41	577	228
Other food retailers	51	442	251
Confectioners, tobacconists, newsagents	21	192	75
Clothing group	43	597	240
Hardware group	11	107	49
Booksellers, stationers	*	*	*
Chemists' goods, photographic goods group	8	120	56
Furniture group	11	111	52
Jewellery, leather and sports goods group	12	5	26
General group (including department stores, variety and other general stores)	*	*	*
Coal, builders' materials, corn group	10	185	60
Other goods	9	125	66
Service Trades			
Services (including hairdressers, funeral furnishers, portrait photographers and repair trades)	21	35	66
Catering group	32	125	166
Motor group	11	298	134
Total all groups	281	2,965	1,469

*Included in other goods.

149

served a population of 20,000. If this ratio of one shop to seventy persons were maintained in the future, then a population of 120,000 would need more than 1,700 shops; alternatively, on the basis of the Reith Committee's estimate for the new towns, between 800 and 1,200 would be required for a population of this size. However, to estimate the 'right' number of shops, either from an economic or a social viewpoint, is an exceedingly difficult thing to do. For one thing, it involves discovering what exactly the consumer wants from his shopping facilities. If he is solely concerned with obtaining his purchases at minimum cost possible and regards the function of a retailer purely and simply as a mechanical intermediary between himself and the manufacturer, then it may be argued, on economic grounds, that the 'right' number of shops is that number which allows each retailer to make a competitive (but no more than competitive) profit while maintaining the costs of retailing at their lowest possible level.

But the snag is that the consumer is very rarely indifferent towards the shopping facilities provided. In many cases the functions of a retailer extend far beyond that of a selling intermediary: thus, where expensive durable goods (e.g. items of furniture) are being bought the consumer will value expert advice as to the relative merits of particular products, while the scope and efficiency of after-sales service may be of crucial importance in influencing the sales of certain mechanical and electrical goods, e.g. cars and television sets. Some consumers value efficiency and courtesy of service very highly; others, such factors as the availability of delivery facilities, the provision of credit, the range of goods offered for sale, the amenities and atmosphere of the shop itself. All these services and others may be offered for sale by retail outlets, quite apart from the products which are the basis for such sales.

Again, shopping facilities are not homogeneous, because they are very rarely sited at equal distances from a consumer's home. Convenience is a vitally important ingredient of most kinds of day-to-day shopping, and clearly a grocer's shop a quarter mile from home is not considered in the same light as a retail outlet a 6d. bus ride away, even if the products and services provided are otherwise identical. And lastly, as Ford and Thomas and others have pointed out,[1] shopping is a personal and social event as well as an economic activity; it is part of community life and as such is of some sociological significance.

[1] P. Ford and C. J. Thomas, *Shops and Planning*, Blackwell, 1953, pp. 35 ff. W. Burns, *British Shopping Centres*, pp. 44 ff.

The shop or shopping centre is not only a place where goods are bought but a meeting-place and a focus of interest, particularly for women shoppers. To this extent the optimum size of a retail outlet or the best location for a shopping centre, looked at from a social viewpoint, may be quite different from the answers arrived at on economic grounds.

For such services as these the consumer may be prepared to pay a certain price. But the question is, how much?—and how are the planners to find out? One solution, of course, is to leave it to private enterprise to undertake the responsibility of estimating consumer requirements. But this may result in a pattern of shopping facilities which, though perhaps desirable from the economic viewpoint, may be undesirable from the wider social and planning aspect, e.g. in respect of the layout and attractiveness of a housing estate. Unless one abolishes the conception of town planning altogether, some overall guidance of each element of planning would seem necessary, so that the whole may be integrated.

Once again, a range of alternatives is open to the planners. Some will be of purely economic significance, some influenced by social and aesthetic considerations, and some decided by street geography and the physical characteristics of the town itself. These and other planning decisions or ideals, e.g. the basic concepts underlying a particular Town Development scheme, the economist has largely to take as given. He may, however, be permitted to offer some guidance from his knowledge of general principles of resource allocation; and to attempt to 'get order, reasonableness and consistency into the broad ideas out of which such rules must spring'.[1]

In a sense the problem of securing the right kind of shopping balance is similar to that of industrial balance dealt with in the previous chapter. But here the purpose is more to satisfy the inhabitants of the town in their role as consumers rather than as suppliers of labour, while ensuring that the shopkeepers earn a reasonable rate of profit. The reconciliation of the interests of buyers and sellers is not an easy thing to plan for in advance. However, it does imply that on the one hand there should be sufficient competition (or potential competition) between retailers to preclude any possibility of monopolistic exploitation (unless this latter can be controlled by other means); on the other, that there must not be too much competition (i.e. too many selling points) to cause permanent instability in the retail trade with

[1] Ford and Thomas, op. cit., p. 9.

all (or most) firms earning below normal profits.[1] Clearly, with a limited market and large number of possible variables influencing consumer choice, some kind of partial monopoly (e.g. in respect of geographical convenience, delivery facilities, range of goods offered for sale, etc.) is inevitable–although its extent and its power can be minimised with careful planning.

A further dilemma in this connection arises as a result of the growing economies of large-scale retailing. The profit or mark-up required by a retailer is often a function of the size of his turnover. A large self-service grocery store finds that it pays to cut the profit margin on certain commodities where the demand is responsive to a fall in price or to attract custom into the shop. Split such a store into ten or twelve small retail units and average retailing costs will be higher–due to duplication of overheads, less specialisation of labour, and so on. But there may be room for only one large self-service store or supermarket in the town, which means it has a degree of monopoly power to charge more than is necessary to render a reasonable profit. It may be argued that the consumer is prepared to accept this providing he obtains his product more cheaply. But, with the additional cost and time involved in travelling into the town centre (where usually such stores must be located), how far, in fact, is this the case?

LOCATION OF SHOPS

This brings us to our second point. The number of shopping outlets required is also related to their geographical distribution throughout the town, and this in turn will partly depend on the type of shop and the overall conception of town layout.

The new towns differ considerably in their shopping layouts and the distribution of shops between the town centre and surrounding neighbourhood. Thus, while Stevenage and Harlow are building a cluster of main neighbourhood shopping centres with twenty-five to sixty shops in each and a group of sub-centres of five to eight shops situated at strategic points–a three-tier system–Crawley and Hemel Hempstead have preferred to divide their residential areas into a larger number of smaller zones with ten to thirty shops each–a two-tier system. Basildon has adopted this latter policy, but has also recognised the additional need for groups of three to four 'pantry'

---

[1] Or, as the New Towns Committee put it, 'a balanced development of shopping centres with all classes of business represented and with sufficient but not excessive competition within each class'. *New Towns Committee*, Final report, para. 135; 1945–46 Cmd. 6876, xiv.

shops in those parts of each neighbourhood farthest removed from the main shopping centre. Bracknell is deliberately leaving a number of corner sites vacant in case the demand for extra individual shops should arise. In Newton Aycliffe there will eventually be six neighbourhoods or wards, each containing about 3,000 people and from ten to fifteen shops. The new towns control the building and occupancy of shops as a matter of policy, to provide for all ordinary needs and to avoid undue duplication or redundancy.[1] Some of the larger expanding towns, e.g. Swindon, are four-tiered in their shopping patterns – with town centre, district centres (of 100 or more shops), neighbourhood centres (twenty to thirty shops) and sub-centres (four to eight shops) or isolated corner shops.

W. Burns in his book *British Shopping Centres*[2] suggests that shopping facilities can best be grouped into a three-tier system:

(i) a town centre, where the great bulk of shops will be more or less specialist in character and mostly departmental or variety chain stores or shops catering for occasional rather than day-to-day shopping needs;

(ii) a suburban shopping centre comprising a large group of retail outlets (one would think between fifty and eighty for a town of 100,000 population) comprising most of the larger convenience shops, e.g. grocers, butchers, etc., and also some outlets selling occasional goods, e.g. hardware, furniture, electrical goods, etc.

(iii) A large number of individual shops mainly in the form of isolated or 'corner' shops, although sometimes it may be found desirable to group these in twos or threes. Such shops are fitted into residential areas as conveniently as possible and flexibility is essential for town planning purposes. Convenience is the predominating factor for the 'general' store which, on the sociological side, satisfies real needs of the housewife for contact with other people. However, there is a case for shops such as chemists, boot and shoe repairers, newsagents and even hairdressers, to be fitted into this category.

THE PHASING OF SHOPPING DEVELOPMENT

In general, it is only to be expected that shopping facilities should, to begin with, lag behind the growth of population. The policy of

[1] J. Nicholson, *New Communities in Britain*, National Council of Social Service, 1961.
[2] Leonard Hill, 1959.

most Development Corporations has been to develop one neighbourhood at a time with its complement of day-to-day shops, and only when the population of the town has reached about one-half of its projected size, to embark on developing the town centre, which will contain the familiar multiples and those shops selling durable goods and/or those less frequently purchased. In any event, the experience of the new towns has already shown that most of the larger chain stores will not consider opening a new branch unless it is situated in the centre of a town which has a potential market of at least 30,000–40,000 people. Partly, at least, this reflects the rising costs of equipping a new branch; one multiple provision merchant estimated in 1959 his own outlay in this respect to be around £70,000. The net result is, however, that the early settlers in the new towns are often forced not only to readjust their existing shopping habits but also to shop outside the new town for certain types of clothing, furniture, hardware, electrical goods, etc. The lack of variety in the neighbourhood shopping facilities has also caused some dissatisfaction: for example, though these are usually provided with two or three grocers, butchers, bakers, etc., other shops and particularly all types of 'service' points, e.g. shoe repairers, hairdressers, dry cleaners, etc., are conspicuous by their absence. In Basingstoke there is, of course, an already established town centre comprising four main shopping streets. It would seem likely that this will remain substantially unchanged in the first phase of town expansion and that the development of shopping facilities will be concentrated in the new neighbourhoods to be built up. By the late 1960s, however, one imagines that the ambitious redevelopment of the Town Centre, laid down in the scheme, will be well under way.

Concerning the number of people likely to be engaged in retail distribution in 1981, given the expenditure calculated in the previous sections, much depends on productivity trends in retailing and, in particular, the extent to which automation develops. Thus comparative figures of numbers employed in retailing in other towns with a population around 100,000 are little guide, unless we can also estimate the likely trends in productivity per head between now and 1981. Between 1950 and 1961, during which time the national income of the U.K. (at constant prices) rose by 52 per cent, the numbers employed in the distributive trades rose by 24·2 per cent. Assuming the same relationship will exist in the future as in the past and that the national income rises by an average of 3 per cent per annum, then by

1981 the numbers likely to be employed in retailing (for a population about 10 per cent greater) will be 3·25 million or nine more per 1,000 of population. This would mean that a town of 120,000, which today employs 5,000 in retailing, would employ 6,080 in 1981. Thus, on the basis of our thirty-six town study, we should expect almost 6,500 to be employed in retailing in Basingstoke (compared with the 1,813 of today). However, assuming that productivity in retailing in Basingstoke will rise faster than in the rest of the country, this figure of 6,500 can probably be reduced to 6,000.

CONCLUSIONS

This chapter has tried to do three things: first, to give some idea of the likely structure of the tertiary industries in Basingstoke in 1981; second, to estimate the total expenditure on retail goods and services in 1981 and then convert this figure into the floor space required; and third, to suggest a possible breakdown of the shopping needs of consumers in Basingstoke and productivity trends.

CHAPTER VII

SUMMARY OF CONCLUSIONS

CHAPTER I

I.1. In the course of the last century, and particularly during the last two decades, the town of Basingstoke has evolved from an agricultural and market centre serving the surrounding countryside into a light industrial town of increasing importance in the economy of Southern England. Three factors in particular have contributed to this widening of the town's activities; first, the coming of the railway in the mid-nineteenth century and the resulting improvement of communications between Basingstoke and the rest of the country; second, the development of the Midlands and Southern England as important industrial regions consequent upon changing locational forces, making for market- rather than raw material-oriented siting of industry; and third, the passing of the Town Development Act, 1952, and the scheduling of Basingstoke as an expanded town to take some of the overspill of industry and people from the congested London area.

I.2. The results of these developments are shown first in a rise of population which has accelerated particularly since 1956 and second, in the growing proportion of the labour force employed in secondary industry.

CHAPTER II

II.1. In October 1961 an agreement was signed between the Basingstoke Borough Council, Hampshire County Council and London County Council by which some 11,500 houses were to be built in Basingstoke by 1976 to accommodate some 37,000 Londoners. This agreement superseded an earlier and less ambitious scheme to accommodate 12,000 Londoners and, with complementary expansion schemes at Tadley and Andover, replaced the plan for establishing a new town at Hook. On the basis of the scheme and the likely natural increase in population, a new Town Map and Development Plan were drawn up giving details of the proposed increase in population to 86,000 in 1982 and the provision for accommodating this increase.

156

II.2. The planning of the expansion has been entrusted to a Joint Development Committee consisting of representatives of the three sponsoring authorities, with the day-to-day work being undertaken by the Basingstoke Development Group headed by a Director of Town Development and a Chief Planning Officer.

II.3. Financial assistance in respect of the scheme is provided from two main sources, (i) central Government funds under the Town Development Act 1952 and (ii) the Hampshire County Council and London County Council in accordance with the agreement signed. As regards the former, Exchequer assistance is given chiefly in the form of an annual housing subsidy (at present £28 per dwelling) payable for sixty years, while the Ministry of Housing is empowered to give financial aid for acquiring land for development and the provision of certain utilities, e.g. water and drainage. The main form of assistance provided by the London County Council is in the form of a capital loan for land purchased outside the comprehensive development area and the development of industrial estates, and contributions to the deficiency on housing revenue account and to the development costs of the scheme. Most of these latter costs, however, will be shared between the Hampshire County Council and Basingstoke Borough Council.

CHAPTER III

III.1. Most of the undertakings supplying the basic services appear confident that they will be able to meet the demands of Basingstoke over the next twenty years, and the experience of the new towns confirms this. Sewerage is well catered for; present water supplies are sufficient at least until 1970 and new boring is now being undertaken. With gas supplies, it would seem likely that on the basis of national trends some 4 m. therms will be consumed in 1981, not 5.9 m. as estimated by the Southern Gas Board. Electricity demand has probably been underestimated at 90 kW. Taking the national trend figures of the past ten years and projecting forwards to 1981 it would seem that consumption will rise to between 120 and 150 kW.

III.2. The cost of the sewerage scheme will be shared between the Ministry of Housing and Basingstoke Borough Council; the gas, electricity and water undertakings will finance their own expansions and they anticipate little difficulty in raising the capital as and when it is needed. There is no guarantee, however, that national economic conditions will be such that labour will be easy to obtain as and when it is needed.

CHAPTER IV

IV.1. The population of Basingstoke Borough at April 1961 was 25,940, an increase of 52·8 per cent over the figure for 1951. Of this increase, net immigration accounted for nearly two-thirds and natural increase for the balance. The population of the new Town Map Area was estimated at 30,130 in 1959, of which 23,130 lived within and 7,000 outside the Borough, i.e. in the 'new intake' area. Since no figures are available of the age distribution of the population in this latter area we excluded it from our more detailed projections.

IV.2. The population estimated for 1982 by the authors of the Town Map is 86,230 (77,980 excluding population in new intake area). In our opinion this estimate errs on the low side in that it makes insufficient allowance both for the natural rate of increase of the existing and migrant population, and the extent of private migration.[1] A more realistic projection for 1982 would be 93,430 (85,180 excluding population in the 'new intake' area) although our subsequent analysis proceeds on the more general basis that the population of Basingstoke will rise to 80,000 between 1976 and 1981.

IV.3. The implications of this target are examined in the light of five alternative patterns of migration build-up and natural increase. The accommodation of 44,000 people (36,400 Londoners and up to 8,000 private migrants) within fifteen years is seen to involve a much larger annual intake of migrants than that achieved by any other new or expanded town since the war, although conditions for such growth in the next twenty years may well be considerably more favourable than in the period since 1947. According to the rates of net immigration assumed, population in Basingstoke will rise to between 67,000 and 80,192 in 1976 and 88,959 and 95,129 in 1981.

IV.4. In computing the likely natural increase of population in this period, we have taken into consideration first the (estimated) existing age structure of Basingstoke and second the pattern of the natural national increase in the new towns. Computed deviations from the natural rate of increase are seen to vary from 7 per thousand in 1962–5 to 11 per thousand in 1971–5 and 10 per thousand in 1976–81.

---

[1] The author is apparently not alone in this view. See discussion on Messrs. Steel and McCulloch's paper, 'Principles and Practice of Town Development. The Basingstoke Scheme', *Journal of the Town Planning Institute*, Jan. 1963, p. 13.

IV.5. To calculate the likely age-distribution of the population in the future, it is necessary first to estimate the age-pattern in 1961. Forward projections are then based on (*a*) the rates of migration assumed in the five projections and (*b*) estimated births and deaths computed from national life tables adjusted to Basingstoke's age-structure and the (cumulative) age-pattern of migrants based on the experience of new towns. The resulting distribution is broken down into four age groups, 0–4, 5–14, 15–64 and 65 and over. Due to a very rapid increase expected in the (younger) child population, both the *proportion* of the working labour force and those over 65 is likely to fall slightly from its present level to between 61·5 and 61·8 per cent by 1971. By 1981, however, when there is little or no sponsored immigration, these proportions will have risen again to 63·3 per cent. It is also likely there will be three times the number of schoolchildren in 1976 as at present and in 1981 four times the number.

CHAPTER V

V.1. The working labour force of Basingstoke is likely to rise to approximately 40,000 by 1976 and 50,000 in 1981. These figures are based on the population projections in Chapter IV and national studies which relate the size of the working labour force to the total population and more particularly to those in the 15–64 age group.

V.2. Apart from agriculture, which within the Town Map Area[1] is likely to remain of minor significance, the distribution of the labour force will be entirely between the manufacturing (or secondary) and service (or tertiary) industries. We consider the main sources of income and spending power in Basingstoke will be derived from three lines of activity: (i) manufacturing industry, (ii) certain service industries and trades, e.g. insurance, research and development, and office activities of construction companies, and (iii) retailing and marketing industries designed to serve the hinterland around Basingstoke (as well as the town itself). Of these it is suggested that employment in (i) will remain of dominant importance, as secondary industry both helps fulfil one of the main purposes of urban decentralisation and also is the major job provider of almost all the other towns in England and Wales similar in size, general character and structure to that envisaged for Basingstoke by 1981. Nevertheless, we assume the proportion of those engaged in manufacturing industry, although

[1] Excluding the new intake area.

159

rising slightly at first, will later fall to about 46 per cent, partly in keeping with the national trend of recent years and the possibility of increased office decentralisation from the London area and partly because the development of most service industries usually follows a little behind that of manufacturing.

V.3. In deciding the future pattern of secondary industry, emphasis should be placed on securing and maintaining a proper industrial and occupational balance. Thus, on the *demand* side, encouragement should be given to industries with an adequate and stable rate of growth, high productivity and/or rate of increase in productivity and likely to do well in international markets. On the *supply* side, care is necessary to attract industries, the labour requirements of which will ensure sufficient and varied employment opportunities for the work force available: in particular, it is important to secure an equilibrium between the supply of different forms of labour, e.g. male and female, juvenile and adult, skilled and unskilled, operative and non-operative, and so on. It is also important to avoid the dangers inherent in one firm dominating the local economy by its size and in an undue preponderance of production units which are branches of firms in other parts of the country. Some of these measures influencing choice are assessed in an Appendix.

CHAPTER VI

VI.1. The numbers employed in the various service trades in Basingstoke are likely to rise to between 25,500 and 26,700 in 1981. On the basis of (i) a study of the new towns and other towns similar in size and economic structure to that envisaged for Basingstoke and (ii) national (and international) trends in the pattern of growth in the service trades, an approximate distribution of employment in these trades can be built up.

VI.2. In estimating requirements for retail distribution in 1980 a calculation is first made of the *per capita* expenditure in Basingstoke on the various categories of retail trade in the years 1950, 1961 and 1980. The figure for this latter year is then related to the estimated size of the shopping population, and from data of retailing efficiency (measured in terms of sales per sq. ft.) the area of shopping space required is computed. By 1980 it is thought that shopping expenditure at 1959 prices might rise to between £45,000,000 and £50,000,000 and that, if trends in the increase of sales per sq. ft. follow the pattern of the past ten years, this would require a shopping area of 37 acres.

VI.3. Care must be taken to obtain a correct balance of shops. On the one hand it is desirable to benefit from all the modern devices of retailing and on the other, to take into consideration the varying requirements of the consumer, while avoiding the creation of monopolies which might be exploited against the public interests.

VI.4. Studies of new towns show diverse patterns of neighbourhood shopping: it is difficult to generalise as to which is the ideal pattern, as so much depends on planning conceptions of town layout, road patterns, topography and social structure.

VI.5. The number likely to be employed in retailing in Basingstoke in 1981 is between 5,000 and 5,250 on the basis of the numbers now employed by comparable size towns, but nearer 6,000 taking into consideration national income and productivity trends.

SELECTED BIBLIOGRAPHY

BOOKS AND REPORTS

Ashworth, W. *The genesis of Modern British town planning*. Routledge & Kegan Paul, 1953.

Burns, W. *British shopping centres*. Leonard Hill, 1959.

Chisholm, M. 'Location of industry', *Planning* (P.E.P.), No. 466, 29.10.62.

Cullingworth, J. B. *Housing needs and planning policy*. Routledge & Kegan Paul, 1960.

Cumbernauld Development Corporation. *Report on the Central Area*. The Corporation, 1960.

Distribution of the Industrial Population. Royal Commission Report; 1939–40. Cmd. 6153, iv.

Economist Intelligence Unit. *Britain and Europe*, E.I.U., 1957.

Estall, R. and Buchanan, R. O. *Industrial activity and economic geography*. Hutchinson University Library, 1961.

Florence, P. Sargant. *Investment, location and size of plant*. C.U.P., 1948.

Ford, P., and Thomas, C. J. *The industrial prospects of the Southampton region*. Basil Blackwell, 1952.

Shops and planning. Basil Blackwell, 1953.

Goss, A. *British industry and town planning*. Fountain Press, 1962.

Greater London Plan 1944; 1945 Non-Parl. Min. of Town and Country Planning.

Gutkind, E. A. (ed.) *Creative demobilisation: case studies in national planning*. Kegan Paul, Trench, Trubner & Co., 1943.

Hampshire County Council. *Hampshire Development Plan 1950*, 1952. *Development Plan 1955*, 1957.

Basingstoke Town Map, Amendment Map and Comprehensive Development Area Map, 1961, 1962.

Basingstoke Town Map, Amendment Map and Comprehensive Development Area Map, 1961. (Report of Survey & Conclusions), 1962.

Logie, G. *Industry in towns*. Allen & Unwin, 1952.

London County Council. *The planning of a new town*. 1961.

Moser, C. A., and Scott, W. *British towns*. Oliver & Boyd, 1961.

Munby, D. L. *Industry and planning in Stepney*. O.U.P., 1951.
New Towns. Committee Reports; 1945–46. Cmd. 6759, Cmd. 6794, Cmd. 6876, xiv.
Nicholson, J. H. *New communities in Britain*. National Council of Social Service, 1961.
Self, P. *Cities in flood*. Faber & Faber, 1961.
Town and Country Planning Association. *Office location in the London region*. Town and Country Planning Association, 1958.
  The paper metropolis. Town and Country Planning Association, 1962.

ARTICLES

Aucott, J. V. 'Dispersal of offices from London.' *The Town Planning Review*, Vol. XXXI, April 1960.
Chisholm, M. 'The Common Market and British manufacturing industry and transport.' *Journal of the Town Planning Institute*, Jan. 1962.
Cullingworth, J. B. 'Housing formation in England & Wales.' *Town Planning Review*, April 1960.
Diamond, D. R., and Gibb, E. B. 'Development of new shopping areas: area estimation.' *Scottish Journal of Political Economy*, Vol. IX, June 1962, No. 2.
Dunning, J. H. 'Manufacturing industry in the new towns.' *Manchester School of Economic and Social Studies*. Vol. XXVIII, No. 2, May 1960.
'First results of the 1961 Census of Distribution.' *Board of Trade Journal*, 8 Feb. 1963.
Fleming, J. B. 'An analysis of shops and service trades in Scottish towns.' *Scottish Geographical Magazine*, Vol. 70, No. 3, Dec. 1954.
Florence, P. Sargant. 'The selection of industries for dispersal into rural areas.' *Journal of the Royal Statistical Society*, 1944, Vol. CVII, Pt. II.
  'The future of the conurbations.' *Town and Country Planning Summer School*, Sep. 1961.
Glass, R. 'Household structure and housing need.' *Population Studies*, Vol. 4, 1950–51.
Green, R. J., and Beaumont, R. M. 'The size of shopping centres. A method of projection in retail floor area.' *Journal of the Town Planning Institute*, Dec. 1962.

Liesner, H. 'The European Common Market and British industry.' *Economic Journal*, Vol. LXVIII, June 1958.

Luttrell, W. F. 'Industry and small town expansion.' *Official Architectural and Planning Review*, Vol. 8, Aug. 1955.

McLelland, W. G. 'Economics of the supermarket.' *Economic Journal*, Vol. LXXII, March 1962.

McMahon, C. W., and Worswick, G. D. N. 'The growth of services in the economy.' *District Bank Review*, Dec. 1960 and March 1961.

Manners, G. 'The crisis of regional planning.' *Westminster Bank Review*, Aug. 1962.

Powell, A. G. 'The recent development of Greater London.' *Advancement of Science*, May 1960.

Schneider, J. R. L. 'Local population projections in England and Wales.' *Population Studies*, Vol. X, No. 1, 1956.

Self, P. 'Congested London–can employment be dispersed.' *Times Review of Industry*, July 1962.

Smailes, A. E. 'Balanced Towns, their bases and occurrence in England & Wales.' *Journal of the Town Planning Institute*, Nov./Dec. 1945.

Steel, R., and McCulloch, A. 'Principles and practice of town development: the Basingstoke Scheme.' *Journal of the Town Planning Institute*, Jan. 1963.

Stone, P. A. 'Urban development and cost prediction.' *Town Planning Review*, Oct. 1959.

INDEX

East Kilbride, 51, 92, 105
Eastleigh, effect of railway works, 17, 105
economic activity, classification of, 85–87
  distribution of, 22–24
  future pattern of, 84–85
Economist Intelligence Unit, 109, 162
education, implications of population age structure, 68–69
electricity supplies, 39–41
  New Towns, 39n., 40–41
employment, 77–128
  distribution of, 17, 19, 23, 77–84
  see also labour force
Estall, R. C., 80n., 96n., 111n., 162
European Economic Community, possible impact of entry, 100, 108–10
expanding towns, 21, 22, 24–25, 28, 29, 51
  age composition of migrants, 59, 68
  differences from New Towns, 55
  shops, 153
  see also Town Development Act, 1952 and individual towns
'exporting' industries, definition of, 85–86

factory/site proportion, 101
Farnborough, 105
Federation of British Industries, 102
female labour force, 87, 88, 89, 103–4
female/male employment ratio, 103–4
financial arrangements, 30–31, 55, 157
firms, administrative structure, 106–7
  size of, 105–6
Fleming, J. B., 137n., 163
Florence, P. S., 114, 116, 162, 163
Ford, P., 150, 151n., 162

garage accommodation, 70
gas supplies, 35–39
  New Towns, 37–38
Gibb, E. B., 135n., 144n., 163
Gillingham, Kent, 89, 90
Glass, R., 163
Goss, A., 21n., 162
Greater London Plan, 1944, 21–22, 24, 162
Green, R. J., 144n., 163
growth industries, 100, 102
Gutkind, E. A., 162

Hague, D. C., 114n.
Hall, M., 141n.
Hampshire County Council, role in expansion scheme, 22, 27–28, 29, 30–31, 55

Hampshire Development Plan, 1950, 23–24, 28, 162
Hampshire Development Plan, 1955, 24, 28, 162
Harlow, 37, 38, 39n., 40, 51, 52, 53, 54, 58, 65, 70, 92, 98, 113, 148, 152
Hatfield, 38, 98, 105
Hemel Hempstead, 37, 38, 40, 51, 52, 53, 54, 56, 58, 59, 65, 70, 113, 148, 152
Hook, projected L.C.C. new town, 28, 156
household, average size of, 27, 49
houses, allocation of, 127
  type needed, 70
housing, 25, 27, 49, 51, 56, 57
  implications of population growth, 69–70
  local authority, 45
  New Towns, 51, 52–53
  private enterprise, 45
  subsidies, 30, 31
Housing and Local Government, Ministry of, 30, 31–32, 35, 126

immigrants, age structure of, 48, 59, 64–66
  natural increase of, 49, 57–60, 158
immigration, 26, 43–44, 45, 50–57, 158–9
  five projections, 56–57
  New Towns, 51–54
  private, 45, 48, 49, 55–56
  New Towns, 53
'importing' industries, definition of, 86
Index of Retail Sales, 139
industrial balance, concept of, 97–108
  influence of demand on, 99–102
  influence of supply on, 102–7
  present position in Basingstoke, 99
industrial development, procedure for, 125–6
Industrial Development Certificates, 82, 84, 125
Industrial Selection Scheme, 27n., 53, 126
industry, 77–128
investment industries, 88, 89
  definition of, 85–86

job ratio, definition of, 87n.
juvenile/adult employment ratio, 27, 104
juveniles, employment of, 88, 104

King's Lynn, retail sales survey, 144–6

Labour, Ministry of, 126, 127
labour force, distribution of, 89–90

population, projections A–E, 50, 61
  total planned, 27, 48
  see also immigration
Powell, A. G., 21n., 164
Power, Ministry of, 39
primary industry *see* agriculture
professional services, employment in, 80, 82, 132
public utilities, 34–42, 157–8

railways, as employers of labour, 16, 17
  effect of works on Eastleigh, 17, 105
  effect on Basingstoke, 15, 16, 22, 156
Reading, 15, 16, 45, 91, 105, 130, 136, 137n., 138
Reith Committee, 86n., 134, 150, 152n., 163
retail sales and population size, 136–44
retail trade, 133–55, 160–1
  employment in, 154–5
  see also shops
Retailers' Advisory Committee on Town Planning, 134
roads, development of, 18, 32, 34, 128
Rochdale Committee, 110
Rotary Club of Basingstoke, 18

Salisbury, 16
Schneider, J. R. L., 164
schools *see* education
Scott, W., 87n., 88, 89–90, 162
season-ticket holders, 45
secondary industry *see* manufacturing industry
Self, P., 163, 164
self-service stores, 135, 152
selling point, definition of, 134n.
service trades, 129–33, 160
  definition of, 86
  employment in, 77, 89–90, 129–31
    and living standards, 91
sewerage, 30, 34–35
shopping unit, size of, 134–5
shops, area needed and consumers' expenditure, 135–49
  location of, 152–3
  New Towns, 53, 92, 134, 148–9, 150
  phasing of development, 153–5
  present provisions, 145–6, 154
  ratio to population, 134, 150
  size of, 134–5
  type of, 149–52
  see also retail trade; service trades
site allocation in Basingstoke, 98
skilled labour, concentration of, 98
Slough industrial area, pattern of industry, 117
Smailes, A. E., 164
South Eastern Electricity Board, 40

Southampton, 84, 91, 110, 136, 138
Southern Electricity Board, 39
Southern Gas Board, 35, 39
space requirements, individual industries, 101–2
Steel, R. W., 158n., 164
Stevenage, 38, 39n., 51, 52, 58, 100, 113, 148
Stone, P. A., 164
supermarkets, 135, 152
Swindon, 24, 51, 59, 89n., 153

Tadley, 28, 28n., 156
technological education, New Towns, 98
tertiary industry *see* service trades
Thetford, 59
Thomas, C. J., 20n., 150, 151n., 162
Thomas, W., 21n.
Town and Country Planning Act, 1947, 23
Town and Country Planning Act, 1954, 31
Town and Country Planning Association, 93–94, 163
Town and Country Planning (Development Plans) Direction, 1954, 46
Town Development Act, 1952, 21, 22, 24, 28–29, 30, 51, 55, 84, 88–89, 156
  1959 Basingstoke Agreement, 25, 28, 45, 48, 48n., 156
  1961 Basingstoke Agreement, 27, 28, 31, 48, 156
Town Map 1962, 27, 31–32, 85, 162
  land allocated for shops, 146
  site allocation for industry, 101, 127–8
  total population envisaged, 47, 158
  underestimate of natural increase of population, 48–49, 61, 158
Trade, Board of, 25, 82, 125, 126
Trades Council of Basingstoke, 18
transport *see* communications

unemployment rate, 19, 84

water supplies, 30, 41
Welwyn Garden City, 38, 51, 58, 59
White, H. P., 17n.
Winchester, 15, 91
women, employment, 87, 88, 89, 103–4
workers, procedure for selection by firms, 127
workers per acre, density of, 93n., 101
working population *see* labour force
Worswick, G. D. N., 164

Young, T., 54n.